Treatment Strategy and Mechanism of Acute Ischemic Stroke

Treatment Strategy and Mechanism of Acute Ischemic Stroke

Guest Editor

Sonu M. M. Bhaskar

MDPI

Basel • Beijing • Wuhan • Barcelona • Belgrade • Novi Sad • Cluj • Manchester

Guest Editor
Sonu M. M. Bhaskar
Department of Neurology
National Cerebral and
Cardiovascular Center (NCVC)
Osaka
Japan

Editorial Office
MDPI AG
Grosspeteranlage 5
4052 Basel, Switzerland

This is a reprint of the Special Issue, published open access by the journal *Neurology International* (ISSN 2035-8377), freely accessible at: https://www.mdpi.com/journal/neurolint/special_issues/56DSP9J7Y5.

For citation purposes, cite each article independently as indicated on the article page online and as indicated below:

Lastname, A.A.; Lastname, B.B. Article Title. *Journal Name* **Year**, *Volume Number*, Page Range.

ISBN 978-3-7258-5645-9 (Hbk)
ISBN 978-3-7258-5646-6 (PDF)
https://doi.org/10.3390/books978-3-7258-5646-6

Cover image courtesy of Sonu M. M. Bhaskar

Contents

About the Editor

Sonu M. M. Bhaskar

Sonu M. M. Bhaskar, MD, PhD, PHF, FANA, is an internationally recognized neurologist and physician-scientist specializing in vascular neurology, stroke, and neuroimaging. He is the Founding Director of the Global Health Neurology Lab and the NSW Brain Clot Bank in Sydney, Australia. Currently, he is on sabbatical as a Visiting Director (2023–2025) at the National Cerebral and Cardiovascular Center in Osaka, Japan, supported by the Japan Society for the Promotion of Science (JSPS) and the Australian Academy of Science. Dr. Bhaskar's work lies at the intersection of cerebrovascular research, neurotechnology, health systems, and translational science. He leads multidisciplinary teams working across acute stroke mechanisms, health inequities, and biomarker discovery, with a strong emphasis on global health impact. His scholarly output includes over 170 peer-reviewed publications, with an H-index of 56 and over 37,000 citations—placing him among the top 2–3% of most-cited physicians and researchers in neurology worldwide. His work has been featured in journals such as *Lancet Neurology*, *Annals of Clinical and Translational Neurology*, and the *European Journal of Neurology*. Dr. Bhaskar has delivered more than 150 invited talks, chaired sessions at leading conferences, and contributed to training programs in neurology and public health leadership. His research and leadership have informed and shaped healthcare policy not only across the Asia–Pacific region but also globally. His numerous accolades include the European Academy of Neurology Investigator Award (2019, two-time recipient), the Rotary Vocational Excellence Award (2020), the Fellowship of the American Neurological Association (2022), the Australian Global Talent Award (2023), the Paul Harris Fellow citation (PHF, four-time recipient), and the TouchNEUROLOGY Future Leader Award (2025). He was elected as a full member of Sigma Xi, one of the world's oldest and most prestigious scientific research honor societies, in 2023.

Preface

This Special Issue of *Neurology International, Treatment Strategy and Mechanism of Acute Ischemic Stroke,* was conceived to address a growing need: to bridge the translational divide between rapidly evolving stroke research and the realities of clinical care. With acute ischemic stroke continuing to be a leading cause of death and disability worldwide, the motivation for this Reprint lies in the urgent demand for integrated, evidence-based approaches to diagnosis, treatment, and recovery.

This Special Issue assembles contributions from a diverse group of clinicians, neuroscientists, and health system researchers across the globe. Together, these papers reflect the state of the art advances in multimodal imaging, interventional strategies, respiratory and motor rehabilitation, prognostic markers, and mechanistic insights—including vascular, immunological, and genomic perspectives.

This Reprint is intended for stroke neurologists, interventionalists, rehabilitation specialists, and scholars of neurovascular science. It is also relevant for public health experts and policymakers who are concerned with the delivery and equity of stroke care.

As Guest Editor, I extend my sincere thanks to the authors for their rigorous contributions and to the editorial staff of *Neurology International* for their support. I also acknowledge the collaborative spirit of our research community, which continues to drive innovation in the face of complex clinical challenges.

Sonu M. M. Bhaskar
Guest Editor

neurology international

MDPI

Editorial

Bridging the Gap Between Research and Clinical Practice in Treatment Strategies and Mechanisms of Acute Ischemic Stroke

Sonu M. M. Bhaskar [1,2,3,4,5]

1 Department of Neurology, Division of Cerebrovascular Medicine and Neurology, National Cerebral and Cardiovascular Center (NCVC), Suita 564-8565, Osaka, Japan; sonu.bhaskar@globalhealthneurolab.org
2 Global Health Neurology Lab, Sydney, NSW 2150, Australia
3 Ingham Institute for Applied Medical Research, Clinical Sciences Stream, Liverpool, NSW 2170, Australia
4 NSW Brain Clot Bank, NSW Health Pathology, Sydney, NSW 2170, Australia
5 Department of Neurology and Neurophysiology, Liverpool Hospital, South Western Sydney Local Health District and South West Sydney Clinical Campuses, UNSW Medicine, Liverpool, NSW 2170, Australia

Acute ischemic stroke (AIS) remains a formidable global health challenge, with approximately 11.9 million incident cases and 7 million deaths annually [1]. As the second leading cause of death worldwide and the third leading cause of death and disability combined, stroke imposes a staggering economic burden of over US$890 billion annually, projected to nearly double by 2050 [2]. This Special Issue, "*Treatment Strategy and Mechanism of Acute Ischemic Stroke*", presents a timely update addressing critical aspects of AIS management, from innovative diagnostics to novel therapeutic strategies and mechanistic insights. By integrating findings, this Special Issue aims to bridge the gap between research and clinical practice, ultimately improving outcomes for the millions affected by stroke globally. This collection of studies represents a significant step towards a more personalized, mechanism-driven, and equitable approach to stroke care.

Imaging plays a pivotal role in AIS diagnosis, treatment selection, and prognostication [3,4]. Lakatos et al. demonstrated that pretreatment cranial computed tomography perfusion (CTP) can predict dynamic cerebral autoregulation changes post-recanalization therapy with 83% accuracy. Their findings highlight the potential of CTP-derived metrics, such as penumbra and infarct core volumes, to guide blood pressure management and predict outcomes. Similarly, Huang et al. proposed redefining infarction size for small-vessel occlusion using advanced imaging, offering a more precise classification that distinguishes small-vessel disease from branch atheromatous disease (BAD) with a sensitivity of 92% and specificity of 88%.

Understanding the mechanisms underlying AIS is essential for developing targeted therapies [5]. Wiszniewska et al. explored the role of Factor XIII in thrombolysis-treated patients, revealing its potential as a biomarker for stroke severity and prognosis. Sablić et al. investigated the synergistic effects of communicating arteries and leptomeningeal collaterals on patient outcomes. This study reinforces the important role of collateral circulation in determining stroke severity and recovery potential [6,7]. The recently proposed *Evolucollateral Dynamics* hypothesis, by our group, suggests that the genetic and environmental factors influencing collateral circulation have evolutionary underpinnings, which may impact stroke outcomes and therapeutic strategies [7]. Integrating this framework into future research could enhance our understanding of collateral remodeling and optimize treatment approaches for AIS. These mechanistic studies provide valuable insights into the complex interplay of vascular, hemostatic, and neuroprotective factors in AIS.

 https://doi.org/10.3390/neurolint17040054

The development of novel therapeutic strategies is a cornerstone of AIS management. Drozdowska et al. evaluated the usability of a remote ischemic conditioning (RIC) device for pre-hospital stroke management, highlighting its potential to protect brain tissue during prolonged transport times. This study exemplifies the growing emphasis on early intervention and neuroprotection in AIS care. Additionally, Latacz et al. demonstrated the safety and efficacy of low-dose eptifibatide in patients with tandem occlusions, offering a promising pharmacological adjunct to endovascular therapy. Baki et al. highlight the life-saving role of decompressive surgery for malignant cerebellar infarction, reporting a 90-day mortality rate of 27.6% and good functional outcomes (mRS 0–3) in 41.4% of patients. Christidi et al. explore the potential of resting-state fMRI as a prognostic tool, demonstrating that increased amplitude of low-frequency fluctuation (ALFF) in bilateral M1 regions during subacute stroke is associated with better motor recovery. Another review by Crispino focuses on the dual complexities of hemorrhagic and ischemic events in patients with coagulation disorders, noting that direct oral anticoagulants can reduce major bleeding rates to as low as 2.13% annually compared to 3.09% with warfarin. Drakopanagiotakis et al. emphasize the critical need for respiratory rehabilitation, showing that respiratory muscle training can improve maximal inspiratory pressure (MIP) by 20–45 cm H_2O and forced vital capacity (FVC) by 200–400 mL in stroke survivors. These improvements are associated with reduced respiratory complications and better functional outcomes. Together, these studies underscore the importance of personalized, mechanism-driven, and multidisciplinary approaches to optimize stroke care across the continuum [8].

Beyond acute settings, post-stroke rehabilitation remains a critical component of stroke management [8]. Baroni et al. reviewed the effectiveness of paired associative stimulation (PAS) in motor recovery, emphasizing its potential to enhance neuroplasticity and functional outcomes. Similarly, Lee and colleagues explored the application of muscle synergies for gait rehabilitation, providing a framework for personalized rehabilitation strategies. These studies highlight the importance of integrating advanced neurorehabilitation techniques into the continuum of stroke care [9].

Addressing disparities in stroke care delivery is essential for achieving equitable outcomes [10,11]. The studies in this issue underscore the need for systemic reforms to ensure access to advanced imaging, mechanical thrombectomy, and rehabilitation services, particularly in underserved regions [11]. Collaborative efforts between researchers, clinicians, and policymakers are crucial for overcoming these challenges.

To bridge the gap between research and clinical practice, an evidence-based, integrated framework for AIS management is essential. This framework emphasizes personalized diagnostics, mechanism-driven therapies, and comprehensive rehabilitation. Advanced imaging techniques, such as resting-state fMRI, can identify neuroplasticity patterns and guide individualized rehabilitation strategies [12], while biomarkers can inform surgical decision-making, as indicated by studies on decompressive surgery outcomes [13]. Mechanistically, understanding the interplay between coagulation disorders and ischemia [14,15], as highlighted by Crispino, facilitates targeted therapies that balance anticoagulation risks in complex patients. Comprehensive rehabilitation, including RMT, has demonstrated improvements in pulmonary function and functional mobility, as shown by Drakopanagiotakis et al. Moreover, equity in access to advanced diagnostics and therapies remains critical, particularly for underserved populations [10]. By integrating these dimensions, this framework aims to optimize outcomes and reduce disparities in stroke care.

This Special Issue highlights the urgent need for a paradigm shift in stroke care. While challenges such as standardizing methodologies and conducting larger-scale studies remain, the potential impact of these advancements is profound. With stroke burden expected to rise, particularly in low- and middle-income countries where 87% of stroke deaths occur [1], implementing these evidence-based strategies is not just a public health or scientific imperative but a moral one. In conclusion, by fostering global collaboration, prioritizing patient-centered approaches, and addressing healthcare disparities [16], we can translate these research insights into tangible improvements in stroke outcomes worldwide. The time for action is now—building on past consensus efforts, including the European Stroke Action Plan (2018–2030), which sets clear targets to reduce stroke incidence, improve care, and implement evidence-based strategies, with millions of lives relying on our collective commitment [17].

Funding: This study was not supported by any funding. The financial support for a separate project was received through the Grant-in-Aid for Scientific Research (KAKENHI) (PI: S.M.M.B.), funded by the Japan Society for the Promotion of Science (JSPS), Japanese Ministry of Education, Culture, Sports, Science and Technology (MEXT), Japan (Grant ID: 23KF0126). S.B. was awarded the JSPS International Fellowship, supported by MEXT and the Australian Academy of Science, for the period 2023–2025 (Grant ID: P23712).

Conflicts of Interest: The author reports leadership or a fiduciary role in other board, society, committee, or advocacy group, paid or unpaid, with the National Cerebral and Cardiovascular Center (Osaka, Japan) as Visiting Director (2023–2025); *Rotary District 9675* (Sydney, Australia) as District Chair for Diversity, Equity, and Inclusion; the Global Health and Migration Hub Community, *Global Health Hub Germany* (Berlin, Germany) as Chair, Founding Member, and Manager; and editorial board memberships at *PLOS One, BMC Neurology, Frontiers in Neurology, Frontiers in Stroke, Frontiers in Public Health, Journal of Aging Research, Neurology International, Diagnostics, and BMC Medical Research Methodology*. Additionally, S.M.M.B. serves as a Member of the *College of Reviewers* for the Canadian Institutes of Health Research (CIHR), Government of Canada; Director of Research for the *World Headache Society* (Bengaluru, India); a member of the Scientific Review Committee at *Cardiff University Biobank* (Cardiff, UK); Chair of the *Rotary Reconciliation Action Plan* (RAP), Rotary District 9675 (NSW, Australia), Healthcare and Medical Adviser for the *Japan Connect* (Osaka, Japan), and as an Expert Adviser/Reviewer for the *Cariplo Foundation* (Milan, Italy). These roles are unrelated to the submitted work. The funding body has no role in the study design, data collection, analysis, interpretation of findings, and manuscript preparation. The content is solely the responsibility of the authors and does not necessarily represent the official views of the affiliated/funding organization/s.

List of Contributions:

1. Baki, E.; Baumgart, L.; Kehl, V.; Hess, F.; Wolff, A.W.; Wagner, A.; Hernandez Petzsche, M.R.; Boeckh-Behrens, T.; Hemmer, B.; Meyer, B.; et al. Functional Outcomes After Decompressive Surgery in Patients with Malignant Space-Occupying Cerebellar Infarction. *Neurol. Int.* **2024**, *16*, 1239–1246.
2. Baroni, A.; Antonioni, A.; Fregna, G.; Lamberti, N.; Manfredini, F.; Koch, G.; D'Ausilio, A.; Straudi, S. The Effectiveness of Paired Associative Stimulation on Motor Recovery after Stroke: A Scoping Review. *Neurol. Int.* **2024**, *16*, 567–589.
3. Christidi, F.; Orgianelis, I.; Merkouris, E.; Koutsokostas, C.; Tsiptsios, D.; Karavasilis, E.; Psatha, E.A.; Tsiakiri, A.; Serdari, A.; Aggelousis, N.; et al. A Comprehensive Review on the Role of Resting-State Functional Magnetic Resonance Imaging in Predicting Post-Stroke Motor and Sensory Outcomes. *Neurol. Int.* **2024**, *16*, 189–201.
4. Crispino, P. Hemorrhagic Coagulation Disorders and Ischemic Stroke: How to Reconcile Both? *Neurol. Int.* **2023**, *15*, 1443–1458.
5. Drakopanagiotakis, F.; Bonelis, K.; Steiropoulos, P.; Tsiptsios, D.; Sousanidou, A.; Christidi, F.; Gkantzios, A.; Serdari, A.; Voutidou, S.; Takou, C.-M.; et al. Pulmonary Function Tests Post-Stroke. Correlation between Lung Function, Severity of Stroke, and Improvement after Respiratory Muscle Training. *Neurol. Int.* **2024**, *16*, 139–161.
6. Drozdowska, B.A.; Lam, K.; Doolan, C.; Violato, E.; Ganesh, A. Evaluating the Usability of a Remote Ischemic Conditioning Device for Pre-Hospital Stroke Management: Insights from Paramedic Simulations. *Neurol. Int.* **2024**, *16*, 1405–1420.

7. Huang, Y.-C.; Weng, H.-H.; Lin, L.-C.; Lee, J.-D.; Yang, J.-T.; Tsai, Y.-H.; Chen, C.-H. Redefining Infarction Size for Small-Vessel Occlusion in Acute Ischemic Stroke: A Retrospective Case–Control Study. *Neurol. Int.* **2024**, *16*, 1164–1174.
8. Kinzhebay, A.; Salybekov, A.A. The Role of Somatic Mutations in Ischemic Stroke: CHIP's Impact on Vascular Health. *Neurol. Int.* **2025**, *17*, 19.
9. Lakatos, L.-B.; Bolognese, M.; Österreich, M.; Müller, M.; Karwacki, G.M. Pretreatment Cranial Computed Tomography Perfusion Predicts Dynamic Cerebral Autoregulation Changes in Acute Hemispheric Stroke Patients Having Undergone Recanalizing Therapy: A Retrospective Study. *Neurol. Int.* **2024**, *16*, 1636–1652.
10. Latacz, P.; Popiela, T.; Brzegowy, P.; Lasocha, B.; Kwiecień, K.; Simka, M. Safety and Efficacy of Low-Dose Eptifibatide for Tandem Occlusions in Acute Ischemic Stroke. *Neurol. Int.* **2024**, *16*, 253–262.
11. Lee, J.; Kim, K.; Cho, Y.; Kim, H. Application of Muscle Synergies for Gait Rehabilitation After Stroke: Implications for Future Research. *Neurol. Int.* **2024**, *16*, 1451–1463.
12. Nishimura, T.; Matsugaki, R.; Matsuda, S. Physical Rehabilitation and Post-Stroke Pneumonia: A Retrospective Observational Study Using the Japanese Diagnosis Procedure Combination Database. *Neurol. Int.* **2023**, *15*, 1459–1468.
13. Sablić, S.; Dolić, K.; Budimir Mršić, D.; Čičmir-Vestić, M.; Matana, A.; Lovrić Kojundžić, S.; Marinović Guić, M. Communicating Arteries and Leptomeningeal Collaterals: A Synergistic but Independent Effect on Patient Outcomes after Stroke. *Neurol. Int.* **2024**, *16*, 620–630.
14. Wiszniewska, M.; Włodarczyk, U.; Sury, M.; Słomka, A.; Piekuś-Słomka, N.; Żdanowicz, A.; Żekanowska, E. The Usefulness of Factor XIII Concentration Assessment in Patients in the Acute Phase of Ischaemic Stroke Treated with Thrombolysis. *Neurol. Int.* **2024**, *16*, 551–560.

References

1. GBD 2021 Stroke Risk Factor Collaborators. Global, regional, and national burden of stroke and its risk factors, 1990–2021: A systematic analysis for the Global Burden of Disease Study 2021. *Lancet Neurol.* **2024**, *23*, 973–1003. [CrossRef] [PubMed]
2. Feigin, V.L.; Brainin, M.; Norrving, B.; Martins, S.O.; Pandian, J.; Lindsay, P.; Grupper, M.F.; Rautalin, I. World Stroke Organization: Global Stroke Fact Sheet 2025. *Int. J. Stroke* **2025**, *20*, 132–144. [CrossRef] [PubMed]
3. Katyal, A.; Bhaskar, S.M.M. Value of pre-intervention CT perfusion imaging in acute ischemic stroke prognosis. *Diagn. Interv. Radiol.* **2021**, *27*, 774–785. [CrossRef] [PubMed]
4. Shaban, S.; Huasen, B.; Haridas, A.; Killingsworth, M.; Worthington, J.; Jabbour, P.; Bhaskar, S.M.M. Digital subtraction angiography in cerebrovascular disease: Current practice and perspectives on diagnosis, acute treatment and prognosis. *Acta Neurol. Belg.* **2022**, *122*, 763–780. [CrossRef] [PubMed]
5. Qin, C.; Yang, S.; Chu, Y.-H.; Zhang, H.; Pang, X.-W.; Chen, L.; Zhou, L.-Q.; Chen, M.; Tian, D.-S.; Wang, W. Signaling pathways involved in ischemic stroke: Molecular mechanisms and therapeutic interventions. *Signal Transduct. Target. Ther.* **2022**, *7*, 215. [CrossRef] [PubMed]
6. Ravindran, A.V.; Killingsworth, M.C.; Bhaskar, S. Cerebral collaterals in acute ischaemia: Implications for acute ischaemic stroke patients receiving reperfusion therapy. *Eur. J. Neurosci.* **2021**, *53*, 1238–1261. [CrossRef] [PubMed]
7. Sinha, A.; Gupta, M.; Bhaskar, S.M.M. Evolucollateral dynamics in stroke: Evolutionary pathophysiology, remodelling and emerging therapeutic strategies. *Eur. J. Neurosci.* **2024**, *60*, 6779–6798. [CrossRef] [PubMed]
8. Lip, G.Y.H.; Lane, D.A.; Lenarczyk, R.; Boriani, G.; Doehner, W.; Benjamin, L.A.; Fisher, M.; Lowe, D.; Sacco, R.L.; Schnabel, R.; et al. Integrated care for optimizing the management of stroke and associated heart disease: A position paper of the European Society of Cardiology Council on Stroke. *Eur. Heart J.* **2022**, *43*, 2442–2460. [CrossRef] [PubMed]
9. Li, X.; He, Y.; Wang, D.; Rezaei, M.J. Stroke rehabilitation: From diagnosis to therapy. *Front. Neurol.* **2024**, *15*, 1402729. [CrossRef] [PubMed]
10. Biswas, R.; Wijeratne, T.; Zelenak, K.; Huasen, B.B.; Iacobucci, M.; Killingsworth, M.C.; Beran, R.G.; Gebreyohanns, M.; Sekhar, A.; Khurana, D.; et al. Disparities in Access to Reperfusion Therapy for Acute Ischemic Stroke (DARTS): A Comprehensive Meta-Analysis of Ethnicity, Socioeconomic Status, and Geographical Factors. *CNS Drugs* **2025**, *39*, 417–442. [CrossRef] [PubMed]
11. Steinmetz, J.D.; Seeher, K.M.; Schiess, N.; Nichols, E.; Cao, B.; Servili, C.; Cavallera, V.; Cousin, E.; Hagins, H.; Moberg, M.E.; et al. Global, regional, and national burden of disorders affecting the nervous system, 1990–2021: A systematic analysis for the Global Burden of Disease Study 2021. *Lancet Neurol.* **2024**, *23*, 344–381. [CrossRef]
12. Reid, L.B.; Boyd, R.N.; Cunnington, R.; Rose, S.E. Interpreting Intervention Induced Neuroplasticity with fMRI: The Case for Multimodal Imaging Strategies. *Neural Plast.* **2016**, *2016*, 2643491. [CrossRef] [PubMed]

13. Guimarães de Almeida Barros, A.; Roquim, E.S.L.; Pessoa, A.; Eiras Falcão, A.; Viana Magno, L.A.; Valadão Freitas Rosa, D.; Aurelio Romano Silva, M.; Marques de Miranda, D.; Nicolato, R. Use of biomarkers for predicting a malignant course in acute ischemic stroke: An observational case-control study. *Sci. Rep.* **2023**, *13*, 16097. [CrossRef] [PubMed]
14. Sun, M.Y.; Bhaskar, S.M.M. When Two Maladies Meet: Disease Burden and Pathophysiology of Stroke in Cancer. *Int. J. Mol. Sci.* **2022**, *23*, 15769. [CrossRef]
15. Sun, M.Y.; Bhaskar, S.M.M. Bridging the Gap in Cancer-Related Stroke Management: Update on Therapeutic and Preventive Approaches. *Int. J. Mol. Sci.* **2023**, *24*, 7981. [CrossRef] [PubMed]
16. Rajeswaran, P.; Huasen, B.B.; Stanwell, P.; Killingsworth, M.C.; Bhaskar, S.M.M. Prevalence and outcomes of mild stroke patients undergoing reperfusion therapy: A meta-analysis and SAFE recommendations for optimal management. *J. Cent. Nerv. Syst. Dis.* **2025**, *17*, 11795735251314881. [CrossRef] [PubMed]
17. Norrving, B.; Barrick, J.; Davalos, A.; Dichgans, M.; Cordonnier, C.; Guekht, A.; Kutluk, K.; Mikulik, R.; Wardlaw, J.; Richard, E.; et al. Action Plan for Stroke in Europe 2018–2030. *Eur. Stroke J.* **2018**, *3*, 309–336. [CrossRef] [PubMed]

neurology international

MDPI

Review

Hemorrhagic Coagulation Disorders and Ischemic Stroke: How to Reconcile Both?

Pietro Crispino

Medicine Unit, Santa Maria Goretti Hospital, Via Scaravelli Snc, 04100 Latina, Italy; picrispino@libero.it

Abstract: Coagulation and fibrinolytic system disorders are conditions in which the blood's ability to clot is impaired, resulting in an increased risk of thrombosis or bleeding. Although these disorders are the expression of two opposing tendencies, they can often be associated with or be a consequence of each other, contributing to making the prognosis of acute cerebrovascular events more difficult. It is important to recognize those conditions that are characterized by dual alterations in the coagulation and fibrinolytic systems to reduce the prognostic impact of clinical conditions with difficult treatment and often unfortunate outcomes. Management of these individuals can be challenging, as clinicians must balance the need to prevent bleeding episodes with the potential risk of clot formation. Treatment decisions should be made on an individual basis, considering the specific bleeding disorder, its severity, and the patient's general medical condition. This review aims to deal with all those forms in which coagulation and fibrinolysis represent two sides of the same media in the correct management of patients with acute neurological syndrome. Precision medicine, personalized treatment, advanced anticoagulant strategies, and innovations in bleeding control represent future directions in the management of these complex pathologies in which stroke can be the evolution of two different acute events or be the first manifestation of an occult or unknown underlying pathology.

Keywords: stroke; emocoagulation; fibrinolysis

1. Background

In the last decade, many steps forward have been made in the search for new anticoagulant drugs capable of acting on the pro-coagulative activity of some pathological conditions such as atrial fibrillation or congenital anomalies of the blood coagulation system (factor V deficiency Leiden, protein C, and protein S deficiency, hyperhomocysteinemia) [1,2]. New drugs have therefore been introduced, including direct anticoagulants, which are now key therapies in numerous pathological conditions and are added to antifibrinolytics, platelet inhibitors, and vitamin-K-dependent anticoagulants [1–4]. Their therapeutic benefits are indisputable; however, all evidence for these treatments must be balanced against the risk of bleeding complications, which are relatively frequent for some of these drugs. Warfarin, the oral anticoagulant that until a few years ago was the most commonly used, substantially increases the risk of major bleeding complications, including hemorrhagic stroke and gastrointestinal bleeding, by up to 3–8% per year [5,6]. Similarly, major bleeding is a serious and frequent complication of all inhibitors of platelet activity, including adenosine diphosphate 3–5 receptor antagonists such as clopidogrel and prasugrel [7,8]. We have therefore focused our attention more on the risk of ischemic stroke in hemorrhagic patients as it always represents a critical issue in the management of patients, especially the most fragile patients. This represents a condition in which it is always difficult to make a choice, with a consequent increase in mortality rates and health care costs in terms of worsening frailty if the patient were to survive. This review analyzes all the aspects linked to these antithetical conditions in order to better think about the appropriate therapies to apply to these patients as categories at greater risk of complications.

2. Stratification of Bleeding Risk in Patients Treated for Vascular Events Prevention

The occurrence of bleeding, major or minor, attributable to the use of direct oral anticoagulants (DOACs) varies based on the introductory trials and, for this reason, the need to stratify patients based on the possible risk associated with this therapy has become widespread. In the RE-LY study [9], which compared dabigatran with warfarin for stroke prevention in atrial fibrillation, the annual rate of major bleeding with dabigatran 150 mg twice daily was approximately 3.11%, while with dabigatran 110 mg twice daily it was approximately 2.71%. Ultimately, the RE-LY study concluded that dabigatran administered at a dose of 110 mg was associated with an incidence of stroke and systemic embolism similar to that found in patients in the warfarin group, compared to a lower incidence of major bleeding. At the 150 mg dosage, it was associated with a lower incidence of stroke and systemic embolism with the same incidence of major bleeding. In the ROCKET AF trial [10], which compared rivaroxaban with warfarin for atrial fibrillation, the annual rate of major bleeding with rivaroxaban was approximately 3.6%, whereas with warfarin it was approximately 3.4%. In the ARISTOTLE study [11], which compared apixaban with warfarin for the prevention of stroke in atrial fibrillation, the annual rate of major bleeding with apixaban was approximately 2.13%, while with warfarin it was approximately 3.09%. In the ENGAGE AF-TIMI 48 study [12], which compared edoxaban with warfarin for the prevention of stroke in atrial fibrillation, the annual rate of major bleeding with edoxaban was approximately 2.75%, whereas with warfarin it was approximately 3.43%. Clinical studies that establish their effectiveness provide valuable information; although these drugs have declared low rates of major bleeding, it has been repeatedly highlighted that the patients tested often belong to a class of individuals at a lower risk of adverse effects compared to the real ones in practice [13]. This does not offer the opportunity to verify the real weight of drug-related adverse events, especially in more complex patients where drug–drug or drug–disease interactions must always be considered [14,15]. With the introduction of some scores such as, for example, the HAS-Bleed in the context of atrial fibrillation, an attempt has been made to classify patients at high or low risk of major bleeding. However, in the case of multipathological and complex patients, these systems often fail to prevent complications [16–20]. A further problem is the misrecognition of conditions with a high risk of bleeding in all patients who should undergo invasive procedures, whose theoretical calculated risk of bleeding may be lower than when corresponding to reality [21]. These are often fragile patients with a precarious state of health who often, as already mentioned, escape the introductory trials of various drugs [22,23]. Therefore, in patients treated with NOACs, a greater stratification of the fragile patient at risk of bleeding necessarily requires knowledge of all the possible clinical conditions at risk of bleeding even if this is not always feasible [24].

3. Hemorrhagic Patients and Risk of Ischemic Stroke

As we have already reported, the greater the degree of frailty of the patient, the greater the risk of adverse events in patients at high cardiovascular risk or already undergoing secondary prevention for atherosclerotic or cardioembolic ischemic stroke or with both associated conditions (Figure 1) [21–24]. The treatment of choice to prevent another ischemic stroke is a long-term oral anticoagulant or antiplatelet strategy. However, many patients cannot take such drugs due to long-term contraindications, comorbidities, cognitive impairment, severe walking difficulties with frequent falls, and patient refusal [25,26]. The risk of death or addiction, as well as the risk of stroke recurrence in this group, is also not well known but must certainly be considered [25,26]. The presence of ongoing bleeding causes a series of problems for patients who, at risk of thromboembolic events, take antiplatelet or anticoagulant therapy [27,28]. However, there is evidence to support the hypothesis that the suspension of these drugs in patients at high cardiovascular and cerebral risk can cause both early and medium-term acute ischemic events related to the hemorrhagic event [29–31]. Even if hemorrhagic events and thromboembolic events are antithetical to each other, they show some shared risk factors. Among these, we consider in particular increasing age, frailty, comorbidity, atherosclerosis, genetic predisposition to blood coagulation disorders, and platelet pathology [30,31]. Doctors are therefore faced with a therapeutic dilemma in the management of patients with concomitant hemorrhagic and ischemic pathologies, especially because, at present, there are no systematic reviews and randomized or observational studies that address this problem [32,33]. Most studies conducted in the past did not mention individual modifiable and non-modifiable risk factors, nor the degree of health complexity and fragility of each patient [34–36]. It is estimated that the recurrence of a stroke during the secondary prevention period is approximately 20% [37]. Among the first causes, there is certainly the impossibility of carrying out the best treatment or the non-optimal coverage of all the risk factors that can contribute to relapse [38,39]. Patients undergoing secondary prevention after an acute cerebral circulation disorder and with high adherence to therapy show a lower incidence of new events [39,40]. Some studies have hypothesized that low adherence to the optimal therapeutic scheme for the prevention of a further cerebrovascular event is linked to the fear of prescribing drugs with important adverse events, uncertainty about long-term clinical benefits and risks, lack of knowledge and experience in managing such therapies, or competing medical issues such as the cost of medications [41]. Others hypothesize that effective information is lacking in how to limit low drug use in populations with different levels of education or cultural traditions adverse to the use of such important drugs [26,42,43]. However, it is the complexity of the patients that represents the fundamental cause of the unfavorable outcome of the disease [44]. We do not yet have adequate knowledge of this complexity in the field of cerebrovascular diseases. We know that there is already a pre-stroke complexity that affects approximately 25% of the affected population, to which is then added the fragility that follows one or more acute cerebrovascular events [44]. It must be said that stroke is one of the most common acute manifestations observed in the elderly and is related to the majority of systemic cardiac, pulmonary, renal, and degenerative pathologies whose synergistic effect is not yet well known [45]. This is why, despite having cutting-edge therapies for the prevention and treatment of stroke, we have a persistently high mortality and disability rate [45,46].

Stroke		
Ischemic	**Hemorrhagic**	**Strategy to balance complications**
• Atherotrombotic • Cardioembolic • Cerebral venous stroke • Miscellaneous	*Intracerebral Hemorrhage* • Hypertension • Arteriovenous Malformation • Head Injury *Subarachnoid Hemorrhage* • Rupture of cerebral aneurism • Arteriovenous malformation • Head injury	• Conduct a thorough medical history and risk factor assessment to identify and manage conditions that may increase the risk of ischemic or hemorrhagic stroke. Address modifiable risk factors such as hypertension, diabetes, and hyperlipidemia. • Research and possible correction of all vascular alterations or anomalies that can be complicated by an ischemic or hemorrhagic stroke • Antiplatelet and antithrombotic agents are commonly used in the prevention of ischemic stroke. However, in hemorrhagic stroke, antiplatelet therapy may increase the risk of bleeding. Adjust the use of these medications based on the patient's risk profile.

Figure 1. Risk factors for ischemic and hemorrhagic strokes.

4. Clinical Outcome of Frail Patients

We know that, together with atherosclerosis, atrial fibrillation (AF) is the most important cause of stroke. All three of these conditions become increasingly common in old age and where frailty is greater, characterized by a decline in biological reserves and the deterioration of physiological mechanisms, making the elderly vulnerable to a series of adverse outcome factors [47–50]. There are little data available on the safety of treatments used in the frail population useful for the primary or secondary prevention of acute cerebrovascular events. In frail patients, a greater number of adverse events related to anticoagulant therapy have been observed, such as bleeding and a greater difficulty in choosing the best therapy for these patients [51]. This leads us to think that the presence of adverse events and the high mortality of these patients is linked precisely to the condition of fragility [52,53]. According to Wilkinson et al. [51], it is precisely bleeding linked to anticoagulant therapy that can be subjected to review in light of the concept of frailty, representing an objective of improving safety and mortality outcomes for this condition. Modifiable hemorrhagic risk factors should be sought and subjected to an optimization process, also in light of the polytherapies often present in these patients [54,55]. Indeed, a dose adjustment of anticoagulant therapy was able to satisfactorily prevent stroke, without a significantly higher incidence of major bleeding compared to placebo [56]. In the following parts, we will discuss what the main factors are that can influence the occurrence of adverse events during chronic treatment for the prevention of stroke or a subsequent event.

4.1. Cerebrovascular Anatomical Condition and Hemorrhage and Ischemic Stroke Risk

Alterations in the blood–brain barrier (BBB) underlie the manifestations that can result from an intracranial hemorrhage or an ischemic stroke associated with several modifiable cardiovascular risk factors (Figure 2) [57]. In pathological conditions, the loss of the homeostatic microenvironment leads to the impairment of normal neuronal function and the alteration in the permeability of the vascular endothelium [58,59]. During and after ischemic stroke, this phenomenon increases the risk of hemorrhage by creating gaps through the tight junctions of the endothelium [60–63] (Figure 3). Diabetes and cigarette smoking have been identified as among the risk factors that mostly cause this catastrophic sequence of events, and are associated with the presence of coagulopathy during chronic liver disease and anticoagulant therapy [64,65]. Obviously, patients undergoing anticoagulant therapy are strongly exposed to the risk of intracerebral hemorrhage due to the increased risk of bleeding associated with the use of these drugs [66,67]. Chronic liver disease with associated coagulation disorders is also associated with cerebral hemorrhage but the definitive role of this condition in the hemorrhagic transformation of ischemic stroke has not yet been fully clarified [68]. The latter study in fact associated liver cirrhosis with a high risk of ischemic stroke, venous thromboembolism, and major hemorrhagic complications only in patients suffering from chronic atrial fibrillation [68]. Another study [69] considered that left ventricular dysfunction is also an independent risk factor in the atrial fibrillation of the evolution of cerebral microvascular damage toward lacunar infarction cerebral hemorrhage or a combined form of both conditions. It is very interesting to underline that one study [70] considered how arterial hypertension contributes more to ischemic stroke in patients up to 65 years of age while, in older patients, it contributes more to intracranial hemorrhage.

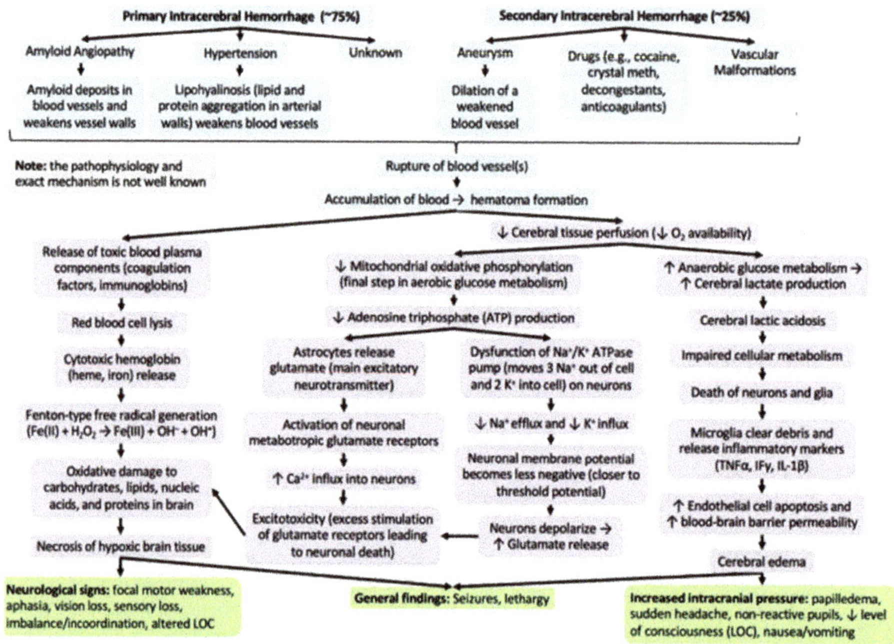

Figure 2. Difference in hemorrhagic stroke.

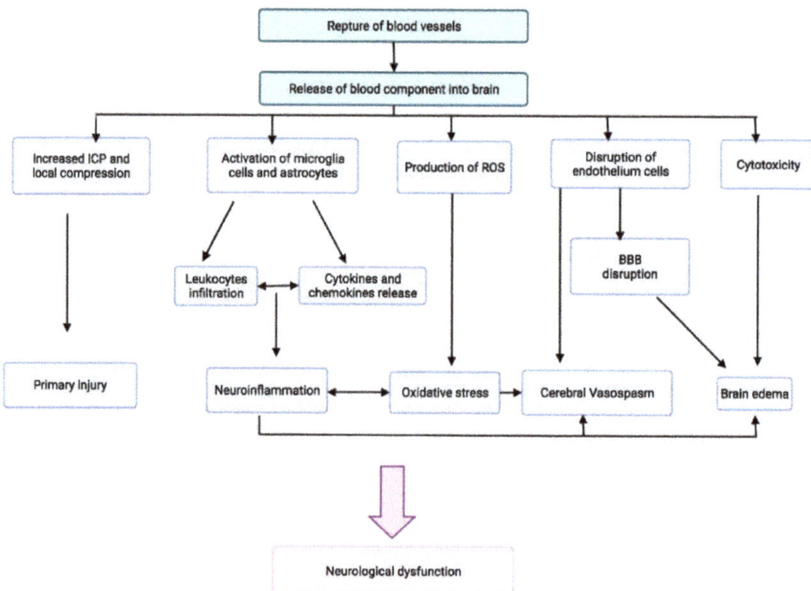

Figure 3. Pathogenesis of rupture of blood–brain barrier *(Adapted)*.

4.2. Size and Severity of the Hemorrhage and Ischemic Stroke Risk

The size and severity of the hemorrhage can influence the risk of complications such as ischemic stroke. Larger hemorrhages involving critical areas of the brain may increase the risk of ischemia [71]. Reperfusion therapies can lead to hemorrhagic transformation in between 3.2 and 43.3% of cases, with a more significant impact on prognosis [72,73]. Etiologically, hemorrhagic transformation is a multifactorial phenomenon, and previous studies, in addition to atrial fibrillation, advanced age, and delay in thrombolytic treatment, have highlighted how the severity of the symptoms and the extension of the ischemic territory are equally important risk factors [74–76]. In particular, it should be underlined that the importance of the vessel affected by the occlusion is also crucial in determining the risk of post-procedure hemorrhage [72,77]. Large vessel occlusion worsens stroke severity and is associated with a worse NIHSS score and an involvement of a larger area of brain tissue [78,79]. In the case of a cardioembolic stroke, the numerousness of the clots was predictive of the overlap of a hemorrhagic complication considering the possibility that, in this case, there could be a greater extension and a multiplicity of the infarcted brain areas [72,80]. Iancu et al. [81] looked for risk factors and hemorrhagic transformation in patients with an initial diagnosis of ischemic stroke. Among these, the presence of male sex, high glycemic values of arterial hypertension at baseline, and high values of NIHSS and ASPECTS scores were all associated with a significantly elevated risk of hemorrhagic transformation [81]. In this study as in others, the assessment of stroke severity and early detection of tomographic characteristics of the infarcted ischemic lesion were the main predictive factors of hemorrhagic transformation [82,83]. In particular, it has been seen that a score above 20 is considered to be at high risk of hemorrhagic transformation regardless of whether the cerebral arterial occlusion has undergone revascularization or not. Another study, however, does not recommend mechanical thrombectomy in the presence of an NIHSS score ≥20 due to the high possibility of hemorrhagic events [84]. Another interesting study found a correlation between an increased NIHSS score and the concomitant presence of atrial fibrillation and tumors [85]. These data regarding the concomitant presence of atrial fibrillation have not been confirmed in other studies [81], while, with regard to the diagnosis of a concomitant neoplasm, it is difficult to find case series that evaluate patients also for the purposes of researching a neoplasm occult. To conclude, a review demonstrated that the occurrence of hemorrhage in the context of an ischemic area depends on many factors, including the location and size of the infarct and poor collateral vascularization, confirming what has been said previously [86]. Therefore, by reflecting on these factors, in the future, one will be able to better stratify the risks before treating a patient with a stroke to avoid worsening the prognosis.

4.3. *Individual Risk Factors in Patients with Hemorrhagic Transformation*

The presence of independent risk factors such as arterial hypertension and diabetes mellitus strongly contributes to the worsening of the prognosis in these patients. The main studies [87,88] that have correlated diabetes mellitus with a greater severity of the stroke and therefore a greater risk of hemorrhagic transformation have highlighted that glycemic decompensation has a negative effect on the BBB and is associated with negative prognostic outcomes. The impact of glycemic imbalance on the metabolism of neurons affected by the stroke is biphasic: first, it is a protective factor; then, it represents the optimal condition for the formation of oxygen free radicals that alter the balance of the barrier, leading to a reduced reperfusion of the microvascular circulation, edema of endothelial and cerebral cells, and therefore hemorrhagic transformation, complicating the clinical picture [89–91]. The same destabilizing effect on the BBB has been observed for blood pressure as the latter is also capable of causing its permeability [92–96]. In patients with a long history of arterial hypertension, the small vessels of the cerebral circulation exhibit stable endothelial damage that further aggravates the inflammatory damage caused by circulating oxygen free radicals [92–96]. It must also be said that the ischemic stroke itself causes an increase in blood pressure values that essentially derive from the type of stroke, the state of recanalization, and the reorganization of the flows through the collateral vessels [97]. In the case of a hemorrhagic stroke, the reduction in blood pressure values can help to limit the extent of the blood share but does not influence functional outcomes [98]. It is therefore possible that blood pressure is not sufficiently controlled alone, but may influence the outcome of treatment in synergy with other measures [95]. AF is associated with an increased risk of intracranial hemorrhage, although this is largely related to concomitant anticoagulant therapy. In a limited number of cases, it has been shown that patients with atrial fibrillation can develop extracranial hemorrhage (ECH) [99]. However, it is unclear whether patients with AF may have a better or worse prognosis than those without arrhythmia [100]. Patients with AF certainly benefit from thrombolytic treatment within the established timeframes, with some associated bleeding risk. Furthermore, it is reiterated that the treatment of all predisposing conditions is truly associated with the favorable outcome of treatments in ischemic stroke. One study [100] also considers that, generally, AF patients suffering from ischemic stroke are significantly older and have a higher NIHSS score than the non-AF group. Therefore, it is necessary to consider that patients with AF are at greater risk of adverse events and poor prognosis, regardless of concomitant anticoagulant therapy [100]. Atherosclerosis is to be considered as a condition that can alternatively cause ischemic or hemorrhage, so its progression is strongly associated with possible dual manifestations of thrombosis and hemorrhage [101]. Intracranial hemorrhage is the only pathological form that unites some rather rare conditions, including cerebral amyloid angiopathy, reversible cerebral vasoconstriction syndrome, and reversible posterior leukoencephalopathy syndrome [102]. The coexistence of intracranial hemorrhage is often described as a complication of acute ischemic stroke without this association being fully understood [103–105]. It is known that, in these cases, half of the patients had severe stenosis or occlusion of the large arteries [103–107]. All these conditions are secondary to degeneration at the level of the stenosis, which causes rupture of the fragile dilated collateral vessels, accompanied by emboli that enter the marginally perfused vessels, leading to necrosis and rupture [108,109].

4.4. Pathological Condition and Risk of Ischemic or Hemorrhagic Stroke

Several studies gathered in a meta-analysis [110] have reported that cirrhosis, especially cirrhosis in the decompensation phase, can increase the risk of stroke. In general, an increase in total strokes was reported, which were largely hemorrhagic strokes. It is known that liver cirrhosis, especially decompensated cirrhosis aggravated by liver failure, can cause defects in the blood coagulation structure, exposing patients to the risk of bleeding, largely linked to the lengthening of clotting times and the sequestration of platelets in the hepatosplenic circulation, but also thrombotic complications due to the defect in the production of proteins with a natural anticoagulant function (protein C, protein S, and antithrombin III) [111]. A meta-analysis, subsequent to that cited previously, aimed to establish the potential association between liver cirrhosis and the various forms of acute cerebrovascular accidents such as hemorrhagic stroke, ischemic stroke, intracranial hemorrhage, and subarachnoid hemorrhage. The study highlighted how, in cirrhotic patients, there was a significant increase in intracranial and subarachnoid hemorrhage compared to both forms of stroke [111]. In this regard, the authors concluded that it was necessary on a case-by-case basis to follow the blood coagulation status over time and subject the patient to suitable prophylaxis [111].

Patients with hematological diseases often experience cerebrovascular complications including ischemic stroke, intracerebral and subarachnoid hemorrhage, and microbleeds [112]. Hemorrhagic stroke is a very rare occurrence in chronic myeloproliferative diseases but when it occurs it can be fatal [113,114]. Particularly in forms such as polycythemia vera and essential thrombocytosis, ischemic stroke is frequent. In these cases, cytoreductive therapy and antiaggregants may be taken into consideration in high-risk patients [115,116]. The leukemic forms, especially the acute ones, can cause ischemic or hemorrhagic stroke, sometimes presenting themselves as the onset of the hematological pathology [117]. In these cases, stroke is linked to the state of hypercoagulability, disseminated intravascular coagulopathy (DIC), leukocyte stasis linked to the proliferation of white blood cells, and forms of endocarditis with platelet vegetations associated with paradoxical embolism [117]. The AF and other traditional vascular risk factors, infections, and antitumor therapies (radiotherapy, chemotherapy, immunotherapy, and transplantation) can be additional risk factors causing stroke. Intracranial bleeding is a serious, potentially fatal cerebrovascular complication in patients with acute leukemia [112]. Bleeding can occur in one or more locations, mainly linked to thrombocytopenia. The presence of multiple hemorrhagic foci of hemorrhage, thrombocytopenia, leukocytosis, and the early presence of a comatose state has been considered as the most unfavorable prognostic factors for survival [118]. The leukemic form with the highest hemorrhagic risk is acute promyelocytic leukemia, which is often associated with disseminated vascular coagulation and fibrinolysis (with alternating sectoral forms of a thrombotic or hemorrhagic type at the same time) [119]. The same mechanisms listed above are common to all other forms of stroke that also occur in lymphomas. In these latter forms, the presence of vasculitis, thrombotic microangiopathy, and blood stasis induced by compression and ab extrinsic of the lymphatic packets can be further factors causing hypercoagulability and therefore stroke [120]. An increase in ischemic or hemorrhagic risk is also reported in patients with multiple myeloma linked in particular to hypercoagulability, hyperviscosity, thrombotic microangiopathy, and embolism [121–123]. Furthermore, many drugs used for these pathological forms can have thrombogenic side effects [124]. Thrombotic and thrombocytopenic purpura can cause ischemic or hemorrhagic strokes even at a young age, usually with multifocal onset with involvement of the large arteries [125]. Approximately half of the patients presented neurological forms associated with hemolysis and thrombocytopenia [125].

4.5. Therapies and Treatments Influences in Patients with Hemorrhagic Transformation

The bleeding risk is known as the main problem of all anticoagulant therapies, despite their proven effectiveness [99]. The introduction of direct oral anticoagulants (DOACs) into clinical practice means that major hemorrhages, especially intracranial, associated with therapy are certainly reduced; however, the risk of bleeding, even with these drugs, is not eliminated. Hence, there is a need to develop new drugs that are ideally capable of overcoming the current limitations of therapies in use (Figure 4) [99]. In particular, it would seem that the risk increases in conjunction with acute renal failure as all direct and indirect anticoagulant drugs have a predominantly renal excretion mechanism and are therefore subject to hyperaccumulation with the possibility of an increase in major and minor hemorrhages [126]. It has been shown that the frequency of cerebral hemorrhage is significantly increased in patients with chronic renal failure, with a significant increase in deaths during dialysis [127]. Although the data available to us showed that DOACs reduced the risk of major bleeding compared to VKAs in patients with chronic renal failure, especially in patients with atrial fibrillation, the risk of major hemorrhagic events is concrete [128,129]. The heterogeneity of the various studies makes it difficult to understand whether in the presence of renal failure the prevention of stroke has, in terms of safety results, entailed the same risk of bleeding compared to not carrying out anticoagulant therapy. An increase in intracranial hemorrhage with the use of anticoagulants was a notable finding. For patients with atrial fibrillation and end-stage renal disease, there is a clinical equilibrium in which the risk of bleeding with any anticoagulant may outweigh any potential benefit [130,131].

Patients at risk of ischemic stroke related to cardioembolic situations may be at risk of stroke despite the use of correct anticoagulant therapy [132]. In most cases, this phenomenon, which involves both large and small arteries, is linked to the atherosclerotic process and to the various nuances that characterize cryptogenic forms [133]. However, if anticoagulation is correctly performed, the optimization of secondary prevention therapy may require integration with antiplatelet therapy to make the treatment more effective in preventing major adverse cardiovascular events [134,135]. However, on the contrary, other data suggest that the addition of antiplatelets to anticoagulants provides no benefit in patients with atrial fibrillation and stroke and increases the long-term risks associated with combined anticoagulant and antiplatelet treatment in patients with atrial fibrillation [136].

Timeline for starting NAO in patients with major bleeding risk and ischemic stroke	Recommendation of timeline for NAO reintroduction in patients with ischemic stroke and risk bleeding
In patients at high risk of ischaemic stroke, (re-)initiation of OAC, with preference for NOACs over VKAs in NOAC-eligible patients, should be considered in consultation with a neurologist/stroke specialist after: • A trauma-related ICH • Acute spontaneous ICH (which includes subdural, subarachnoid, or intracerebral hemorrhage), after careful consideration of risks and benefits	In AF patients with an ischaemic stroke or TIA, long-term secondary prevention of stroke using OAC is recommended if there is no strict contraindication to OAC use, with a preference for NOACs over VKAs in NOAC-eligible patients.
The optimal time of anticoagulation after ICH is unknown, but should be delayed beyond the acute phase, probably at least 4 weeks.	Patients presenting with acute ischaemic stroke, very early anticoagulation (<48 hours) is not recommended.
	In patients with stroke at high risk of hemorrhagic conversion it is reasonable to delay the initiation of oral anticoagulation beyond 14 days to reduce the risk of ICH (AHA/ASA)

Figure 4. The optimal strategy to reintroduce NAO in therapy according to standard guidelines 2020 ESC Guidelines for the diagnosis and management of atrial fibrillation (European Heart Journal 2020-doi/10.1093/eurheartj/ehaa612).

5. Ischemic and Hemorrhagic Stroke in COVID-19 Patients

During the peak of the COVID-19 pandemic, an increase in acute neurological disorders of cerebral circulation was observed, especially in patients hospitalized in intensive care units [136–138]. These events were mainly of a thrombotic nature but a non-negligible number of cerebral hemorrhagic manifestations was also highlighted [139,140]. The establishment of a state of hypercoagulability during COVID-19 infection seems to play a decisive role in these manifestations so, in these patients, it was carried out early with pharmacological prophylaxis of venous thromboembolism (VTE) and directed against the excessive tendency to induce platelet aggregation [141–145]. Intracranial hemorrhage (ICH) is less frequent than ischemic strokes in COVID-19 patients. However, ICH is associated with a high mortality rate [146]. At the beginning of the pandemic, Sharifi-Razavi et al. [147] wondered whether ICH was a direct effect of COVID-19 infection or the consequence of prophylactic anticoagulant treatment, and other studies have shown that these events are associated with a loss of the integrity of the BBB and with an increase in inflammatory cytokines (IL-1, IL-6, TNF-α) [148,149]. Finally, as previously mentioned, a state of hypercoagulation can also produce thrombotic microangiopathy of small vessels, loss of endothelial integrity, and subsequent rupture [150].

6. Conclusions

When treating patients with neurological manifestations related to an ischemic or hemorrhagic stroke, any underlying conditions or possible consequences related to the modulation of the hemocoagulatory and fibrinolytic system must always be taken into consideration. Patients with AF and a higher NIHSS score, diabetics, and elderly patients should be considered in the highest risk category for the probable evolution of ischemic stroke to hemorrhagic stroke. Hemorrhagic transformation in the different contexts in which it can occur is one of the complications that most influence the good outcome in terms of functional prognosis and in terms of survival. It is mainly due to an alteration in the BBB with a consequent alteration in endothelial permeability and inflammatory damage. The same etiopathogenesis is also common to those cases related to COVID-19 infection, where the inflammatory cascade linked to the circulating virus plays an essential role in modifying the endothelial balance. Future studies should be aimed at improving knowledge on the management of anticoagulant therapy in fragile patients, perhaps with the introduction of more selective anticoagulant drugs with less hemorrhagic impact than those still available. Another direction of study is to acquire further knowledge on the correct timing of the suspension and reactivation of anticoagulant therapy in high-risk patients. In this field, there are still gray areas, especially when it comes to recovery after a major hemorrhagic event and when it comes to any factors that can favor the appearance of complications.

Funding: This research received no external funding.

Institutional Review Board Statement: Not applicable.

Informed Consent Statement: Not applicable.

Conflicts of Interest: The author declares no conflict of interest.

References

1. Poddar, R. Hyperhomocysteinemia is an emerging comorbidity in ischemic stroke. *Exp. Neurol.* **2020**, *336*, 113541. [CrossRef] [PubMed]
2. Martinelli, I.; Passamonti, S.M.; Bucciarelli, P. Thrombophilic states. *Handb. Clin. Neurol.* **2014**, *120*, 1061–1071. [CrossRef] [PubMed]
3. Ciarambino, T.; Crispino, P.; Minervini, G.; Giordano, M. Cerebral Sinus Vein Thrombosis and Gender: A Not Entirely Casual Relationship. *Biomedicines* **2023**, *11*, 1280. [CrossRef] [PubMed]
4. Ciarambino, T.; Crispino, P.; Mastrolorenzo, E.; Viceconti, A.; Giordano, M. Stroke and Etiopathogenesis: What Is Known? *Genes* **2022**, *13*, 978. [CrossRef] [PubMed]
5. Beyth, R.J.; Landefeld, C.S. Anticoagulants in older patients. A safety perspective. *Drugs Aging* **1995**, *6*, 45–54. [CrossRef]
6. Cunningham, A.; Stein, C.M.; Chung, C.P.; Daugherty, J.R.; Smalley, W.E.; Ray, W.A. An automated database case definition for serious bleeding related to oral anticoagulant use. *Pharmacoepidemiol. Drug Saf.* **2011**, *20*, 560–566. [CrossRef] [PubMed]
7. Yildirim, M.; Mueller-Hennessen, M.; Milles, B.R.; Biener, M.; Hund, H.; Frey, N.; Giannitsis, E.; Salbach, C. Real-World Evidence on Disparities on the Initiation of Ticagrelor Versus Prasugrel in Patients with Acute Coronary Syndrome. *J. Am. Heart Assoc.* **2023**, *12*, e030879. [CrossRef] [PubMed]
8. Wiviott, S.D.; Braunwald, E.; McCabe, C.H.; Montalescot, G.; Ruzyllo, W.; Gottlieb, S.; Neumann, F.-J.; Ardissino, D.; De Servi, S.; Murphy, S.A.; et al. Prasugrel versus Clopidogrel in Patients with Acute Coronary Syndromes. *N. Engl. J. Med.* **2007**, *357*, 2001–2015. [CrossRef]
9. Reilly, P.A.; Lehr, T.; Haertter, S.; Connolly, S.J.; Yusuf, S.; Eikelboom, J.W.; Ezekowitz, M.D.; Nehmiz, G.; Wang, S.; Wallentin, L.; et al. The effect of dabigatran plasma concentrations and patient characteristics on the frequency of ischemic stroke and major bleeding in atrial fibrillation patients: The RE-LY Trial (Randomized Evaluation of Long-Term Anticoagulation Therapy). *J. Am. Coll. Cardiol.* **2014**, *63*, 321–328. [CrossRef]
10. Bansilal, S.; Bloomgarden, Z.; Halperin, J.L.; Hellkamp, A.S.; Lokhnygina, Y.; Patel, M.R.; Becker, R.C.; Breithardt, G.; Hacke, W.; Hankey, G.J.; et al. Efficacy and safety of rivaroxaban in patients with diabetes and nonvalvular atrial fibrillation: The Rivaroxaban Once-daily, Oral, Direct Factor Xa Inhibition Compared with Vitamin K Antagonism for Prevention of Stroke and Embolism Trial in Atrial Fibrillation (ROCKET AF Trial). *Am. Heart J.* **2015**, *170*, 675–682.e8. [CrossRef]
11. Guimarães, P.O.; Pokorney, S.D.; Lopes, R.D.; Wojdyla, D.M.; Gersh, B.J.; Giczewska, A.; Carnicelli, A.; Lewis, B.S.; Hanna, M.; Wallentin, L.; et al. Efficacy and safety of apixaban vs warfarin in patients with atrial fibrillation and prior bioprosthetic valve replacement or valve repair: Insights from the ARISTOTLE trial. *Clin. Cardiol.* **2019**, *42*, 568–571. [CrossRef] [PubMed]
12. Kato, E.T.; Giugliano, R.P.; Ruff, C.T.; Koretsune, Y.; Yamashita, T.; Kiss, R.G.; Nordio, F.; Murphy, S.A.; Kimura, T.; Jin, J.; et al. Efficacy and Safety of Edoxaban in Elderly Patients with Atrial Fibrillation in the ENGAGE AF–TIMI 48 Trial. *J. Am. Heart Assoc.* **2016**, *5*, e003432. [CrossRef] [PubMed]
13. Herrera, A.P.; Snipes, S.A.; King, D.W.; Torres-Vigil, I.; Goldberg, D.S.; Weinberg, A.D. Disparate Inclusion of Older Adults in Clinical Trials: Priorities and Opportunities for Policy and Practice Change. *Am. J. Public Health* **2010**, *100* (Suppl. 1), S105–S112. [CrossRef] [PubMed]
14. Ray, W.A.; Griffin, M.R.; Avorn, J. Evaluating Drugs after Their Approval for Clinical Use. *N. Engl. J. Med.* **1993**, *329*, 2029–2032. [CrossRef] [PubMed]
15. Cho, H.J.; Chae, J.; Yoon, S.H.; Kim, D.S. Factors related to polypharmacy and hyper-polypharmacy for the elderly: A nationwide cohort study using National Health Insurance data in South Korea. *Clin. Transl. Sci.* **2022**, *16*, 193–205. [CrossRef] [PubMed]
16. Holbrook, A.M.; Pereira, J.A.; Labiris, R.; McDonald, H.; Douketis, J.D.; Crowther, M.; Wells, P.S. Systematic Overview of Warfarin and Its Drug and Food Interactions. *Arch. Intern. Med.* **2005**, *165*, 1095–1106. [CrossRef] [PubMed]
17. Zhou, Z.; Yu, J.; Carcel, C.; Delcourt, C.; Shan, J.; Lindley, R.I.; Neal, B.; Anderson, C.S.; Hackett, M.L. Resuming anticoagulants after anticoagulation-associated intracranial haemorrhage: Systematic review and meta-analysis. *BMJ Open* **2018**, *8*, e019672. [CrossRef] [PubMed]
18. Ballestri, S.; Romagnoli, E.; Arioli, D.; Coluccio, V.; Marrazzo, A.; Athanasiou, A.; Di Girolamo, M.; Cappi, C.; Marietta, M.; Capitelli, M. Risk and Management of Bleeding Complications with Direct Oral Anticoagulants in Patients with Atrial Fibrillation and Venous Thromboembolism: A Narrative Review. *Adv. Ther.* **2023**, *40*, 41–66. [CrossRef]

19. Shorr, R.I.; Ray, W.A.; Daugherty, J.R.; Griffin, M.R. Concurrent Use of Nonsteroidal Anti-inflammatory Drugs and Oral Anticoagulants Places Elderly Persons at High Risk for Hemorrhagic Peptic Ulcer Disease. *Arch. Intern. Med.* **1993**, *153*, 1665–1670. [CrossRef]
20. Penner, L.S.; Gavan, S.P.; Ashcroft, D.M.; Peek, N.; Elliott, R.A. Does coprescribing nonsteroidal anti-inflammatory drugs and oral anticoagulants increase the risk of major bleeding, stroke and systemic embolism? *Br. J. Clin. Pharmacol.* **2022**, *88*, 4789–4811. [CrossRef]
21. Proietti, M.; Cesari, M. Describing the relationship between atrial fibrillation and frailty: Clinical implications and open research questions. *Exp. Gerontol.* **2021**, *152*, 111455. [CrossRef]
22. Proietti, M.; Romiti, G.F.; Vitolo, M.; Harrison, S.L.; Lane, D.A.; Fauchier, L.; Marin, F.; Näbauer, M.; Potpara, T.S.; Dan, G.-A.; et al. Epidemiology and impact of frailty in patients with atrial fibrillation in Europe. *Age Ageing* **2022**, *51*, afac192. [CrossRef] [PubMed]
23. Borre, E.D.; Goode, A.; Raitz, G.; Shah, B.; Lowenstern, A.; Chatterjee, R.; Sharan, L.; Allen LaPointe, N.M.; Yapa, R.; Davis, J.K.; et al. Predicting Thromboembolic and Bleeding Event Risk in Patients with Non-Valvular Atrial Fibrillation: A Systematic Review. *Thromb. Haemost.* **2018**, *118*, 2171–2187. [CrossRef] [PubMed]
24. Guidoux, C.; Meseguer, E.; Ong, E.; Lavallée, P.C.; Hobeanu, C.; Monteiro-Tavares, L.; Charles, H.; Cabrejo, L.; Martin-Bechet, A.; Rigual, R.; et al. Twelve-month outcome in patients with stroke and atrial fibrillation not suitable to oral anticoagulant strategy: The WATCH-AF registry. *Open Heart* **2019**, *6*, e001187. [CrossRef] [PubMed]
25. Yang, M.; Cheng, H.; Wang, X.; Ouyang, M.; Shajahan, S.; Carcel, C.; Anderson, C.; Kristoffersen, E.S.; Lin, Y.; Sandset, E.C.; et al. Antithrombotics prescription and adherence among stroke survivors: A systematic review and meta-analysis. *Brain Behav.* **2022**, *12*, e2752. [CrossRef] [PubMed]
26. Lowres, N.; Giskes, K.; Hespe, C.; Freedman, B. Reducing Stroke Risk in Atrial Fibrillation: Adherence to Guidelines Has Improved, but Patient Persistence with Anticoagulant Therapy Remains Suboptimal. *Korean Circ. J.* **2019**, *49*, 883–907. [CrossRef] [PubMed]
27. Wani, M.; Nga, E.; Navaratnasingham, R. Should a patient with primary intracerebral haemorrhage receive antiplatelet or anticoagulant therapy? *BMJ* **2005**, *331*, 439–442. [CrossRef] [PubMed]
28. Flynn, R.; Doney, A. Antithrombotic medicines following intracerebral haemorrhage: Where's the evidence? *Ther. Adv. Drug Saf.* **2011**, *2*, 205–211. [CrossRef]
29. Vermeer, S.E.; Algra, A.; Franke, C.L.; Koudstaal, P.J.; Rinkel, G.J. Long-term prognosis after recovery from primary intracerebral hemorrhage. *Neurology* **2002**, *59*, 205–209. [CrossRef]
30. Li, L.; Murthy, S.B. Cardiovascular Events After Intracerebral Hemorrhage. *Stroke* **2022**, *53*, 2131–2141. [CrossRef]
31. Wong, Y.-S.; Tsai, C.-F.; Ong, C.-T. Risk factors for stroke recurrence in patients with hemorrhagic stroke. *Sci. Rep.* **2022**, *12*, 17151. [CrossRef]
32. Flynn, R.W.; MacDonald, T.M.; Murray, G.D.; Doney, A.S. Systematic Review of Observational Research Studying the Long-Term use of Antithrombotic Medicines Following Intracerebral Hemorrhage. *Cardiovasc. Ther.* **2010**, *28*, 177–184. [CrossRef]
33. Wardlaw, J.M.; Keir, S.L.; Seymour, J.; Lewis, S.; Sandercock, P.A.G.; Dennis, M.S.; Cairns, J. What is the best imaging strategy for acute stroke? *Int. J. Technol. Assess. Health Care* **2005**, *21*, 148. [CrossRef]
34. Bailey, R.D.; Hart, R.G.; Benavente, O.; Pearce, L.A. Recurrent brain hemorrhage is more frequent than ischemic stroke after intracranial hemorrhage. *Neurology* **2001**, *56*, 773–777. [CrossRef] [PubMed]
35. Li, L.; Luengo-Fernandez, R.; Zuurbier, S.M.; Beddows, N.C.; Lavallee, P.; Silver, L.E.; Kuker, W.; Rothwell, P.M. Ten-year risks of recurrent stroke, disability, dementia and cost in relation to site of primary intracerebral haemorrhage: Population-based study. *J. Neurol. Neurosurg. Psychiatry* **2020**, *91*, 580–585. [CrossRef] [PubMed]
36. Viswanathan, A.; Rakich, S.M.; Engel, C.; Snider, R.; Rosand, J.; Greenberg, S.M.; Smith, E.E. Antiplatelet use after intracerebral hemorrhage. *Neurology* **2006**, *66*, 206–209. [CrossRef] [PubMed]
37. Benjamin, E.J.; Virani, S.S.; Callaway, C.W.; Chamberlain, A.M.; Chang, A.R.; Cheng, S.; Chiuve, S.E.; Cushman, M.; Delling, F.N.; Deo, R.; et al. American Heart Association Council on Epidemiology and Prevention Statistics Committee and Stroke Statistics Subcommittee. Heart Disease and Stroke Statistics-2018 Update: A Report from the American Heart Association. *Circulation* **2018**, *137*, e67–e492. [CrossRef] [PubMed]
38. Broderick, J.P.; Bonomo, J.B.; Kissela, B.M.; Khoury, J.C.; Moomaw, C.J.; Alwell, K.; Woo, D.; Flaherty, M.L.; Khatri, P.; Adeoye, O.; et al. Withdrawal of Antithrombotic Agents and Its Impact on Ischemic Stroke Occurrence. *Stroke* **2011**, *42*, 2509–2514. [CrossRef] [PubMed]
39. Kim, J.; Lee, H.S.; Nam, C.M.O.; Heo, J.H. Effects of Statin Intensity and Adherence on the Long-Term Prognosis after Acute Ischemic Stroke. *Stroke* **2017**, *48*, 2723–2730. [CrossRef] [PubMed]
40. Rijkmans, M.; de Jong, G.; Van Den Berg, J.S.P. Non-persistence in ischaemic stroke: Risk of recurrent vascular events. *Acta Neurol. Scand.* **2018**, *137*, 288–292. [CrossRef]
41. Kirley, K.; Rao, G.; Bauer, V.; Masi, C. The Role of NOACs in Atrial Fibrillation Management: A Qualitative Study. *J. Atr. Fibrillation* **2016**, *9*, 1416. [CrossRef] [PubMed]
42. Hsu, J.C.; Maddox, T.M.; Kennedy, K.F.; Katz, D.F.; Marzec, L.N.; Lubitz, S.A.; Gehi, A.K.; Turakhia, M.P.; Marcus, G.M. Oral Anticoagulant Therapy Prescription in Patients with Atrial Fibrillation Across the Spectrum of Stroke Risk. *JAMA Cardiol.* **2016**, *1*, 55–62. [CrossRef] [PubMed]

43. Guo, J.S.; He, M.; Gabriel, N.; Magnani, J.W.; Kimmel, S.E.; Gellad, W.F.; Hernandez, I. Underprescribing vs underfilling to oral anticoagulation: An analysis of linked medical record and claims data for a nationwide sample of patients with atrial fibrillation. *J. Manag. Care Spéc. Pharm.* **2022**, *28*, 1400–1409. [CrossRef] [PubMed]
44. Burton, J.K.; Stewart, J.; Blair, M.; Oxley, S.; Wass, A.; Taylor-Rowan, M.; Quinn, T.J. Prevalence and implications of frailty in acute stroke: Systematic review & meta-analysis. *Age Ageing* **2022**, *51*, afac064. [CrossRef] [PubMed]
45. Conroy, S.; Carpenter, C.; Banerjee, J. Silver Book II: Quality Care for Older People with Urgent Care Needs. 2021. Available online: https://www.bgs.org.uk/resources/resource-series/silver-book-ii (accessed on 1 November 2023).
46. Quinn, T.J.; Taylor-Rowan, M.; Coyte, A.; Clark, A.B.; Musgrave, S.D.; Metcalf, A.K.; Day, D.J.; Bachmann, M.O.; Warburton, E.A.; Potter, J.F.; et al. Pre-Stroke Modified Rankin Scale: Evaluation of Validity, Prognostic Accuracy, and Association with Treatment. *Front. Neurol.* **2017**, *8*, 275. [CrossRef] [PubMed]
47. Rockwood, K.; Howlett, S.E. Fifteen years of progress in understanding frailty and health in aging. *BMC Med.* **2018**, *16*, 220. [CrossRef] [PubMed]
48. Clegg, A.; Young, J.; Iliffe, S.; Rikkert, M.O.; Rockwood, K. Frailty in elderly people. *Lancet* **2013**, *381*, 752–762. [CrossRef]
49. Wilkinson, C.; Clegg, A.; Todd, O.; Rockwood, K.; Yadegarfar, M.E.; Gale, C.P.; Hall, M. Atrial fibrillation and oral anticoagulation in older people with frailty: A nationwide primary care electronic health records cohort study. *Age Ageing* **2021**, *50*, 772–779. [CrossRef]
50. Wilkinson, C.; Todd, O.; Clegg, A.; Gale, C.P.; Hall, M. Management of atrial fibrillation for older people with frailty: A systematic review and meta-analysis. *Age Ageing* **2019**, *48*, 196–203. [CrossRef]
51. Wilkinson, C.; Wu, J.; Searle, S.D.; Todd, O.; Hall, M.; Kunadian, V.; Clegg, A.; Rockwood, K.; Gale, C.P. Clinical outcomes in patients with atrial fibrillation and frailty: Insights from the ENGAGE AF-TIMI 48 trial. *BMC Med.* **2020**, *18*, 401. [CrossRef]
52. Farooqi, M.A.M.; Gerstein, H.; Yusuf, S.; Leong, D.P. Accumulation of Deficits as a Key Risk Factor for Cardiovascular Morbidity and Mortality: A Pooled Analysis of 154,000 Individuals. *J. Am. Heart Assoc.* **2020**, *9*, e014686. [CrossRef] [PubMed]
53. Stow, D.; Matthews, F.E.; Hanratty, B. Frailty trajectories to identify end of life: A longitudinal population-based study. *BMC Med.* **2018**, *16*, 171. [CrossRef] [PubMed]
54. Hindricks, G.; Potpara, T.; Dagres, N.; Arbelo, E.; Bax, J.J.; Blomström-Lundqvist, C.; Boriani, G.; Castella, M.; Dan, J.-A.; Dilaveris, P.E.; et al. The Task Force for the diagnosis and management of atrial fibrillation of the European Society of Cardiology (ESC) Developed with the special contribution of the European Heart Rhythm Association (EHRA) of the ESC. *Eur. Heart J.* **2020**, *42*, 373–498. [CrossRef] [PubMed]
55. Rodríguez, L.A.G.; Martín-Pérez, M.; Vora, P.; Roberts, L.; Balabanova, Y.; Brobert, G.; Fatoba, S.; Suzart-Woischnik, K.; Schaefer, B.; Ruigomez, A. Appropriateness of initial dose of non-vitamin K antagonist oral anticoagulants in patients with non-valvular atrial fibrillation in the UK. *BMJ Open* **2019**, *9*, e031341. [CrossRef] [PubMed]
56. Okumura, K.; Akao, M.; Yoshida, T.; Kawata, M.; Okazaki, O.; Akashi, S.; Eshima, K.; Tanizawa, K.; Fukuzawa, M.; Hayashi, T.; et al. Low-Dose Edoxaban in Very Elderly Patients with Atrial Fibrillation. *N. Engl. J. Med.* **2020**, *383*, 1735–1745. [CrossRef]
57. Jiang, X.; Andjelkovic, A.V.; Zhu, L.; Yang, T.; Bennett, M.V.L.; Chen, J.; Keep, R.F.; Shi, Y. Blood-brain barrier dysfunction and recovery after ischemic stroke. *Prog. Neurobiol.* **2018**, *163–164*, 144–171. [CrossRef] [PubMed]
58. Mozaffarian, D.; Benjamin, E.J.; Go, A.S.; Arnett, D.K.; Blaha, M.J.; Cushman, M.; Das, S.R.; de Ferranti, S.; Despres, J.-P.; Fullerton, H.J.; et al. Stroke Statistics Subcommittee (2016). Heart Disease and Stroke Statistics-2016 Update: A Report from the American Heart Association. *Circulation* **2016**, *133*, e38–e60. [CrossRef]
59. Liu, H.; Wang, Y.; Xiao, Y.; Hua, Z.; Cheng, J.; Jia, J. Hydrogen Sulfide Attenuates Tissue Plasminogen Activator-Induced Cerebral Hemorrhage Following Experimental Stroke. *Transl. Stroke Res.* **2016**, *7*, 209–219. [CrossRef]
60. Jin, X.; Liu, J.; Liu, W. Early Ischemic Blood Brain Barrier Damage: A Potential Indicator for Hemorrhagic Transformation Following Tissue Plasminogen Activator (tPA) Thrombolysis? *Curr. Neurovasc. Res.* **2014**, *11*, 254–262. [CrossRef]
61. Jickling, G.C.; Liu, D.; Stamova, B.; Ander, B.P.; Zhan, X.; Lu, A.; Sharp, F.R. Hemorrhagic Transformation after Ischemic Stroke in Animals and Humans. *J. Cereb. Blood Flow Metab. Off. J. Int. Soc. Cereb. Blood Flow Metab.* **2013**, *34*, 185–199. [CrossRef]
62. Wang, W.; Li, M.; Chen, Q.; Wang, J. Hemorrhagic Transformation after Tissue Plasminogen Activator Reperfusion Therapy for Ischemic Stroke: Mechanisms, Models, and Biomarkers. *Mol. Neurobiol.* **2015**, *52*, 1572–1579. [CrossRef] [PubMed]
63. Zhang, Y.; Wang, Y.; Zuo, Z.; Wang, Z.; Roy, J.; Hou, Q.; Tong, E.; Hoffmann, A.; Sperberg, E.; Bredno, J.; et al. Effects of tissue plasminogen activator timing on blood–brain barrier permeability and hemorrhagic transformation in rats with transient ischemic stroke. *J. Neurol. Sci.* **2014**, *347*, 148–154. [CrossRef] [PubMed]
64. Lioutas, V.-A.; Beiser, A.; Himali, J.; Aparicio, H.; Romero, J.R.; DeCarli, C.; Seshadri, S. Lacunar Infarcts and Intracerebral Hemorrhage Differences: A Nested Case-Control Analysis in the FHS (Framingham Heart Study). *Stroke* **2017**, *48*, 486–489. [CrossRef] [PubMed]
65. Lioutas, V.-A.; Beiser, A.S.; Aparicio, H.J.; Himali, J.J.; Selim, M.H.; Romero, J.R.; Seshadri, S. Assessment of Incidence and Risk Factors of Intracerebral Hemorrhage Among Participants in the Framingham Heart Study Between 1948 and 2016. *JAMA Neurol.* **2020**, *77*, 1252–1260. [CrossRef] [PubMed]
66. Da Silva, I.R.F.; Provencio, J.J. Intracerebral Hemorrhage in Patients Receiving Oral Anticoagulation Therapy. *J. Intensiv. Care Med.* **2015**, *30*, 63–78. [CrossRef] [PubMed]

67. Murphy, M.P.; Kuramatsu, J.B.; Leasure, A.; Falcone, G.J.; Kamel, H.; Sansing, L.H.; Kourkoulis, C.; Schwab, K.; Elm, J.J.; Gurol, M.E.; et al. Cardioembolic Stroke Risk and Recovery After Anticoagulation-Related Intracerebral Hemorrhage. *Stroke* **2018**, *49*, 2652–2658. [CrossRef] [PubMed]
68. Riahi, E.B.; Adelborg, K.; Pedersen, L.; Kristensen, S.R.; Hansen, A.T.; Sørensen, H.T. Atrial fibrillation, liver cirrhosis, thrombosis, and bleeding: A Danish population-based cohort study. *Res. Pract. Thromb. Haemost.* **2022**, *6*, e12668. [CrossRef]
69. Muscari, A.; Masetti, G.; Faccioli, L.; Ghinelli, M.; Trossello, M.P.; Puddu, G.M.; Spinardi, L.; Zoli, M. Association of Left Ventricular Hypertrophy and Atrial Fibrillation with Hemorrhagic Evolution of Small Vessel Disease. *J. Stroke Cerebrovasc. Dis. Off. J. Natl. Stroke Assoc.* **2021**, *30*, 105946. [CrossRef]
70. Stanton, R.; Demel, S.L.; Flaherty, M.L.; Antzoulatos, E.; Gilkerson, L.A.; Osborne, J.; Behymer, T.P.; Moomaw, C.J.; Sekar, P.; Langefeld, C.; et al. Risk of intracerebral haemorrhage from hypertension is greatest at an early age. *Eur. Stroke J.* **2021**, *6*, 28–35. [CrossRef]
71. Neuberger, U.; Kickingereder, P.; Schönenberger, S.; Schieber, S.; Ringleb, P.A.; Bendszus, M.; Pfaff, J.; Möhlenbruch, M.A. Risk factors of intracranial hemorrhage after mechanical thrombectomy of anterior circulation ischemic stroke. *Neuroradiology* **2019**, *61*, 461–469. [CrossRef]
72. Sun, J.; Lam, C.; Christie, L.; Blair, C.; Li, X.; Werdiger, F.; Yang, Q.; Bivard, A.; Lin, L.; Parsons, M. Risk factors of hemorrhagic transformation in acute ischaemic stroke: A systematic review and meta-analysis. *Front. Neurol.* **2023**, *14*, 1079205. [CrossRef] [PubMed]
73. Jaillard, A.; Cornu, C.; Durieux, A.; Moulin, T.; Boutitie, F.; Lees, K.R.; Hommel, M. Hemorrhagic transformation in acute ischemic stroke. The MAST-E study. MAST-E Group. *Stroke* **1999**, *30*, 1326–1332. [CrossRef] [PubMed]
74. Hassan, A.E.; Kotta, H.; Shariff, U.; Preston, L.; Tekle, W.; Qureshi, A. There Is No Association Between the Number of Stent Retriever Passes and the Incidence of Hemorrhagic Transformation for Patients Undergoing Mechanical Thrombectomy. *Front. Neurol.* **2019**, *10*, 818. [CrossRef] [PubMed]
75. Yu, X.; Pan, J.; Zhao, X.; Hou, Q.; Liu, B. Predicting hemorrhagic transformation after thrombectomy in acute ischemic stroke: A multimodal score of the regional pial collateral. *Neuroradiology* **2021**, *64*, 493–502. [CrossRef] [PubMed]
76. Zhang, X.; Xie, Y.; Wang, H.; Yang, D.; Jiang, T.; Yuan, K.; Gong, P.; Xu, P.; Li, Y.; Chen, J.; et al. Symptomatic intracranial hemorrhage after mechanical thrombectomy in Chinese ischemic stroke patients: The Asian Score. *Stroke* **2020**, *51*, 2690–2696. [CrossRef] [PubMed]
77. Mokin, M.; Kass-Hout, T.; Kass-Hout, O.; Dumont, T.M.; Kan, P.; Snyder, K.V.; Hopkins, L.N.; Siddiqui, A.H.; Levy, E.I. Intravenous thrombolysis and endovascular therapy for acute ischemic stroke with internal carotid artery occlusion: A systematic review of clinical outcomes. *Stroke* **2012**, *43*, 2362–2368. [CrossRef] [PubMed]
78. Monteiro, A.; Khan, S.; Waqas, M.; Dossani, R.H.; Ruggiero, N.; Siddiqi, N.M.; Baig, A.A.; Rai, H.H.; Cappuzzo, J.M.; Levy, E.I.; et al. Mechanical thrombectomy versus intravenous alteplase alone in acute isolated posterior cerebral artery occlusion: A systematic review. *J. NeuroInterv. Surg.* **2022**, *14*, 564–567. [CrossRef] [PubMed]
79. Waqas, M.; Kuo, C.C.; Dossani, R.H.; Monteiro, A.; Baig, A.A.; Alkhaldi, M.; Cappuzzo, J.M.; Levy, E.I.; Siddiqui, A.H. Mechanical thrombectomy versus intravenous thrombolysis for distal large-vessel occlusion: A systematic review and meta-analysis of observational studies. *Neurosurg. Focus* **2021**, *51*, E5. [CrossRef]
80. Molina, C.A.; Montaner, J.; Abilleira, S.; Ibarra, B.; Romero, F.; Arenillas, J.F.; Alvarez-Sabín, J. Timing of Spontaneous Recanalization and Risk of Hemorrhagic Transformation in Acute Cardioembolic Stroke. *Stroke* **2001**, *32*, 1079–1084. [CrossRef]
81. Iancu, A.; Buleu, F.; Chita, D.S.; Tutelca, A.; Tudor, R.; Brad, S. Early Hemorrhagic Transformation after Reperfusion Therapy in Patients with Acute Ischemic Stroke: Analysis of Risk Factors and Predictors. *Brain Sci.* **2023**, *13*, 840. [CrossRef]
82. NINDS t-PA Stroke Study Group. Intracerebral hemorrhage after intravenous t-PA therapy for ischemic stroke. *Stroke* **1997**, *28*, 2109–2118. [CrossRef] [PubMed]
83. Kidwell, C.S.; Saver, J.L.; Carneado, J.; Sayre, J.; Starkman, S.; Duckwiler, G.; Gobin, Y.P.; Jahan, R.; Vespa, P.; Villablanca, J.P.; et al. Predictors of hemorrhagic trans-formation in patients receiving intra-arterial thrombolysis. *Stroke* **2002**, *33*, 717–724. [CrossRef] [PubMed]
84. Luo, A.; Bandela, S.; Mozumder, A.; Lee, B. Predictors of Hemorrhagic Transformation After Mechanical Thrombectomy in Acute Ischemic Stroke (1644). *Neurology* **2021**, *96*, 1644.
85. Shen, Z.; Jin, H.; Lu, Y.; Sun, W.; Liu, R.; Li, F.; Shu, J.; Tai, L.; Li, G.; Chen, H.; et al. Predictors and Prognosis of Symptomatic Intracranial Hemorrhage in Acute Ischemic Stroke Patients Without Thrombolysis: Analysis of Data from the Chinese Acute Ischemic Stroke Treatment Outcome Registry. *Front. Neurol.* **2021**, *12*, 727304. [CrossRef] [PubMed]
86. Hong, J.M.; Kim, D.S.; Kim, M. Hemorrhagic Transformation After Ischemic Stroke: Mechanisms and Management. *Front. Neurol.* **2021**, *12*, 703258. [CrossRef] [PubMed]
87. McCann, S.K.; Lawrence, C.B. Comorbidity and age in the modeling of stroke: Are we still failing to consider the characteristics of stroke patients? *BMJ Open Sci.* **2020**, *4*, e100013. [CrossRef] [PubMed]
88. Demchuk, A.M.; Morgenstern, L.B.; Krieger, D.W.; Linda Chi, T.; Hu, W.; Wein, T.H.; Hardy, R.J.; Grotta, J.C.; Buchan, A.M. Serum Glucose Level and Diabetes Predict Tissue Plasminogen Activator–Related Intracerebral Hemorrhage in Acute Ischemic Stroke. *Stroke* **1999**, *30*, 34–39. [CrossRef] [PubMed]
89. Jurcau, A.; Ardelean, A.I. Oxidative Stress in Ischemia/Reperfusion Injuries following Acute Ischemic Stroke. *Biomedicines* **2022**, *10*, 574. [CrossRef]

90. Janaszak-Jasiecka, A.; Płoska, A.; Wierońska, J.M.; Dobrucki, L.W.; Kalinowski, L. Endothelial dysfunction due to eNOS uncoupling: Molecular mechanisms as potential therapeutic targets. *Cell. Mol. Biol. Lett.* **2023**, *28*, 21. [CrossRef]
91. Martini, S.R.; Kent, T.A. Hyperglycemia in Acute Ischemic Stroke: A Vascular Perspective. *J. Cereb. Blood Flow Metab.* **2007**, *27*, 435–451. [CrossRef]
92. Yeo, L.L.L.; Paliwal, P.; Teoh, H.L.; Seet, R.C.; Chan, B.P.L.; Liang, S.; Venketasubramanian, N.; Rathakrishnan, R.; Ahmad, A.; Ng, K.W.P.; et al. Timing of Recanalization After Intravenous Thrombolysis and Functional Outcomes After Acute Ischemic Stroke. *JAMA Neurol.* **2013**, *70*, 353–358. [CrossRef] [PubMed]
93. Tsivgoulis, G.; Saqqur, M.; Sharma, V.K.; Lao, A.Y.; Hill, M.D.; Alexandrov, A.V.; CLOTBUST Investigators. Asso-ciation of pretreatment blood pressure with tissue plasminogen activator-induced arterial recanalization in acute ischemic stroke. *Stroke* **2007**, *38*, 961–966. [CrossRef]
94. Van den Berg, S.A.; Uniken Venema, S.M.; LeCouffe, N.E.; Postma, A.A.; Lycklama, À.; Nijeholt, G.J.; Rinkel, L.A.; Treurniet, K.M.; Kappelhof, M.; Bruggeman, A.E.; et al. Admission blood pressure and clinical outcomes in patients with acute ischaemic stroke treated with intravenous alteplase and endovascular treatment versus endovascular treatment alone: A MR CLEAN-NO IV substudy. *Eur. Stroke J.* **2023**, *8*, 647–654. [CrossRef] [PubMed]
95. Sandset, E.C.; Anderson, C.S.; Bath, P.M.; Christensen, H.; Fischer, U.; Gąsecki, D.; Lal, A.; Manning, L.S.; Sacco, S.; Steiner, T.; et al. European Stroke Organisation (ESO) guidelines on blood pressure management in acute ischaemic stroke and intracerebral haemorrhage. *Eur. Stroke J.* **2021**, *6*, XLVIII–LXXXIX. [CrossRef] [PubMed]
96. Berg, S.A.v.D.; Venema, S.M.U.; Mulder, M.J.; Treurniet, K.M.; Samuels, N.; Lingsma, H.F.; Goldhoorn, R.-J.B.; Jansen, I.G.; Coutinho, J.M.; Roozenbeek, B.; et al. Admission Blood Pressure in Relation to Clinical Outcomes and Successful Reperfusion After Endovascular Stroke Treatment. *Stroke* **2020**, *51*, 3205–3214. [CrossRef] [PubMed]
97. Mistry, E.A.; Mehta, T.; Mistry, A.; Arora, N.; Starosciak, A.K.; Rosa, F.D.L.R.L.; Siegler, J.E.; Chitale, R.; Anadani, M.; Yaghi, S.; et al. Blood Pressure Variability and Neurologic Outcome After Endovascular Thrombectomy: A Secondary Analysis of the BEST Study. *Stroke* **2020**, *51*, 511–518. [CrossRef] [PubMed]
98. Krishnan, K.; Scutt, P.; Woodhouse, L.; Adami, A.; Becker, J.L.; Cala, L.A.; Casado, A.M.; Chen, C.; Dineen, R.A.; Gommans, J.; et al. Continuing versus Stopping Prestroke Antihypertensive Therapy in Acute Intracerebral Hemorrhage: A Subgroup Analysis of the Efficacy of Nitric Oxide in Stroke Trial. *J. Stroke Cerebrovasc. Dis. Off. J. Natl. Stroke Assoc.* **2016**, *25*, 1017–1026. [CrossRef]
99. Zhou, E.; Lord, A.; Boehme, A.; Henninger, N.; de Havenon, A.; Vahidy, F.; Ishida, K.; Torres, J.; Mistry, E.A.; Mac Grory, B.; et al. Risk of Ischemic Stroke in Patients with Atrial Fibrillation After Extracranial Hemorrhage. *Stroke* **2020**, *51*, 3592–3599. [CrossRef]
100. Dang, H.; Ge, W.-Q.; Zhou, C.-F.; Zhou, C.-Y. The Correlation between Atrial Fibrillation and Prognosis and Hemorrhagic Transformation. *Eur. Neurol.* **2019**, *82*, 9–14. [CrossRef]
101. Chai, J.; Nie, H.; Wu, X.; Guan, Y.; Dai, T.; Shen, Y. The clinical and neuroradiological features of patients of coexisting atraumatic convexity subarachnoid hemorrhage and large artery atherosclerosis stroke: A retrospective observational study. *Medicine* **2021**, *100*, e28155. [CrossRef]
102. Kumar, S.; Goddeau, R.P.; Selim, M.H.; Thomas, A.; Schlaug, G.; Alhazzani, A.; Searls, D.E.; Caplan, L.R. Atraumatic convexal subarachnoid hemorrhage: Clinical presentation, imaging patterns, and etiologies. *Neurology* **2010**, *74*, 893–899. [CrossRef]
103. Usmani, N.; Ahmad, F.; Koch, S. Convexity subarachnoid hemorrhage in ischemic stroke. *J. Neurol. Sci.* **2015**, *348*, 259–261. [CrossRef] [PubMed]
104. Nakajima, M.; Inatomi, Y.; Yonehara, T.; Hirano, T.; Ando, Y. Nontraumatic convexal subarachnoid hemorrhage concomitant with acute ischemic stroke. *J. Stroke Cerebrovasc. Dis. Off. J. Natl. Stroke Assoc.* **2014**, *23*, 1564–1570. [CrossRef] [PubMed]
105. Safan, A.S.; Imam, Y.; Akhtar, N.; Al-Taweel, H.; Zakaria, A.; Quateen, A.; Own, A.; Kamran, S. Acute ischemic stroke and convexity subarachnoid hemorrhage in large vessel atherosclerotic stenosis: Case series and review of the literature. *Clin. Case Rep.* **2022**, *10*, e5968. [CrossRef] [PubMed]
106. Takamiya, S.; Yoshimoto, T.; Maruichi, K. Subarachnoid Hemorrhage with Progressive Cerebral Steno-Occlusive Disease: Report of 2 Cases. *J. Stroke Cerebrovasc. Dis. Off. J. Natl. Stroke Assoc.* **2019**, *28*, e14–e16. [CrossRef] [PubMed]
107. Graff-Radford, J.; Fugate, J.E.; Klaas, J.; Flemming, K.D.; Brown, R.D.; Rabinstein, A.A. Distinguishing clinical and radiological features of non-traumatic convexal subarachnoid hemorrhage. *Eur. J. Neurol.* **2016**, *23*, 839–846. [CrossRef] [PubMed]
108. Introna, A.; Mezzapesa, D.M.; Petruzzellis, M.; Savarese, M.; Chiumarulo, L.; Zimatore, D.S.; Dicuonzo, F.; Simone, I.L. Convexal subarachnoid hemorrhage and acute ischemic stroke: A border zone matter? *Neurol. Sci. Off. J. Ital. Neurol. Soc. Ital. Soc. Clin. Neurophysiol.* **2019**, *40*, 1419–1424. [CrossRef] [PubMed]
109. Autieri, V.; Gaillard, N.; Mourand, I.; Laurent-Chabalier, S.; Mura, T.; Trandafir, C.; Wacongne, A.; de Champfleur, N.M.; Thouvenot, E.; Pereira, F.; et al. Primary acute convexity subarachnoid hemorrhage in older patients: Analysis of baseline, clinical and MRI characteristics including quantitative surface study and topographical probabilistic mapping of convexity subarachnoid hemorrhage. *Acta Neurol. Belg.* **2023**, *123*, 519–527. [CrossRef] [PubMed]
110. Xiong, J.; Xu, W.; Huang, H.; Bian, J.; Wang, A.; Bai, Y.; Lin, J.; Xu, Y.; Lu, X.; Zhao, H. Cirrhosis and risk of stroke: A systematic review and meta-analysis. *Atherosclerosis* **2018**, *275*, 296–303. [CrossRef]
111. Zheng, K.; Yoshida, E.M.; Tacke, F.; Li, Y.; Guo, X.; Qi, X. Risk of Stroke in Liver Cirrhosis: A Systematic Review and Meta-Analysis. *J. Clin. Gastroenterol.* **2020**, *54*, 96–105. [CrossRef]
112. Ferro, J.M.; Infante, J. Cerebrovascular manifestations in hematological diseases: An update. *J. Neurol.* **2021**, *268*, 3480–3492. [CrossRef] [PubMed]

113. Momozaki, A.; Masuoka, J.; Furukawa, T.; Koguchi, M.; Ito, H.; Yoshioka, F.; Inoue, K.; Ogata, A.; Nakahara, Y.; Abe, T. Hemorrhagic stroke associated with essential thrombocythemia: Case report and literature review. *J. Stroke Cerebrovasc. Dis.* **2020**, *29*, 105069. [CrossRef] [PubMed]
114. Sugiyama, M.; Ueno, Y.; Kamo, H.; Edahiro, Y.; Miyamoto, N.; Yamashiro, K.; Tanaka, R.; Shimo, Y.; Komatsu, N.; Hattori, N. Specific mechanisms of subarachnoid hemorrhage accompanied by ischemic stroke in essential thrombocythemia: Two case reports and a literature review. *J. Neurol.* **2019**, *266*, 1869–1878. [CrossRef] [PubMed]
115. Tefferi, A.; Vannucchi, A.M.; Barbui, T. Polycythemia vera treatment algorithm 2018. *Blood Cancer J.* **2018**, *8*, 3. [CrossRef] [PubMed]
116. Tefferi, A.; Vannucchi, A.M.; Barbui, T. Essential thrombocythemia treatment algorithm 2018. *Blood Cancer J.* **2018**, *8*, 2. [CrossRef] [PubMed]
117. Del Prete, C.; Kim, T.; Lansigan, F.; Shatzel, J.; Friedman, H. The Epidemiology and Clinical Associations of Stroke in Patients with Acute Myeloid Leukemia: A Review of 10,972 Admissions From the 2012 National Inpatient Sample. *Clin. Lymphoma Myeloma Leuk.* **2018**, *18*, 74–77.e1. [CrossRef] [PubMed]
118. Raghavan, A.; Wright, C.H.; Wright, J.M.; Jensen, K.; Malloy, P.; Elder, T.; Burant, C.; Sajatovic, M.; Hoffer, A. Outcomes and Clinical Characteristics of Intracranial Hemorrhage in Patients with Hematologic Malignancies: A Systematic Literature Review. *World Neurosurg.* **2020**, *144*, e15–e24. [CrossRef]
119. Sanz, M.A.; Montesinos, P. Advances in the management of coagulopathy in acute promyelocytic leukemia. *Thromb. Res.* **2020**, *191*, S63–S67. [CrossRef]
120. Zaorsky, N.G.; Zhang, Y.; Tchelebi, L.T.; Mackley, H.B.; Chinchilli, V.M.; Zacharia, B.E. Stroke among cancer patients. *Nat. Commun.* **2019**, *10*, 5172. [CrossRef]
121. Smith, N.; Kimberger, K.; Parrish, C.; Currie, S.; Butterworth, S.; Alty, J. Multiple myeloma with multiple neurological presentations. *Pract. Neurol.* **2019**, *19*, 511–517. [CrossRef]
122. Pinto, M.J.; Medeiros, P.B.; Príncipe, F.; Carvalho, M. Cerebral Venous Thrombosis in Hematological Malignancy: Balancing the Risks. *J. Stroke Cerebrovasc. Dis.* **2020**, *29*, 104683. [CrossRef] [PubMed]
123. Fotiou, D.; Gavriatopoulou, M.; Terpos, E. Multiple Myeloma and Thrombosis: Prophylaxis and Risk Prediction Tools. *Cancers* **2020**, *12*, 191. [CrossRef] [PubMed]
124. Osafehinti, D.; Zivari, K. Case of Stroke from Cerebral Vasculitis following Carfilzomib, Lenalidomide, and Dexamethasone Therapy in a Patient with Relapsing Multiple Myeloma. *Case Rep. Hematol.* **2019**, *2019*, 5180424. [CrossRef] [PubMed]
125. Tomich, C.; Debruxelles, S.; Delmas, Y.; Sagnier, S.; Poli, M.; Olindo, S.; Renou, P.; Rouanet, F.; Sibon, I. Immune-Thrombotic Thrombocytopenic Purpura is a Rare Cause of Ischemic Stroke in Young Adults: Case Reports and Literature Review. *J. Stroke Cerebrovasc. Dis.* **2018**, *27*, 3163–3171. [CrossRef]
126. Padrini, R. Clinical Pharmacokinetics and Pharmacodynamics of Direct Oral Anticoagulants in Patients with Renal Failure. *Eur. J. Drug Metab. Pharmacokinet.* **2019**, *44*, 1–12. [CrossRef] [PubMed]
127. Parker, K.; Hartemink, J.; Saha, A.; Mitra, R.; Lewis, P.; Power, A.; Choudhuri, S.; Mitra, S.; Thachil, J. A systematic review of the efficacy and safety of anticoagulants in advanced chronic kidney disease. *J. Nephrol.* **2022**, *35*, 2015–2033. [CrossRef]
128. Reed, D.; Palkimas, S.; Hockman, R.; Abraham, S.; Le, T.; Maitland, H. Safety and effectiveness of apixaban compared to warfarin in dialysis patients. *Res. Pract. Thromb. Haemost.* **2018**, *2*, 291–298. [CrossRef]
129. Sarratt, S.C.; Nesbit, R.; Moye, R. Safety Outcomes of Apixaban Compared with Warfarin in Patients With End-Stage Renal Disease. *Ann. Pharmacother.* **2017**, *51*, 445–450. [CrossRef]
130. Kuno, T.; Takagi, H.; Ando, T.; Sugiyama, T.; Miyashita, S.; Valentin, N.; Shimada, Y.J.; Kodaira, M.; Numasawa, Y.; Briasoulis, A.; et al. Oral Anticoagulation for Patients with Atrial Fibrillation on Long-Term Dialysis. *J. Am. Coll. Cardiol.* **2020**, *75*, 273–285. [CrossRef]
131. Parker, K.; Mitra, S.; Thachil, J. Is anticoagulating haemodialysis patients with non-valvular atrial fibrillation too risky? *Br. J. Haematol.* **2018**, *181*, 725–736. [CrossRef]
132. RAF and RENO-EXTEND Investigators; Caliandro, P.; Cancelloni, V.; Marco, M.; Reale, G.; Zauli, A.; Agnelli, G.; Caso, V.; Becattini, C.; Calabresi, P.; et al. Risk of recurrent stroke in patients with atrial fibrillation treated with oral anticoagulants alone or in combination with anti-platelet therapy. *Eur. Stroke J.* **2023**, *8*, 722–730. [CrossRef]
133. Purrucker, J.C.; Hölscher, K.; Kollmer, J.; Ringleb, P.A. Etiology of Ischemic Strokes of Patients with Atrial Fibrillation and Therapy with Anticoagulants. *J. Clin. Med.* **2020**, *9*, 2938. [CrossRef] [PubMed]
134. Sharma, M.; Hart, R.G.; Connolly, S.J.; Bosch, J.; Shestakovska, O.; Ng, K.K.H.; Catanese, L.; Keltai, K.; Aboyans, V.; Alings, M. Stroke Outcomes in the COMPASS Trial. *Circulation* **2019**, *139*, 1134–1145. [CrossRef] [PubMed]
135. Anand, S.S.; Bosch, J.; Eikelboom, J.W.; Connolly, S.J.; Diaz, R.; Widimsky, P.; Aboyans, V.; Alings, M.; Kakkar, A.K.; Keltai, K.; et al. Rivaroxaban with or without aspirin in patients with stable peripheral or carotid artery disease: An international, randomised, double-blind, placebo-controlled trial. *Lancet* **2018**, *391*, 219–229. [CrossRef] [PubMed]
136. So, C.H.; Eckman, M.H. Combined aspirin and anticoagulant therapy in patients with atrial fibrillation. *J. Thromb. Thrombolysis* **2017**, *43*, 7–17. [CrossRef] [PubMed]
137. Deana, C.; Bagatto, D. Severe stroke in patients admitted to intensive care unit after COVID-19 infection: Pictorial essay of a case series. *Brain Hemorrhages* **2022**, *3*, 29–35. [CrossRef] [PubMed]

138. Fara, M.G.; Stein, L.K.; Skliut, M.; Morgello, S.; Fifi, J.T.; Dhamoon, M.S. Macrothrombosis and stroke in patients with mild COVID-19 infection. *J. Thromb. Haemost.* **2020**, *18*, 2031–2033. [CrossRef] [PubMed]
139. Nannoni, S.; de Groot, R.; Bell, S.; Markus, H.S. Stroke in COVID-19: A systematic review and meta-analysis. *Int. J. Stroke* **2021**, *16*, 137–149. [CrossRef]
140. Wang, Z.; Yang, Y.; Liang, X.; Gao, B.; Liu, M.; Li, W.; Chen, Z.; Wang, Z. COVID-19 Associated Ischemic Stroke and Hemorrhagic Stroke: Incidence, Potential Pathological Mechanism, and Management. *Front. Neurol.* **2020**, *11*, 571996. [CrossRef]
141. Pluta, J.; Cieniewicz, A.; Trzebicki, J. COVID-19: Coagulation disorders and anticoagulant treatment in patients hospitalised in ICU. *Anaesthesiol. Intensiv. Ther.* **2021**, *53*, 153–161. [CrossRef]
142. Cuker, A.; Tseng, E.K.; Nieuwlaat, R.; Angchaisuksiri, P.; Blair, C.; Dane, K.; Davila, J.; DeSancho, M.T.; Diuguid, D.L.; Griffin, D.O.; et al. American Society of Hematology living guidelines on the use of anticoagulation for thromboprophylaxis in patients with COVID-19: July 2021 update on postdischarge thromboprophylaxis. *Blood Adv.* **2022**, *6*, 664–671. [CrossRef]
143. Godon, A.; Tacquard, C.A.; Mansour, A.; Garrigue, D.; Nguyen, P.; Lasne, D.; Testa, S.; Levy, J.H.; Albaladejo, P.; Gruel, Y.; et al. Prevention of venous thromboembolism and haemostasis monitoring in patients with COVID-19: Updated proposals (April 2021): From the French working group on perioperative haemostasis (GIHP) and the French study group on thrombosis and haemostasis (GFHT), in collaboration with the French society of anaesthesia and intensive care (SFAR). *Anaesth. Crit. Care Pain Med.* **2021**, *40*, 100919. [CrossRef] [PubMed]
144. Varga, Z.; Flammer, A.J.; Steiger, P.; Haberecker, M.; Andermatt, R.; Zinkernagel, A.S.; Mehra, M.R.; Schuepbach, R.A.; Ruschitzka, F.; Moch, H. Endothelial cell infection and endotheliitis in COVID-19. *Lancet* **2020**, *395*, 1417–1418. [CrossRef] [PubMed]
145. Wijeratne, T.; Crewther, S.G.; Sales, C.; Karimi, L. COVID-19 Pathophysiology Predicts That Ischemic Stroke Occurrence Is an Expectation, Not an Exception—A Systematic Review. *Front. Neurol.* **2021**, *11*, 607221. [CrossRef] [PubMed]
146. Abbas, R.; El Naamani, K.; Sweid, A.; Schaefer, J.W.; Bekelis, K.; Sourour, N.; Elhorany, M.; Pandey, A.S.; Tjoumakaris, S.; Gooch, M.R.; et al. Intracranial Hemorrhage in Patients with Coronavirus Disease 2019 (COVID-19): A Case Series. *World Neurosurg.* **2021**, *154*, e473–e480. [CrossRef] [PubMed]
147. Sharifi-Razavi, A.; Karimi, N.; Rouhani, N. COVID-19 and intracerebral haemorrhage: Causative or coincidental? *New Microbes New Infect.* **2020**, *35*, 100669. [CrossRef] [PubMed]
148. Ackermann, M.; Verleden, S.E.; Kuehnel, M.; Haverich, A.; Welte, T.; Laenger, F.; Vanstapel, A.; Werlein, C.; Stark, H.; Tzankov, A.; et al. Pulmonary Vascular Endothelialitis, Thrombosis, and Angiogenesis in COVID-19. *N. Engl. J. Med.* **2020**, *383*, 120–128. [CrossRef]
149. Quinaglia, T.; Shabani, M.; Breder, I.; Silber, H.A.; Lima, J.A.; Sposito, A.C. Coronavirus disease-19: The multi-level, multi-faceted vasculopathy. *Atherosclerosis* **2021**, *322*, 39–50. [CrossRef]
150. Sagris, M.; Theofilis, P.; Antonopoulos, A.S.; Tsioufis, C.; Oikonomou, E.; Antoniades, C.; Crea, F.; Kaski, J.C.; Tousoulis, D. Inflammatory Mechanisms in COVID-19 and Atherosclerosis: Current Pharmaceutical Perspectives. *Int. J. Mol. Sci.* **2021**, *22*, 6607. [CrossRef]

neurology international

MDPI

Article

Physical Rehabilitation and Post-Stroke Pneumonia: A Retrospective Observational Study Using the Japanese Diagnosis Procedure Combination Database

Takehiro Nishimura [1], Ryutaro Matsugaki [2,*] and Shinya Matsuda [1]

[1] Department of Preventive Medicine and Community Health, University of Occupational and Environmental Health, Kitakyushu 807-8555, Japan; nsmr@med.uoeh-u.ac.jp (T.N.); smatsuda@med.uoeh-u.ac.jp (S.M.)

[2] Department of Work Systems and Health, Institute of Industrial Ecological Sciences, University of Occupational and Environmental Health, Kitakyushu 807-8555, Japan

* Correspondence: pt-matsugaki@med.uoeh-u.ac.jp; Tel.: +81-93-603-1611

Abstract: In this study, the relationship between the duration of physical rehabilitation and occurrence of pneumonia after ischemic stroke was examined. We included 426,508 patients aged ≥75 years with acute ischemic stroke. A multilevel logistic regression analysis nested at the hospital level was conducted to examine the association between the duration of physical rehabilitation and occurrence of pneumonia. The duration of physical rehabilitation refers to the hours of physical rehabilitation performed daily until the 7th day of hospitalization. In the multivariable analysis, the intensity of rehabilitation for durations of 20–39 min/day (adjusted odds ratio [aOR]: 0.78, 95% Confidence Interval [CI]: 0.75–0.81, $p < 0.001$), 40–59 min/day (aOR: 0.68, 95% CI: 0.66–0.71, $p < 0.001$), 60–79 min/day (aOR:0.56, 95% CI: 0.53–0.58, $p < 0.001$), and ≥80 min/day (aOR: 0.46, 95% CI: 0.44–0.48, $p < 0.001$) were significantly associated with a reduced incidence of pneumonia. In addition, the trend identified for duration of rehabilitation was significant ($p < 0.001$). The results of this study suggest the usefulness of high-duration physical rehabilitation for preventing pneumonia in older patients with ischemic stroke.

Keywords: ischemic stroke; Japan; physical rehabilitation; pneumonia

1. Introduction

Pneumonia is one of the most common stroke complications. Its incidence after stroke is estimated to be 3.2–14.3% [1–3], with most cases occurring within 7 days of hospitalization [3]. Older age [4], male sex [4], stroke severity [4,5], comorbidity (including chronic obstructive pulmonary disease, angina, or coronary artery disease) [4], dysphasia [4–8], dependency before admission [4], low mobility on admission [9], mechanical ventilation [5], and nasogastric tube [5,9] use are known to increase the risk of pneumonia and respiratory complications after stroke. Pneumonia prevention is of upmost importance because pneumonia is associated with undesirable outcomes, such as increased length of hospitalization [4,10,11], poor functional outcomes [3,4,10], increased costs of hospitalization [12], reduced frequency of routine home discharge [2], serious complications (such as sepsis), and mortality [2–4,10,11].

Physical rehabilitation is one of the most common treatments for stroke. Rehabilitation programs should be individualized according to the pathophysiology of the stroke within 24–48 h of admission to prevent complications and enhance functional recovery [13–15]. Poststroke physical rehabilitation generally includes swallowing training, early mobilization, and exercise interventions to improve disability by physical, occupational, and speech therapists, which are thought to contribute to the prevention of pneumonia [16–18].

Recently, high-duration physical rehabilitation has been shown to contribute to improved outcomes after ischemic stroke. An observational study using Japanese claims-data reported that higher-duration physical rehabilitation contributed to improved activities of daily living [19]. We hypothesized that this dose-response relationship would also apply to pneumonia as an outcome, although the relationship between rehabilitation intensity and the occurrence of pneumonia after ischemic stroke remains unclear. In this study, we examined the relationship between the duration of physical rehabilitation and occurrence of pneumonia after ischemic stroke using the Japanese Diagnosis Procedure Combination (DPC) database. We hypothesized that physical rehabilitation duration is associated with a lower incidence of pneumonia.

2. Materials and Methods

2.1. Study Design and Data Sources

This observational study utilized data from the Japanese DPC database. DPC is the main medical service reimbursement system for acute inpatient care in Japan, launched by the Ministry of Health, Labour and Welfare of Japan in 2002 [20]. The DPC database consists of information relevant to that system and contains the following information concerning hospitalized patients: (1) patient demographics such as sex, age, and admission-discharge pathways; (2) information pertaining to the diagnosis; and (3) details of the medical procedures performed during hospitalization. Importantly, the DPC data used were collated by the DPC research group and represents approximately 90% of the tertiary emergency hospitals in Japan [21]. In this study, we used case data of patients discharged between April 2014 and March 2020. This study was approved by the Institutional Review Board of the University of Occupational and Environmental Health, Japan (approval number: R4-045).

2.2. Study Participants

We included individuals aged 75 years or older with acute ischemic stroke (International Classification of Diseases, Tenth Revision [ICD-10 code]: I63) who were hospitalized within three days of stroke onset (n = 470,337). Patients who did not undergo rehabilitation during hospitalization and those with an unknown modified Rankin Scale score prior to admission or Japan Coma Scale score at admission were excluded. Of the 470,337 patients identified, 426,508 were included in the analysis (Figure 1).

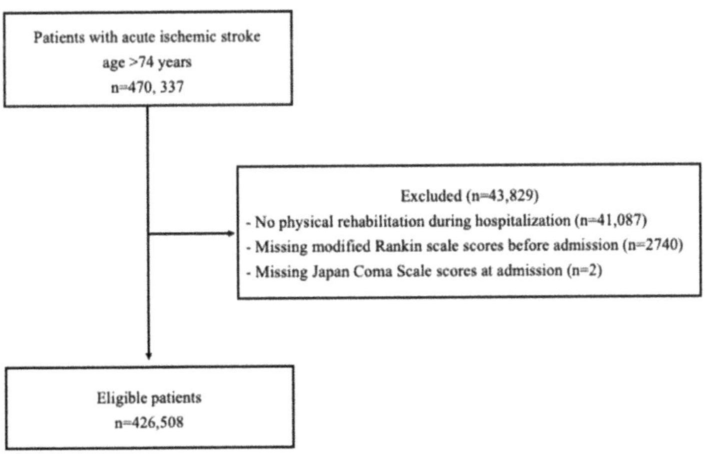

Figure 1. Flowchart of study inclusion.

2.3. Assessment of Physical Rehabilitation Duration

Physical rehabilitation was provided by medical doctors and physical, occupational, and speech therapists under Japan's national healthcare insurance system. Each session lasted 20 min, and patients could undergo up to 180 min of rehabilitation per day. In previous studies, the duration of physical rehabilitation was determined by dividing the total number of hours of rehabilitation provided during a hospital stay by the number of days spent in the hospital [19,22]. Considering that pneumonia after stroke in this study often occurred within seven days [3], the duration of physical rehabilitation was calculated up to the 7th day of hospitalization. We divided the duration of physical rehabilitation into five groups: <20, 20–39, 40–59, 60–79, ≥80 min/d.

2.4. Assessment of Pneumonia

Pneumonia, including aspiration pneumonia, was identified using the ICD-10 codes J12-J18 and J69 for complications during hospitalization in acute care hospital [10,23].

2.5. Assessment of Covariates

The following variables were examined in this study: age, sex, subtype of ischemic stroke (arterial thrombosis, arterial embolism, and others) [23], Charlson comorbidity index (CCI) score [24,25], Japan Coma Scale (JCS) score at admission [26,27], modified Rankin Scale (mRS) score before admission [28], type of acute care implemented (intravenous thrombolysis, and endovascular thrombectomy), fiscal year (2014, 2015, 2016, 2017, 2018, and 2019), and hospital case volume.

The CCI was introduced by Charlson et al. in 1987 as a method to quantify comorbid conditions [24]. It evaluates 17 comorbid conditions, such as myocardial infarction and congestive heart failure, based on ICD-10 codes [25]. The index assigns weights to each condition, with the aggregate of these scores constituting the CCI score. In this study, the CCI scores were categorized into four groups: 0, 1, 2, and ≥3.

The JCS is a metric predominantly utilized in Japan for assessing the level of consciousness in patients [26,27]. It categorizes consciousness impairment into nine levels based on severity, with a score of 0 indicating no impairment. For patients who are awake without any stimuli, a single-digit score (1, 2, 3) is assigned. If the patient can be aroused but reverts to their previous state after stimulation ceases, a two-digit score (10, 20, 30) is used. For patients who cannot be aroused with any forceful mechanical stimuli, a three-digit score (100, 200, 300) is designated. In this study, scores were classified into four categories based on this system: 0 for alert consciousness; 1–3 for wakefulness without stimuli; 10–30 for arousal by some stimuli; and 100–300 for coma. The JCS score can be converted to the internationally used Glasgow Coma Scale (GCS) score. For instance, a JCS score of 3 corresponds to a GCS score of 13, a JCS score of 30 corresponds to a GCS score of 9, and a JCS score of 300 corresponds to a GCS score of 3 [29].

The mRS is a scale used to assess the degree of disability and independence in patients who have suffered a stroke [28]. The mRS is a six-tiered metric ranging from 0 to 5, where 0 indicates no symptoms; 1 signifies no significant disability; 2 represents slight disability; 3 denotes moderate disability; 4 corresponds to moderately severe disability; and 5 reflects severe disability.

2.6. Statistical Analysis

Continuous variables, such as age, were presented as the median and interquartile range. categorical variables were expressed as numbers and percentages.

In the main analysis, a multilevel logistic regression analysis (nested at the hospital level) was conducted, with pneumonia as the dependent variable and the duration of physical rehabilitation as the independent variable. To adjust for potential confounders, we used age, sex, ischemic stroke subtype, CCI score, JCS score at admission, mRS score before admission, acute care type, fiscal year, and hospital case volume as covariates.

As a supplementary analysis, multilevel logistic regression analysis (nested at the hospital level) was conducted by classifying pneumonia into aspiration pneumonia (J69) and other pneumonia (J12-18) [30], with these as the dependent variables and the duration of physical rehabilitation as an independent variable. The covariates were the same as those used in the main analysis.

All statistical analyses were performed using Stata software (Stata Statistical Software: Release 18; StataCorp LLC, College Station, TX, USA). A *p*-value of <0.05 were considered statistically significant.

2.7. Sensitivity Analysis

We conducted the following sensitivity analyses: First, while our primary analysis adopted the duration of physical rehabilitation within the first seven days of hospitalization as the exposure, we performed sensitivity analyses to validate the robustness of our findings. For this purpose, we redefined exposure as the duration of physical rehabilitation up to three and five days post-admission. Second, we hypothesized that the association between the duration of physical rehabilitation and the onset of pneumonia would vary depending on the severity of the patient's consciousness impairment. To examine potential effect modifications, we conducted stratified sensitivity analyses based on the severity of patients' consciousness impairment. Finally, in patients requiring mechanical ventilation shortly after admission, implementing high-duration physical rehabilitation is challenging. Mechanical ventilation is a known risk factor for pneumonia onset. Given that this could introduce bias in our results, we conducted a sensitivity analysis excluding 3227 patients who required mechanical ventilation within the first 2 days of admission, focusing on the remaining 423,281 patients.

3. Results

Table 1 shows the characteristics of the participants according to the duration of physical rehabilitation. The proportion of patients with a Japan Coma Scale score ranging from 100 to 300 decreased as the duration of physical rehabilitation increased: <20 min/d (11.1%), 20–39 min/d (5.9%), 40–59 min/d (3.6%), 60–79 min/d (2.5%), and ≥80 min/d (1.9%). Similarly, the proportion of patients with a mRS score of 5 before admission also declined with increasing duration of physical rehabilitation: <20 min/d (9.3%), 20–39 min/d (5.9%), 40–59 min/d (5.0%), 60–79 min/d (4.6%), and ≥80 min/d (4.4%).

Table 1. Participant Characteristics.

		Duration of Physical Rehabilitation				
	Total (n = 426,508)	<20 min/Day (n = 43,505)	20–39 min/Day (n = 125,126)	40–59 min/Day (n = 123,718)	60–79 min/Day (n = 73,172)	≥80 min/Day (n = 60,987)
Age, median (IQR)	83.0 (79.0, 88.0)	84.0 (79.0, 88.0)	83.0 (79.0, 88.0)	83.0 (79.0, 88.0)	83.0 (79.0, 88.0)	83.0 (79.0, 87.0)
Sex						
men	206,564 (48.4%)	20,208 (46.4%)	60,618 (48.4%)	60,210 (48.7%)	35,860 (49.0%)	29,668 (48.6%)
women	219,944 (51.6%)	23,297 (53.6%)	64,508 (51.6%)	63,508 (51.3%)	37,312 (51.0%)	31,319 (51.4%)
Subtype of ischemic stroke						
Arterial thrombosis	160,613 (37.7%)	14,021 (32.2%)	45,424 (36.3%)	47,311 (38.2%)	29,146 (39.8%)	24,711 (40.5%)
Arterial embolism	134,911 (31.6%)	15,827 (36.4%)	41,543 (33.2%)	38,997 (31.5%)	21,323 (29.1%)	17,221 (28.2%)
Others	130,984 (30.7%)	13,657 (31.4%)	38,159 (30.5%)	37,410 (30.2%)	22,703 (31.0%)	19,055 (31.2%)
Charlson comorbidity index						
0	142,996 (33.5%)	14,669 (33.7%)	42,495 (34.0%)	41,323 (33.4%)	24,559 (33.6%)	19,950 (32.7%)
1	134,435 (31.5%)	13,761 (31.6%)	39,641 (31.7%)	39,556 (32.0%)	23,128 (31.6%)	18,349 (30.1%)
2	85,057 (19.9%)	8519 (19.6%)	24,682 (19.7%)	24,616 (19.9%)	14,558 (19.9%)	12,682 (20.8%)
3≤	64,020 (15.0%)	6556 (15.1%)	18,308 (14.6%)	18,223 (14.7%)	10,927 (14.9%)	10,006 (16.4%)
Japan Coma Scale score at admission						
0	157,335 (36.9%)	14,341 (33.0%)	46,992 (37.6%)	45,397 (36.7%)	27,662 (37.8%)	22,943 (37.6%)
1–3	200,509 (47.0%)	17,503 (40.2%)	54,944 (43.9%)	59,339 (48.0%)	36,803 (50.3%)	31,920 (52.3%)
10–30	48,999 (11.5%)	6816 (15.7%)	15,832 (12.7%)	14,471 (11.7%)	6894 (9.4%)	4986 (8.2%)
100–300	19,665 (4.6%)	4845 (11.1%)	7358 (5.9%)	4511 (3.6%)	1813 (2.5%)	1138 (1.9%)

Table 1. *Cont.*

	Duration of Physical Rehabilitation					
	Total (n = 426,508)	**<20 min/Day (n = 43,505)**	**20–39 min/Day (n = 125,126)**	**40–59 min/Day (n = 123,718)**	**60–79 min/Day (n = 73,172)**	**≥80 min/Day (n = 60,987)**
Modified Rankin Scale score before admission						
0	163,342 (38.3%)	15,442 (35.5%)	47,664 (38.1%)	47,430 (38.3%)	28,490 (38.9%)	24,316 (39.9%)
1	75,758 (17.8%)	7490 (17.2%)	22,369 (17.9%)	22,273 (18.0%)	13,120 (17.9%)	10,506 (17.2%)
2	57,076 (13.4%)	5563 (12.8%)	16,821 (13.4%)	17,057 (13.8%)	9920 (13.6%)	7715 (12.7%)
3	51,049 (12.0%)	4921 (11.3%)	14,808 (11.8%)	14,976 (12.1%)	8896 (12.2%)	7448 (12.2%)
4	55,459 (13.0%)	6016 (13.8%)	15,998 (12.8%)	15,764 (12.7%)	9391 (12.8%)	8290 (13.6%)
5	23,709 (5.6%)	4046 (9.3%)	7430 (5.9%)	6190 (5.0%)	3343 (4.6%)	2700 (4.4%)
Acute care						
Intravenous thrombolysis	29,556 (6.9%)	2502 (5.8%)	8840 (7.1%)	9393 (7.6%)	5118 (7.0%)	3703 (6.1%)
Endovascular thrombectomy	16,480 (3.9%)	1582 (3.6%)	5144 (4.1%)	5226 (4.2%)	2622 (3.6%)	1906 (3.1%)
Mechanical ventilation	3227 (0.8%)	1119 (2.6%)	1280 (1.0%)	518 (0.4%)	199 (0.3%)	111 (0.2%)
Tube feeding	29,162 (6.8%)	3175 (7.3%)	9763 (7.8%)	8766 (7.1%)	4205 (5.7%)	3253 (5.3%)
Fiscal year						
2014	64,972 (15.2%)	9720 (22.3%)	20,345 (16.3%)	17,233 (13.9%)	9511 (13.0%)	8163 (13.4%)
2015	70,889 (16.6%)	8016 (18.4%)	21,715 (17.4%)	19,623 (15.9%)	11,622 (15.9%)	9913 (16.3%)
2016	73,779 (17.3%)	7468 (17.2%)	21,667 (17.3%)	21,136 (17.1%)	12,545 (17.1%)	10,963 (18.0%)
2017	73,113 (17.1%)	6287 (14.5%)	21,232 (17.0%)	21,868 (17.7%)	13,051 (17.8%)	10,675 (17.5%)
2018	73,411 (17.2%)	6501 (14.9%)	20,450 (16.3%)	22,214 (18.0%)	13,324 (18.2%)	10,922 (17.9%)
2019	70,344 (16.5%)	5513 (12.7%)	19,717 (15.8%)	21,644 (17.5%)	13,119 (17.9%)	10,351 (17.0%)
Hospital volume (patients/year)						
−73	142,787 (33.5%)	19,710 (45.3%)	43,445 (34.7%)	40,364 (32.6%)	22,229 (30.4%)	17,039 (27.9%)
74–133	142,124 (33.3%)	13,599 (31.3%)	44,344 (35.4%)	43,642 (35.3%)	23,793 (32.5%)	16,746 (27.5%)
134–	141,597 (33.2%)	10,196 (23.4%)	37,337 (29.8%)	39,712 (32.1%)	27,150 (37.1%)	27,202 (44.6%)

Table 2 shows the incidence of pneumonia according to the duration of physical rehabilitation. The occurrence of pneumonia decreased with increasing duration of rehabilitation: <20 min/d (13.4%), 20–39 min/d (10.1%), 40–59 min/d (8.9%), 60–79 min/d (7.4%), and ≥80 min/d (6.4%). This trend was consistent when pneumonia was divided into non-aspiration and aspiration pneumonia.

Table 2. Occurrence of pneumonia according to physical rehabilitation duration.

	Intensity of Physical Rehabilitation					
	<20 min/Day (n = 43,505)	**20–39 min/Day (n = 125,126)**	**40–59 min/Day (n = 123,718)**	**60–79 min/Day (n = 73,172)**	**≥80 min/Day (n = 60,987)**	***p*-Value**
Pneumonia (ICD-10: J12-J18, J69)	5849 (13.4%)	12,678 (10.1%)	10,972 (8.9%)	5384 (7.4%)	3917 (6.4%)	<0.001
Pneumonia (ICD-10: J12-J18)	2006 (4.6%)	4040 (3.2%)	3473 (2.8%)	1766 (2.4%)	1539 (2.5%)	<0.001
Aspiration pneumonia (ICD-10: J69)	3843 (8.8%)	8638 (6.9%)	7499 (6.1%)	3618 (4.9%)	2378 (3.9%)	<0.001

The adjusted odds ratios (aORs) for pneumonia associated with the physical rehabilitation duration are shown in Table 3. In the multivariable analysis, physical rehabilitation durations of 20–39 min/day (aOR: 0.78, 95% CI: 0.75–0.81, $p < 0.001$), 40–59 min/day (aOR: 0.68, 95% CI: 0.66–0.71, $p < 0.001$), 60–79 min/day (aOR: 0.56, 95% CI: 0.53–0.58, $p < 0.001$), and ≥80 min/day (aOR: 0.46, 95% CI: 0.44–0.48, $p < 0.001$) were significantly associated with a reduced incidence of pneumonia. Furthermore, the analysis indicated a significant trend for the duration of physical rehabilitation (p-value for trend <0.001).

Table 3. Association between physical rehabilitation duration and pneumonia.

	Age-Sex Adjusted					Multivariate Adjusted *				
	OR	95% CI		p-Value	p-Value for Trend	OR	95% CI		p-Value	p-Value for Trend
Duration of physical rehabilitation										
<20 min/day		Reference					Reference			
20–39 min/day	0.69	0.67	0.72	<0.001	<0.001	0.78	0.75	0.81	<0.001	<0.001
40–59 min/day	0.56	0.54	0.58	<0.001		0.68	0.66	0.71	<0.001	
60–79 min/day	0.42	0.40	0.44	<0.001		0.56	0.53	0.58	<0.001	
≥80 min/day	0.33	0.31	0.35	<0.001		0.46	0.44	0.48	<0.001	

* Adjusted for age, sex, subtype of ischemic stroke, Charlson comorbidity index score, Japan Coma Scale score at admission, modified Rankin Scale score before admission, acute care type, fiscal year, and hospital case volume. OR, odds ratio; 95% CI, 95% confidence interval.

Table 4 presents the supplemental analysis results when patients were categorized into the non-aspiration and aspiration pneumonia groups. Consistent with the primary analysis, a higher duration of physical rehabilitation was associated with a decreased incidence of both non-aspiration and aspiration pneumonia ($p < 0.001$).

Table 4. Association between the duration of physical rehabilitation and non-aspiration and aspiration pneumonia.

	Age-Sex Adjusted					Multivariate Adjusted *				
	OR	95% CI		p-Value	p-Value for Trend	OR	95% CI		p-Value	p-Value for Trend
Non-aspiration pneumonia (n = 12,824)										
Duration of physical rehabilitation										
<20 min/day		Reference					Reference			
20–39 min/day	0.70	0.66	0.74	<0.001	<0.001	0.79	0.74	0.83	<0.001	<0.001
40–59 min/day	0.56	0.53	0.59	<0.001		0.68	0.64	0.72	<0.001	
60–79 min/day	0.41	0.38	0.44	<0.001		0.54	0.50	0.58	<0.001	
≥80 min/day	0.35	0.32	0.38	<0.001		0.47	0.43	0.51	<0.001	
Aspiration pneumonia (n = 25,976)										
Duration of physical rehabilitation										
<20 min/day		Reference					Reference			
20–39 min/day	0.71	0.68	0.74	<0.001	<0.001	0.80	0.77	0.84	<0.001	<0.001
40–59 min/day	0.59	0.57	0.62	<0.001		0.72	0.68	0.75	<0.001	
60–79 min/day	0.46	0.43	0.48	<0.001		0.60	0.57	0.64	<0.001	
≥80 min/day	0.35	0.33	0.37	<0.001		0.49	0.46	0.52	<0.001	

* Adjusted for age, sex, subtype of ischemic stroke, Charlson comorbidity index score, Japan Coma Scale score at admission, modified Rankin Scale score before admission, acute care type, fiscal year, and hospital case volume. OR, odds ration; 95% CI, 95% confidence interval.

We performed three sensitivity analyses, and in each analysis, the association between the duration of physical rehabilitation and the occurrence of pneumonia was similar to that in the main analysis (Tables S1–S3).

4. Discussion

We examined the relationship between physical rehabilitation duration and the incidence of pneumonia in older stroke patients using Japanese DPC data. The results revealed that a higher physical rehabilitation duration was associated with a lower incidence of pneumonia, and a dose-response relationship was also confirmed. The results of this study suggest the usefulness of high-duration physical rehabilitation for pneumonia prevention in older patients with stroke.

This study has two main strengths. First, it uses a database that covers approximately 90% of all tertiary emergency hospitals in Japan [21,31]. This may contribute to the generalizability of the results of this study. Second, the study showed a dose-response relationship between the duration of physical rehabilitation and the prevention of pneumonia. Although several findings have suggested that physical rehabilitation is important in preventing the occurrence of pneumonia [32], the novelty of this study lies in demonstrating the significance of the intensity of physical rehabilitation.

Our results suggest that patients with ischemic stroke who receive high-duration physical rehabilitation have a lower incidence of pneumonia. Another observational study in Japan suggested that high-duration physical rehabilitation was associated with independence in activities of daily living [19]. The results of this study suggest that high-duration physical rehabilitation for patients with ischemic stroke not only improves activities of daily living, but may also contribute to the prevention of pneumonia, thus emphasizing the clinical significance of intensive physical rehabilitation. From another perspective, our results complement those of previous studies in that high-duration physical rehabilitation may contribute to improving patients' activities of daily living by reducing the risk of pneumonia.

The mechanism by which rehabilitation suppresses pneumonia has not been elucidated. There may be two possible mechanisms for this phenomenon. One is the prevention of aspiration through early assessment and training of swallowing by a speech-language pathologist. A previous meta-analysis reported a prevalence of 42% for post-stroke dysphagia (PSD) and a 4.35-fold increased risk of pneumonia in the group with PSD compared to the group without PSD [7]. This suggests that the prevention and treatment of dysphagia are important for preventing the development of post-stroke pneumonia. Early dysphagia screening and swallowing therapy have been reported to be useful in preventing pneumonia [18,33]. Although debate continues regarding which method is the best for swallowing screening, recent systematic reviews suggest the usefulness of the Bedside Aspiration test, Gugging Swallowing Screen and Toronto Bedside Swallowing Screening Test [34]. Another possibility is that early mobilization and exercise therapy for disabilities by occupational and physical therapists may have improved low mobility, which is a risk factor for respiratory complications, and that early mobilization may have improved the impaired consciousness and dysphagia [17]. In addition, various interventions, including mobilization, positioning, and breathing exercises, have been recommended to eliminate retained airway secretions in critically ill patients [35]. These interventions may promote the removal of airway secretions, reduce the occurrence of atelectasis, and decrease the risk of pneumonia.

This study has a few limitations. First, in studies using DPC data, the type of physical rehabilitation and timing of pneumonia onset remain unknown. Therefore, although the study demonstrated the benefit of physical rehabilitation, the specific type of physical rehabilitation that is beneficial and the combination of physical rehabilitation that is appropriate, remains unknown. Previous studies have demonstrated the effectiveness of multidisciplinary team interventions [17]. Therefore, a balanced intervention consisting of speech, occupational, and physical therapies is desirable. Second, there are unmeasured confounding factors. A major limitation of this study is the lack of adjustment for stroke severity at onset, as assessed by the NIHSS score, which is associated with functional disability, mortality, and occurrence of pneumonia after ischemic stroke [2,36–38]. Patients with a higher severity of ischemic stroke may receive a shorter duration of physical rehabilitation, and concurrently, exhibit an increased frequency of pneumonia occurrence. This scenario suggests that if the severity of stroke, a potential confounding factor, is not adequately controlled for, it could lead to an overestimation of the protective effect of prolonged physical rehabilitation on the risk of pneumonia. However, we adjusted for variables that are surrogates for stroke severity, such as conscious state at admission and stroke type. In addition to stroke severity, detailed classification of ischemic stroke and the presence of dysphasia are also unmeasured confounding factors. Third, it is known that DPC data have high specificity but low sensitivity for diagnoses [39]. The possibility of misclassification of pneumonia, which is the outcome, cannot be denied. If misclassification of the outcome is occurring, the association between the duration of physical rehabilitation and pneumonia might be underestimated. Finally, the possibility of reverse causality cannot be ruled out, because the timing of pneumonia onset could not be specified in this study. However, we conducted a sensitivity analysis excluding patients who were ventilated soon after hospitalization and a sensitivity analysis using the rehabilitation duration within a very short period of time after hospitalization (within 3 days) as the exposure and confirmed that the relationship between rehabilitation intensity and the occurrence of pneumonia remained the same even when the situations in which reverse causality occurred were controlled as much as possible. The relationship between rehabilitation duration and the occurrence of pneumonia remained unchanged even when situations in which reverse causality occurred were controlled as much as possible.

5. Conclusions

Increasing the duration of rehabilitation in older patients with stroke may reduce the risk of developing pneumonia. This finding underscores the need for active rehabilitation in the acute phase of ischemic stroke. Healthcare providers should intensify their rehabilitation efforts to prevent pneumonia after ischemic stroke.

Supplementary Materials: The following supporting information can be downloaded at: https://www.mdpi.com/article/10.3390/neurolint15040094/s1, Table S1: Association between the intensity of physical rehabilitation within 3 and 5 days of admission and non-aspiration and aspiration pneumonia; Table S2: Association between physical rehabilitation intensity and pneumonia stratified by Japan Coma Scale scores; Table S3: The association between the intensity of physical rehabilitation and pneumonia, excluding patients on mechanical ventilators within 2 days of admission (n = 423,281).

Author Contributions: Conceptualization, T.N., R.M. and S.M.; methodology, T.N., R.M. and S.M.; software, R.M.; validation, T.N., R.M. and S.M.; formal analysis, R.M.; investigation, S.M.; resources, S.M.; data curation, S.M.; writing—original draft preparation, T.N. and R.M.; writing—review and editing, T.N., R.M. and S.M.; supervision, S.M.; project administration, S.M.; funding acquisition, S.M. All authors have read and agreed to the published version of the manuscript.

Funding: This study was funded by the Health Labour Sciences Research Grant from the Ministry of Health, Labour, and Welfare of Japan (H30-Policy-Designation-004).

Institutional Review Board Statement: This study was approved by the Institutional Review Board of the University of Occupational and Environmental Health, Japan (approval number R4-045).

Informed Consent Statement: Patient consent was waived because the study used a database and any personally identifiable information was removed.

Data Availability Statement: The datasets used and/or the analyses conducted in the current study are not publicly available for ethical reasons. The datasets are available from the corresponding author upon reasonable request, with ethical approval.

Conflicts of Interest: The authors declare no conflict of interest.

References

1. Kishore, A.K.; Vail, A.; Chamorro, A.; Garau, J.; Hopkins, S.J.; Di Napoli, M.; Kalra, L.; Langhorne, P.; Montaner, J.; Roffe, C.; et al. How is pneumonia diagnosed in clinical stroke research? A systematic review and meta-analysis. *Stroke* **2015**, *46*, 1202–1209. [CrossRef] [PubMed]
2. Ahmed, R.; Mhina, C.; Philip, K.; Patel, S.D.; Aneni, E.; Osondu, C.; Lamikanra, O.; Akano, E.O.; Anikpezie, N.; Albright, K.C.; et al. Age- and Sex-Specific Trends in Medical Complications After Acute Ischemic Stroke in the United States. *Neurology* **2023**, *100*, e1282–e1295. [CrossRef] [PubMed]
3. de Jonge, J.C.; van de Beek, D.; Lyden, P.; Brady, M.C.; Bath, P.M.; van der Worp, H.B. Temporal Profile of Pneumonia After Stroke. *Stroke* **2022**, *53*, 53–60. [CrossRef] [PubMed]
4. Finlayson, O.; Kapral, M.; Hall, R.; Asllani, E.; Selchen, D.; Saposnik, G.; Canadian Stroke Network; Stroke Outcome Research Canada (SORCan) Working Group. Risk factors, inpatient care, and outcomes of pneumonia after ischemic stroke. *Neurology* **2011**, *77*, 1338–1345. [CrossRef] [PubMed]
5. Yuan, M.; Li, F.; Tian, X.; Wang, W.; Jia, M.; Wang, X.; Liu, G. Risk factors for lung infection in stroke patients: A meta-analysis of observational studies. *Expert Rev. Anti Infect. Ther.* **2015**, *13*, 1289–1298. [CrossRef] [PubMed]
6. Chang, M.C.; Choo, Y.J.; Seo, K.C.; Yang, S. The Relationship Between Dysphagia and Pneumonia in Acute Stroke Patients: A Systematic Review and Meta-Analysis. *Front. Neurol.* **2022**, *13*, 834240. [CrossRef] [PubMed]
7. Banda, K.J.; Chu, H.; Kang, X.L.; Liu, D.; Pien, L.; Jen, H.; Hsiao, S.S.; Chou, K. Prevalence of dysphagia and risk of pneumonia and mortality in acute stroke patients: A meta-analysis. *BMC Geriatr.* **2022**, *22*, 420–425. [CrossRef] [PubMed]
8. Feng, M.; Lin, Y.; Chang, Y.; Chen, C.; Chiang, H.; Huang, L.; Yang, Y.; Hung, C. The Mortality and the Risk of Aspiration Pneumonia Related with Dysphagia in Stroke Patients. *J. Stroke Cerebrovasc. Dis.* **2019**, *28*, 1381–1387. [CrossRef]
9. Brogan, E.; Langdon, C.; Brookes, K.; Budgeon, C.; Blacker, D. Can't swallow, can't transfer, can't toilet: Factors predicting infections in the first week post stroke. *J. Clin. Neurosci.* **2015**, *22*, 92–97. [CrossRef]
10. Teh, W.H.; Smith, C.J.; Barlas, R.S.; Wood, A.D.; Bettencourt-Silva, J.H.; Clark, A.B.; Metcalf, A.K.; Bowles, K.M.; Potter, J.F.; Myint, P.K. Impact of stroke-associated pneumonia on mortality, length of hospitalization, and functional outcome. *Acta Neurol. Scand.* **2018**, *138*, 293–300. [CrossRef]
11. Barlas, R.S.; Clark, A.B.; Bettencourt-Silva, J.H.; Sawanyawisuth, K.; Kongbunkiat, K.; Kasemsap, N.; Tiamkao, S.; Myint, P.K. Pneumonia and Risk of Serious Adverse Outcomes in Hospitalized Strokes in Thailand. *J. Stroke Cerebrovasc. Dis.* **2019**, *28*, 1448–1454. [CrossRef]
12. Katzan, I.L.; Dawson, N.V.; Thomas, C.L.; Votruba, M.E.; Cebul, R.D. The cost of pneumonia after acute stroke. *Neurology* **2007**, *68*, 1938–1943. [CrossRef] [PubMed]
13. Langhorne, P.; Collier, J.M.; Bate, P.J.; Thuy, M.N.; Bernhardt, J. Very early versus delayed mobilisation after stroke. *Cochrane Database Syst. Rev.* **2018**, *10*, CD006187 (1-81). [CrossRef] [PubMed]
14. AVERT Trial Collaboration Group. Efficacy and safety of very early mobilisation within 24 h of stroke onset (AVERT): A randomised controlled trial. *Lancet* **2015**, *386*, 46–55. [CrossRef] [PubMed]
15. Miyamoto, S.; Ogasawara, K.; Kuroda, S.; Itabashi, R.; Toyoda, K.; Itoh, Y.; Iguchi, Y.; Shiokawa, Y.; Takagi, Y.; Ohtsuki, T.; et al. Japan Stroke Society Guideline 2021 for the Treatment of Stroke. *Int. J. Stroke* **2022**, *17*, 1039–1049. [CrossRef]
16. Ingeman, A.; Andersen, G.; Hundborg, H.H.; Svendsen, M.L.; Johnsen, S.P. Processes of care and medical complications in patients with stroke. *Stroke* **2011**, *42*, 167–172. [CrossRef]
17. Fujita, A.; Iwata, M.; Hagii, J.; Metoki, N.; Shiroto, H.; Tsuda, E. The effects of combined early oral health care and early mobilisation on the incidence of stroke-associated pneumonia. *J. Stroke Cerebrovasc. Dis.* **2023**, *32*, 106893. [CrossRef]
18. Bath, P.M.; Lee, H.S.; Everton, L.F. Swallowing therapy for dysphagia in acute and subacute stroke. *Cochrane Database Syst. Rev.* **2018**, *10*, CD000323 (1-142). [CrossRef]
19. Yagi, M.; Yasunaga, H.; Matsui, H.; Morita, K.; Fushimi, K.; Fujimoto, M.; Koyama, T.; Fujitani, J. Impact of Rehabilitation on Outcomes in Patients With Ischemic Stroke: A Nationwide Retrospective Cohort Study in Japan. *Stroke* **2017**, *48*, 740–746. [CrossRef]
20. Hayashida, K.; Murakami, G.; Matsuda, S.; Fushimi, K. History and Profile of Diagnosis Procedure Combination (DPC): Development of a Real Data Collection System for Acute Inpatient Care in Japan. *J. Epidemiol.* **2021**, *31*, 1–11. [CrossRef]
21. Yasunaga, H. Real World Data in Japan: Chapter II The Diagnosis Procedure Combination Database. *Ann. Clin. Epidemiol.* **2019**, *1*, 76–79. [CrossRef]
22. Yamaura, R.; Kaneko, T.; Ishikawa, K.B.; Ikeda, S.; Fushimi, K.; Yamazaki, T. Factors Associated With Time to Achieve Physical Functional Recovery in Patients With Severe Stoke After Inpatient Rehabilitation: A Retrospective Nationwide Cohort Study in Japan. *Arch. Rehabil. Res. Clin. Transl.* **2022**, *4*, 100229. [CrossRef] [PubMed]
23. Wada, T.; Yasunaga, H.; Inokuchi, R.; Horiguchi, H.; Fushimi, K.; Matsubara, T.; Nakajima, S.; Yahagi, N. Effects of edaravone on early outcomes in acute ischemic stroke patients treated with recombinant tissue plasminogen activator. *J. Neurol. Sci.* **2014**, *345*, 106–111. [CrossRef] [PubMed]
24. Charlson, M.E.; Pompei, P.; Ales, K.L.; MacKenzie, C.R. A new method of classifying prognostic comorbidity in longitudinal studies: Development and validation. *J. Chronic Dis.* **1987**, *40*, 373–383. [CrossRef] [PubMed]

25. Quan, H.; Sundararajan, V.; Halfon, P.; Fong, A.; Burnand, B.; Luthi, J.; Saunders, L.D.; Beck, C.A.; Feasby, T.E.; Ghali, W.A. Coding Algorithms for Defining Comorbidities in ICD-9-CM and ICD-10 Administrative Data. *Med. Care* **2005**, *43*, 1130–1139. [CrossRef] [PubMed]
26. Todo, T.; Usui, M.; Takakura, K. Treatment of severe intraventricular hemorrhage by intraventricular infusion of urokinase. *J. Neurosurg.* **1991**, *74*, 81–86. [CrossRef]
27. Ono, Y.; Ono, S.; Yasunaga, H.; Matsui, H.; Fushimi, K.; Tanaka, Y. Clinical characteristics and outcomes of myxedema coma: Analysis of a national inpatient database in Japan. *J. Epidemiol.* **2017**, *27*, 117–122. [CrossRef]
28. van Swieten, J.C.; Koudstaal, P.J.; Visser, M.C.; Schouten, H.J.; van Gijn, J. Interobserver agreement for the assessment of handicap in stroke patients. *Stroke* **1988**, *19*, 604–607. [CrossRef]
29. Nakajima, M.; Okada, Y.; Sonoo, T.; Goto, T. Development and Validation of a Novel Method for Converting the Japan Coma Scale to Glasgow Coma Scale. *J. Epidemiol.* **2023**, *33*, 531–535. [CrossRef]
30. Williams, P.T. Dose-response relationship between exercise and respiratory disease mortality. *Med. Sci. Sports Exerc.* **2014**, *46*, 711–717. [CrossRef]
31. Matsugaki, R.; Muramatsu, K.; Fushimi, K.; Matsuda, S. Dementia and acute care of ischemic stroke in Japan: A retrospective observational study using the Japanese Diagnosis Procedure Combination database. *Geriatr. Gerontol. Int.* **2023**, *23*, 270–274. [CrossRef] [PubMed]
32. Cuesy, P.G.; Sotomayor, P.L.; Piña, J.O.T. Reduction in the Incidence of Poststroke Nosocomial Pneumonia by Using the "Turn-mob" Program. *J. Stroke Cerebrovasc. Dis.* **2010**, *19*, 23–28. [CrossRef] [PubMed]
33. Al-Khaled, M.; Matthis, C.; Binder, A.; Mudter, J.; Schattschneider, J.; Pulkowski, U.; Strohmaier, T.; Niehoff, T.; Zybur, R.; Eggers, J.; et al. Dysphagia in Patients with Acute Ischemic Stroke: Early Dysphagia Screening May Reduce Stroke-Related Pneumonia and Improve Stroke Outcomes. *Cerebrovasc. Dis.* **2016**, *42*, 81–89. [CrossRef] [PubMed]
34. Boaden, E.; Burnell, J.; Hives, L.; Dey, P.; Clegg, A.; Lyons, M.W.; Lightbody, C.E.; Hurley, M.A.; Roddam, H.; McInnes, E.; et al. Screening for aspiration risk associated with dysphagia in acute stroke. *Cochrane Database Syst. Rev.* **2021**, *10*, CD012679. [CrossRef] [PubMed]
35. Gosselink, R.; Bott, J.; Johnson, M.; Dean, E.; Nava, S.; Norrenberg, M.; Schonhofer, B.; Stiller, K.; van de Leur, H.; Vincent, J.L. Physiotherapy for adult patients with critical illness: Recommendations of the European Respiratory Society and European Society of Intensive Care Medicine Task Force on Physiotherapy for Critically Ill Patients. *Intensive Care Med.* **2008**, *34*, 1188–1199. [CrossRef] [PubMed]
36. Bhaskar, S.; Stanwell, P.; Bivard, A.; Spratt, N.; Walker, R.; Kitsos, G.H.; Parsons, M.W.; Evans, M.; Jordan, L.; Nilsson, M.; et al. The influence of initial stroke severity on mortality, overall functional outcome and in-hospital placement at 90 days following acute ischemic stroke: A tertiary hospital stroke register study. *Neurol. India* **2017**, *65*, 1252–1259. [CrossRef] [PubMed]
37. Romano, J.G.; Gardener, H.; Campo-Bustillo, I.; Khan, Y.; Tai, S.; Riley, N.; Smith, E.E.; Sacco, R.L.; Khatri, P.; Alger, H.M.; et al. MaRISS Investigators* Predictors of Outcomes in Patients With Mild Ischemic Stroke Symptoms: MaRISS. *Stroke* **2021**, *52*, 1995–2004. [CrossRef]
38. Henke, C.; Foerch, C.; Lapa, S. Early Screening Parameters for Dysphagia in Acute Ischemic Stroke. *Cerebrovasc. Dis.* **2017**, *44*, 285–290. [CrossRef]
39. Yamana, H.; Moriwaki, M.; Horiguchi, H.; Kodan, M.; Fushimi, K.; Yasunaga, H. Validity of diagnoses, procedures, and laboratory data in Japanese administrative data. *J. Epidemiol.* **2017**, *27*, 476–482. [CrossRef]

neurology international

MDPI

Systematic Review

Pulmonary Function Tests Post-Stroke. Correlation between Lung Function, Severity of Stroke, and Improvement after Respiratory Muscle Training

Fotios Drakopanagiotakis [1], Konstantinos Bonelis [1], Paschalis Steiropoulos [1], Dimitrios Tsiptsios [2,*], Anastasia Sousanidou [2], Foteini Christidi [2], Aimilios Gkantzios [2], Aspasia Serdari [3], Styliani Voutidou [2], Chrysoula-Maria Takou [2], Christos Kokkotis [4], Nikolaos Aggelousis [4] and Konstantinos Vadikolias [2]

1 Department of Respiratory Medicine, Medical School, Democritus University of Thrace, University General Hospital of Alexandroupolis, 68100 Alexandroupolis, Greece; fdrakopanagiotakis@gmail.com (F.D.); kon.bonelis@gmail.com (K.B.); steiropoulos@yahoo.com (P.S.)
2 Department of Neurology, Medical School, Democritus University of Thrace, University General Hospital of Alexandroupolis, 68100 Alexandroupolis, Greece; anastasiasousanidou@gmail.com (A.S.); christidi.f.a@gmail.com (F.C.); aimilios.gk@gmail.com (A.G.); stylvout@med.duth.gr (S.V.); chrytako1@med.duth.gr (C.-M.T.); vadikosm@yahoo.com (K.V.)
3 Department of Child and Adolescent Psychiatry, School of Medicine, Democritus University of Thrace, 68100 Alexandroupolis, Greece; aserntar@med.duth.gr
4 Department of Physical Education and Sport Science, Democritus University of Thrace, 69100 Komotini, Greece; ckokkoti@affil.duth.gr (C.K.); nagelous@phyed.duth.gr (N.A.)
* Correspondence: tsiptsios.dimitrios@yahoo.gr

Abstract: Stroke is a significant cause of mortality and chronic morbidity caused by cardiovascular disease. Respiratory muscles can be affected in stroke survivors, leading to stroke complications, such as respiratory infections. Respiratory function can be assessed using pulmonary function tests (PFTs). Data regarding PFTs in stroke survivors are limited. We reviewed the correlation between PFTs and stroke severity or degree of disability. Furthermore, we reviewed the PFT change in stroke patients undergoing a respiratory muscle training program. We searched PubMed until September 2023 using inclusion and exclusion criteria in order to identify studies reporting PFTs post-stroke and their change after a respiratory muscle training program. Outcomes included lung function parameters (FEV_1, FVC, PEF, MIP and MEP) were measured in acute or chronic stroke survivors. We identified 22 studies of stroke patients, who had undergone PFTs and 24 randomised controlled trials in stroke patients having PFTs after respiratory muscle training. The number of patients included was limited and studies were characterised by great heterogeneity regarding the studied population and the applied intervention. In general, PFTs were significantly reduced compared to healthy controls and predicted normal values and associated with stroke severity. Furthermore, we found that respiratory muscle training was associated with significant improvement in various PFT parameters and functional stroke parameters. PFTs are associated with stroke severity and are improved after respiratory muscle training.

Keywords: pulmonary function tests; MIP; MEP; stroke; respiratory muscle training

1. Introduction

Stroke is one of the most important causes of death and disability. Symptoms are based on the location of the lesion and include minor to major motor and sensory deficits (including hemiplegia and gait disorders) [1]. Respiratory muscles, inspiratory and expiratory, may be affected and lead to changes in muscle distribution, muscle fiber architecture and strength generation. A respiratory muscle deficit, especially affecting the diaphragm, causes respiratory dysfunction affecting the proper contraction of the muscle [2]. Post-stroke lung changes (during the acute, subacute and chronic phases) have been documented. These could include changed ventilatory patterns, increased aspiration risk, sleep difficulties and a high frequency of chest infections [3]. Given that they present a serious risk to patients, most of these clinical problems are recognised and handled in clinical settings [3–5].

Respiratory function can be evaluated by performing pulmonary function tests. Pulmonary function tests (PFTs) include spirometry, which assesses dynamic lung volumes such as forced expiratory volume in 1 s (FEV_1), forced vital capacity (FVC), peak expiratory flow (PEF) and static lung volumes and capacity (total lung capacity (TLC)) and residual volume (RV), performed most commonly by body plethysmography) [6]. The strength of respiratory muscles is assessed via maximum inspiratory pressure (MIP) and maximum expiratory pressure (MEP). PFTs are easily performed but cooperation of the participants is needed [6].

An impaired pulmonary function has been associated with increased incidence of cardiovascular diseases, including stroke, in several epidemiologic studies [7–9]. Duong et al. [10] performed a community-based cohort study, encompassing more than 126,000 patients, which showed that FEV_1 is an independent predictor of impairment and mortality of cardiovascular diseases, including stroke. Although PFTs have been examined as a possible risk factor marker of stroke, evidence is limited regarding the association of PFTs with the prognosis of patients, who have already suffered a stroke. Few studies have examined whether PFTs are a functional marker of stroke severity and prognosis [11–13]. PFTs, however, can provide useful information about the functional status of patients with stroke.

Rehabilitation programs for stroke survivors (depending on the motor defect) comprise a multidisciplinary approach to muscle exercise to improve the motion of the trunk, limbs and coordination of movement along gait [14]. Respiratory muscle training is an important part of these programs to improve respiratory function and reduce possible respiratory complications. Many studies have examined the effect of stroke on cardiopulmonary exercise capacity, measured either with treadmill or bicycle ergospirometers or walking tests after rehabilitation [14]. However, data are limited regarding the association of PFTs with functional improvement after a respiratory muscle training program [15].

The aim of this review is to examine whether PFTs could provide an index of stroke patient severity classification and follow-up. Moreover, we will review PFT changes post-stroke in patients undergoing respiratory muscle training.

2. Materials and Methods

We used the Preferred Reporting Items for Systematic Reviews and Meta-Analyses (PRISMA) checklist (CRD42023466356) as a guide for this study. The study's methods were a priori designed. We searched for studies including stroke survivors and performance of PFTs. These studies were further classified in studies examining baseline PFTs in stroke survivors and PFTs in stroke survivors who underwent a respiratory muscle training program. Therefore, the authors conducted literature research of two databases (MEDLINE and Scopus) for eligible studies. The key search terms were ((pulmonary function tests) AND (stroke)) OR ((pulmonary function tests) AND stroke AND (respiratory muscle training)). A search with the terms ((spirometry) AND (stroke)) OR ((spirometry) AND stroke AND (respiratory muscle training)) was also performed and did not retrieve additional studies. There were no sex restrictions and all articles that were published until September 2023 were retrieved. Observational, case-controlled studies and controlled trials were included. The title and abstract of the articles were screened by two different reviewers. Articles not related to the scope of this review (i.e., PFT measurement at baseline after stroke and how PFTs change after respiratory muscle training in patients with stroke) were removed. In total, 276 articles were retrieved for further analysis and their references were studied in order to search for other relevant studies. If the full-text article could not be retrieved, reference was included only if the abstract described all relevant information. After specific inclusion and exclusion criteria were used (Table 1), 46 articles were eligible for our review. Articles were eligible only if their results included measurements of at least one parameter of lung function tests in association to stroke patients. Primary parameters of interest included measurements of forced vital capacity (FVC), forced expiratory volume in 1 s

(FEV_1), peak expiratory flow (PEF), maximum expiratory pressure (MEP) and maximum inspiratory pressure (MEP) and how they are associated with stroke (Table 2).

Table 1. Inclusion and exclusion criteria.

Inclusion Criteria	*Exclusion Criteria*
Abstracts written in English language Patients with stroke Adult patient population PFTs performed post-stroke as baseline or RCTs after implementation of a respiratory muscle training program	Reviews, meta-analyses, editorials, case reports, Articles in pediatric population Articles in special populations (e.g., pregnancy) Articles with outcomes related exclusively to exercise testing PFTs performed in the setting of rehabilitation but not including a respiratory muscle training program

In the second part of article retrieval, articles regarding correlation of PFTs with respiratory muscle training were reviewed. Only results of randomised controlled trials were included (Table 3). Data extraction was performed using a predefined data form created in Excel. Data regarding the author, year of publication, number of participants, the scale of stroke severity and prognosis, and the main results of each study were captured. We refrained from undertaking a meta-analysis or other statistical analysis, due to the high heterogeneity among the studies. Thus, the data were only descriptively analysed.

The screening and selection process is displayed in flow diagram 1 (Figure 1).

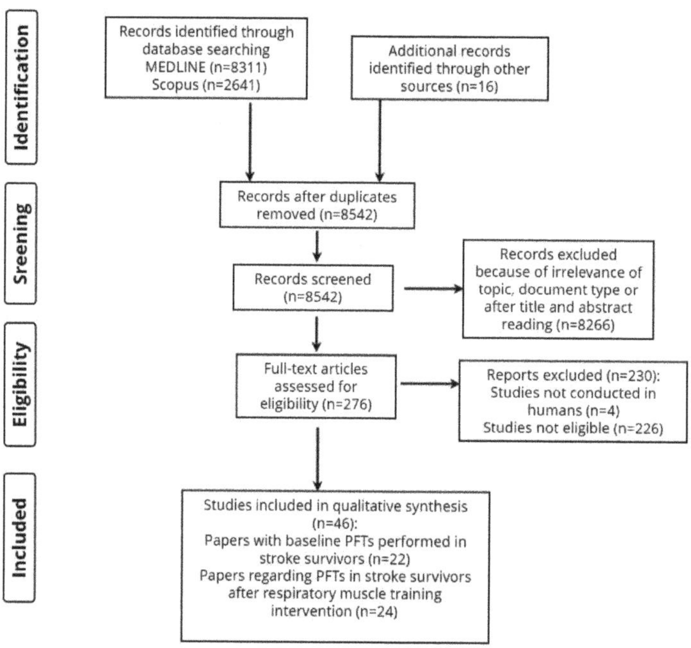

Figure 1. Screening and selection process.

3. Results

After the inclusion/exclusion criteria were evaluated, a total of 46 studies were included in this review. All the studies included stroke survivors and performance of PFTs. These studies were further classified in studies examining baseline PFTs in stroke survivors (n = 22) and PFTs in stroke survivors who underwent a respiratory muscle training program (n = 24). The main characteristics of the included studies are listed in Tables 2 and 3.

Table 2. Studies examining PFTs in post-stroke patients.

Author/Date	Study Population/Mean Age/Gender (M/F)	Type of Study	Type of Stroke	Follow-Up Time	Scale of Stroke Severity	Study Aim	Results
Annoni J.M. et al. 1990 [11]	23 non-smoking hemiplegic patients in the acute phase (53, 12/11)	Case–control	Any type	-	-	Correlation of PFTs with proximal arm function	FIVC and FEVC were reduced with time, independent of motor impairment but related to duration of illness. Patients exhibited a restrictive respiratory pattern. PEF and MEF were reduced by 75%.
De Almeida I.C. et al. [16] 2011	8 patients with right side hemiplegia (51.25 ± 13.8, 4/4), 12 patients with left side hemiplegia (55.33 ± 9.57, 4/8) and 8 controls (52.12 ± 7.28, 5/3)	Case–control	Any type	-	Motor Assessment Scale	Comparison of PFTs and diaphragmatic excursion between groups	MIP significantly better in controls compared to patients with hemiplegia. $FEF_{25–75\%}$ and PEF significantly correlated to left diaphragmatic excursion. IC was not changed. No difference of FVC%, FEV_1%, FEV_1/FVC (small number of patients able to perform these PFT maneuvers)
Ezeugwu et al. 2013 [12]	35 patients with stroke (55.8 ± 8.99, 21/14) and 35 healthy controls (55.6 ± 9.03, 21/14)	Case–control, cross-sectional	Any type	-	-	Comparison of PFTs between stroke patients and controls, correlation with chest excursion	Lower FEV_1, FVC and PEF in stroke patients. Obstructive and restrictive pattern in stroke patients. Lower chest excursion in stroke patients. No correlation between chest excursion and PFTs.
Fugl-Meyer et al. [17] 1983	54 patients with stroke and hemiplegia or hemiparesis	Cross-sectional	Any type	-	-	Correlation of PFTs with stroke severity	PFTs, MIP, MEP, lung compliance and resistance related to the degree of motor impairment and to the interval between stroke and investigation. Lower IC six months after stroke, more evident restrictive disturbance.
Jandt S.R. et al. 2011 [18]	21 patients with stroke (58.9 ± 13.5, 12/9)	Observative, descriptive	Any type	-	-	Correlation of PFTs with trunk impairment scale (TIS)	Significant correlation between TIS and PEF and between TIS and MEF. No correlation of TIS with FEV_1, FVC, FEV_1/FVC and MIF.
Jeong Y. et al. 2020 [13]	52 patients with stroke within six months of onset (34/18)	Prospective	Any type	4 weeks	NIHSS score, Berg Balance Scale	Correlation of PFTs at baseline and 4 weeks after rehabilitation with TIS, Berg Balance Scale and functional independence measure	Baseline FVC, FEV_1 and PEF correlated with initial TIS. Initial PEF significantly associated with Berg Balance Scale and Functional Independence Measure. No correlation with MIP and MEP.
Jung et al. 2014 [19]	10 stroke patients (59.7 ± 12.9, 8/2) and 16 healthy controls (56.1 ± 9.3, 6/10)	Case–control	Any type	-	Korean Modified Barthel Index	Correlation of diaphragmatic excursion with PFTs	Restrictive PFTs in stroke patients. Left diaphragmatic excursion reduction correlated with reduced FEV_1 and FVC in stroke patients.

Table 2. *Cont.*

Author/Date	Study Population/Mean Age/Gender (M/F)	Type of Study	Type of Stroke	Follow-Up Time	Scale of Stroke Severity	Study Aim	Results
Khedr et al. 2000 [20]	34 acute stroke patients (57.23 ± 13.26, 24/10) and 25 healthy volunteers (47.2 ± 22, 17/8)	Case–control, cross-sectional	Ischemic stroke	-	Scandinavian Stroke Scale	Comparison of diaphragmatic excursion and PFTs between groups; no PFTs in healthy controls	41% of the stroke group had decreased diaphragmatic excursion and 70% decreased magnetic evoked potentials. Hemiplegic patients with restrictive PFTs. Negative correlation of FEV_1, FVC and FEV_1/FVC with motor mobility and excitability threshold of affected hemisphere.
Kimura Y et al. 2013 [21]	20 stroke patients without dysphagia (65.7 ± 8.1), 10 stroke patients with dysphagia (74.1 ± 10.2) and 10 healthy controls (68.2 ± 7.2) all male	Case–control, cross-sectional	Any type	-	Brunnstrom's recovery stage	Comparison of peak cough flow and spirometry between groups	Lower peak cough flow and IRV in stroke patients with dysphagia compared to healthy controls. Lower peak cough flow in stroke patients with dysphagia vs. without dysphagia. No differences in ERV or TV between groups.
Kulnik S.T. et al. 2016 [22]	72 patients with stroke (64.6 ± 14.4, 42/30)	Single blind randomised control trial	Any type	4 weeks	NIHSS score	Comparison of peak cough flow in voluntary and reflex cough	Weaker flow in patients' aspiration pneumonia.
Liaw M.Y. et al. 2016 [23]	47 stroke patients with congestive heart failure (65.9 ± 11.5, 24/23)	Cohort	Any type	-	Brunnstrom stage, Barthel Index	Correlation of PFTs with Brunnstrom change	MIP negatively associated with Brunnstrom stage of the proximal and distal parts of the upper extremities and lower extremities, FVC, predicted FVC% and FEV_1%. MEP positively associated with average Brunnstrom stage of the distal area of the upper extremities, FVC, FEV_1, and FEV_1/FVC. FEV_1/FVC negatively associated with the average Brunnstrom stage. Stroke patients had restrictive lung disorder and respiratory muscle weakness, associated with the neurological status of the affected limbs.
Lista Paz A. et al. 2016 [24]	30 chronic stroke patients with a diagnosis of hemiplegia/hemiparesis who were able to walk (55.60 ± 15.84, 22/8) and 30 healthy controls (55.33 ± 14.61, 22/8)	Observational, cross-sectional	Any type	-	Scale Impact of Stroke version 16.0	Comparison of MIP and MEP between groups	Significantly lower MIP and MEP in patients with stroke, <60%. Other spirometry parameters not measured.

Table 2. *Cont.*

Author/Date	Study Population/Mean Age/Gender (M/F)	Type of Study	Type of Stroke	Follow-Up Time	Scale of Stroke Severity	Study Aim	Results
Lista-Paz A. et al. 2023 [25]	33 patients with stroke (56.9 ± 15.7, 24/9) and 33 healthy controls (56.2 ± 15.2, 24/9)	Observational, cross-sectional	Any type	-	Stroke Impact Scale version 16.0	Comparison of PFTs and 6MWT between groups	Stroke patients had significantly lower lung volumes and capacities (VC, FVC, FEV_1, ERV, IC), than controls. Median FVC was 79% and PEF 64% of the reference value. The weak correlation of 6MWD with inspiratory reserve volume and PIF.
Luvizutto G.J. et al. 2017 [26]	32 patients with acute stroke (14/18)	Cross-sectional	Ischemic stroke	-	NIHSS score, mRS score	Correlation of MIP and MEP with anthropometric data and neurologic severity	Lower MIP and MEP than predicted. No association with neurologic severity, positive association with BMI. Other spirometry parameters not measured.
Min S.W. et al. 2018 [27]	57 patients with stroke (69.58 ± 10.29, 34/23)	Cross-sectional	Ischemic stroke	-	-	Correlation of PFTs with dysphagia and aspiration pneumonia	Increased dysphagia associated with worse PCF, FVC and FEV_1 values and aspiration pneumonia.
Nunez Filha M.C. et al. 2020 [28]	53 patients with stroke (55 ± 13.43, 27/26)	Cross-sectional	Any type	-	NIHSS, Modified Barthel Index	Correlation of MIP and MEP and stroke severity with functional mobility	MIP, but not MEP, was independently associated with functional mobility in multivariate analysis. No other spirometry parameters were measured.
Pinheiro M.B. et al. 2014 [29]	89 patients with stroke (56.2 ± 12.0, 48/41)	Cross-sectional, observational	Any type	2 days	-	Correlation of MIP and MEP with stroke population (community vs. non-community ambulators)	Stroke subjects demonstrated decreases of 26.5 and 20% in the MIP and MEP. Significantly worse MIP values seen in non-community ambulators but not statistical significance of MEP, FEV_1 and FVC between community and non-community ambulators.
Santos R.S.D. et al. 2019 [30]	44 patients with stroke (59.4 ± 12.2, 19/25)	Cross-sectional	Any type	-	Functional Independence Measure scale	Correlation of PFTs, MIP and MEP with TIS and Functional Independence Measure	Lower PFTs, MIP and MEP of predicted values, correlation of TIS with FVC, FEV_1 and MIP but not with MEP
Sezer N. et al. 2004 [31]	20 patients with stroke (54.25 ± 11.42, 9/11) and 15 controls (9/6)	Cross-sectional	Any type	-	Brunnstrom classification stage, Barthel Index, Massachusetts General Hospital Functional Ambulation Classification	Comparison of cardiopulmonary response between groups	FEV_1, FVC, VC, PEF and MVV reduced in patients with stroke compared with controls but no correlation with motor disability. $FEF_{25–75\%}$ and FEV_1/FVC no different between groups. Significant respiratory dysfunction in hemiplegic patients.

Table 2. *Cont.*

Author/Date	Study Population/Mean Age/Gender (M/F)	Type of Study	Type of Stroke	Follow-Up Time	Scale of Stroke Severity	Study Aim	Results
Teixeira-Salmela L.F. et al. 2005 [32]	16 community-dwelling stroke survivors (58.37 ± 15.47, 8/8) and 19 age-matched healthy subjects (60.21 ± 4.47, 9/10)	Descriptive case–control	Any type	-	-	Comparison of PFTs, MIP and MEP between stroke patients and controls	Significantly lower MIP and MEP in stroke patients compared to controls and decreased abdominal contribution to tidal volume. Dynamic lung volumes not measured.
Voyvoda et al. 2011 [33]	23 hemiplegic patients (60.5 ± 10.7, 13/10) and 20 controls (61.2 ± 12.1, 13/7)	Descriptive case–control	Ischemic stroke	-	-	Comparison of diaphragm motility with ultrasonography and PFTs between groups	Significantly worse PFTs (FEV_1, FVC, FEV_1/FVC, MIP and MEP) in hemiplegic patients compared to control. No evidence of obstructive disturbance. No significance in diaphragmatic excursion between groups.
Xiao L.J. et al. 2020 [34]	30 patients with stroke and dysphagia (53 ± 11, 20/10), 30 with stroke without dysphagia (59 ± 11, 17/13) and 30 healthy controls (55 ± 18, 18/12)	Descriptive case–control	Any type	-	-	Comparison PFTs between patients with dysphagia after stroke, patients without dysphagia and normal people. Correlation between swallowing function and pulmonary function.	Patients with dysphagia had significantly lower PEF, MIP, MEP FVC, $FEF_{25–75\%}$ and FIV but not FEV_1 compared to those without dysphagia.

3.1. Baseline PFTs in Stroke Survivors

There was high variability in the number of participants included in the studies, as well of the characteristics of the examined patients. As shown in Table 2, only seven studies included more than fifty patients with stroke. Comparisons were made between the PFTs of stroke survivors with healthy controls and those with normal predicted values. The examined PFT parameters were not consistent among studies, particularly regarding the measurement of dynamic lung volumes (Table 2). Xiao et al. and Min et al. compared the PFTs of stroke patients with dysphagia and aspiration pneumonia with those without [27,34]. Stroke severity was assessed using various methods, including Trunk Impairment Score (TIS) [13,18,30], Berg Balance Scale [13] and Functional Independence Measure [30]. Diaphragmatic excursion and diaphragm dysfunction, assessed by ultrasound, were also measured in patients with stroke and correlated to PFTs [12,16,19,20,33].

3.2. Lung Volumes and Flows in Stroke Patients

Patients with stroke exhibited both an obstructive and a restrictive pattern of PFTs. The obstructive pattern was less often as the restrictive pattern and it was not associated with smoking or prior obstructive lung disease, since these were exclusion criteria in almost all the studies (Table 2). Specifically, Ezeugwu et al. compared the PFTs of 35 stroke survivors with 35 age-and sex-matched healthy controls [12]. They found that lung function, including FEV_1, FVC, FEV_1/FVC and PEF was significantly worse in stroke patients. Obstructive pattern was the most common, seen in 46% of the patients, followed by a restrictive pattern in 38% of the stroke patients [12]. Fugl-Meyer at al. reported a restrictive pattern, associated with the degree of hemiplegia [17]. Stroke patients exhibited a restrictive PFT pattern in the study by Jung K. et al. with lower spirometric values compared to controls [19] and in the study by Kimura et al., in which stroke patients with and without dysphagia had lower VCs compared to healthy participants [21]. Lista-Paz et al. showed that the main lung volumes were significantly reduced in people with chronic stroke compared to healthy volunteers matched by age and sex (VC, IRV, IC, FVC, FEV_1 and PEF), as well as to their own reference values (VC, ERV, IC, FVC, FEV_1), results suggestive of a restrictive ventilatory defect in this population [25]. Sezer et al. also found a restrictive pattern with lower FVC and FEV_1 in stroke survivors [31]. Annoni et al. have also reported decreased VC in hemiplegic patients, more evident in patients with severe motor impairment and consistent with a restrictive pattern. The restrictive pattern could not be seen when forced dynamic volumes (FEVC and FIVC) were examined, but this was due to the small number of patients who were able to perform the respiratory maneuvers [11]. Khedr et al. and Liaw et al. also showed that stroke patients exhibited a restrictive PFT pattern [20,23]. Voyvoda et al. found a significantly lower FVC and FEV_1 in hemiplegic patients, suggestive of a restrictive pattern [33]. Liaw et al., who examined the PFTs of patients with stroke and congestive heart failure, found an FVC of 2.0 $\pm$ 0.8 L, with predicted FVC% of 67.9 $\pm$ 18.8%, average FEV_1 of 1.6 $\pm$ 0.7 L, average predicted FEV_1% of 70.6 $\pm$ 20.1% and average FEV_1/FVC of 84.2 $\pm$ 10.5%, all lower than predicted and indicative of a restrictive pattern [23]. When patients with acute stroke were examined, values of FVC and FEV_1 were at 55% of those predicted and FEV_1/FVC values were at 33% of those predicted [20]. In the study by Jandt et al., mean FVC and FEV_1 as well as PEF were all above 80% of those predicted, thus being in the limits of normal [18]. Only in a study of chronic stroke survivors living in the community, who were further classified according to their gait speed, were FVC and FEV_1 above 85% of those values predicted in the total population and FEV_1/FVC was normal [29].

Expiratory flow has also been reported to be reduced in stroke patients: De Almeida et al. reported a significantly reduced PEF in patients with right hemiplegia [16] and so did Sezer et al. [31], Annoni et al. [11] and Ezeugwu et al. [12]. The peak cough flow was also reduced in the study carried out by Kimura et al. [21] and Lista-Paz et al. [25]. These results are supported by the study carried out by Kulnik et al., who showed that peak cough flow was significantly reduced in patients with stroke [22] and in the study by Liaw et al., who showed a maximum mid-expiratory flow of 65.4 ± 29.5% [23].

3.3. MIP and MEP in Stroke Patients

Various studies measured inspiratory and expiratory pressure as the only PFT or in addition to other spirometric parameters. A common finding was the reduction in MIP and MEP in patients with stroke [16,23]. Lista Paz et al. reported decreased MIP and MEP in chronic stroke patients compared with healthy subjects. More importantly, values of MIP and MEP in stroke patients were also below 60% of those predicted in patients with stroke, which is the limit for defining muscle weakness [24]. In another study, the average MIP and MEP were 52.9 ± 33.0 cm H_2O and 60.8 ± 29.0 cm H_2O, respectively, i.e., significantly low [23]. Similar results regarding MIP and MEP were reported from Luvizutto et al.: respiratory pressures were compared with the predicted value, a significant reduction in MIP was observed in the total sample and separately for men and women. When compared with the predicted values, a reduction in MEP was observed in the total sample, in men and in women [26]. MIP and MEP were reduced compared to normal values in the study by Pinheiro et al. [29] and Santos et al. [30].

3.4. Correlation of PFTs with Functional Impairment

Various scales were used in order to assess stroke severity: trunk impairment scale (TIS), the proximal arm function according to the British Medical Research Council, the Motor Assessment Scale, the Berg Balance Scale, the Functional Independence Measure, the Barthel Index, the Scandinavian Stroke Scale, Brunnstrom's recovery rate and function ambulation categories, the Motricity Index, the Scale Impact of Stroke, the NIHSS score, the functional reach test, the International Physical Activity Questionnaire, gait speed and functional ambulation classification (Table 2). Moreover, PFTs were associated in some studies with dysphagia scores or diaphragmatic excursion (Table 2).

Annoni et al. showed that VC was significantly associated with the severity of proximal arm function in hemiplegic patients and FEV_1 and FVC with the duration of stroke illness [11]. Santos et al. reported a significant correlation of FEV_1 and FVC with the TIS [30] and Jeong et al. reported that FVC and FEV_1 values correlated with the TIS scores at admission [13]. FEV_1/FVC was negatively associated with the average Brunnstrom stage over the proximal and distal parts of the upper extremities and lower extremities and Barthel Index [23]. Jung et al. observed a significant positive correlation between left diaphragmatic excursion during deep breathing and FEV_1 (rho = 0.7, p = 0.021) and FVC (rho = 0.86, p = 0.007) in stroke patients. Diaphragmatic excursion did not correlate with the Korean Modified Barthel Index scores. There was found no correlation of these scores with PFTs in hemiplegic patients [19]. Sezer et al. also failed to report a significant correlation of PFTs with motor disability [31].

Abnormal magnetic potentials in the affected hemisphere and central conduction time have also been correlated to PFTs: a significant decline of disability score and higher excitability threshold percentage with lower FEV_1 were found in patients with reduced hemi-diaphragmatic excursion [20].

The expiratory flows PEF and MEF75% also correlated to functional impairment in the study by Annoni et al. [11]. Similar results were reported by Jandt et al. [18]. The latter showed that PEF was significantly associated with TIS. In the prospective study carried out by Jeong et al., the initial peak cough flow correlated with the TIS scores at admission [13]; the initial peak cough flow and FVC were predictive factors for the final TIS score. In linear regression analysis, the initial peak cough flow could predict test scores at discharge for the Berg Balance Scale and Functional Independence Measure [13]. The peak cough flow was also related to function ambulation categories in stroke patients with and without dysphagia [21].

Nunez Filha et al. showed that MIP was associated with stroke severity, as assessed with NIHSS [28] and Pinheiro et al. reported that MIP but not MEP had a negative correlation with gait velocity in stroke survivors [29]. Santos et al. reported a significant correlation of MIP with the TIS [30]. In patients with stroke and heart failure, MIP was negatively associated with the average Brunnstrom stage of the proximal and distal parts of the upper extremities and lower extremities [23].

MEP was positively associated with the average Brunnstrom stage of the distal area of the upper extremities [23]. Jandt et al. found that MEP and not MIP significantly correlated to TIS [18]. In acute stroke patients, however, a correlation of MIP and MEP with functional status could not be found [26]. As expected, a negative correlation of MIP and MEP with BMI was reported [26].

3.5. Correlation of PFTs with Dysphagia and Risk of Aspiration

Kimura Y. et al. compared the PFTs between stroke patients with and without dysphagia and healthy controls: peak cough flow, inspiratory reserve volume and VC were significantly lower in patients with stroke and dysphagia. Peak cough flow values significantly correlated to inspiratory reserve volume [21]. In a secondary analysis of trial data, Kulnik et al. found that increased peak voluntary cough flow and to a lesser extend peak reflex cough flow were associated with a lower possibility of aspiration pneumonia [22]. Min et al. reported similar results regarding peak cough flow, FEV_1 and FVC and their association with the dysphagia score [27]. Finally, Xiao et al. reported that patients with dysphagia had significantly lower PEF, MIP, MEP FVC, $FEF_{25–75\%}$ and FIV compared to those without dysphagia [34].

3.6. Correlation of PFTs with Diaphragmatic Dysfunction

De Almeida et al. measured the diaphragmatic excursion using ultrasound after the acute phase in stroke patients [16]. In their study, right-side hemiplegia affected the respiratory muscles more than left-side hemiplegia, as measured by MIP. Although spirometry was performed only in a few patients, FEV_1, PEF and $FEF_{25–75\%}$ were lower in patients with right-side hemiplegia. In right-side hemiplegia, movement was 4.97 ± 0.78 cm and 4.20 ± 1.45 cm for the right and left diaphragm, respectively, while in left-side hemiplegia, these values were 4.42 ± 0.92 cm and 4.66 ± 1.17 cm [16]. Jung et al. also evaluated the diaphragmatic motion using ultrasound, as mentioned earlier [19]: Stroke patients had a significant unilateral reduction in motion on the hemiplegic side [19]. Diaphragmatic excursion in patients with right-hemiplegia was lower than the one of controls on both sides [19]. On the contrary, in patients with left hemiplegia, diaphragmatic excursion was reduced only on the left side and increased on the right side. Left diaphragmatic motion during deep breathing correlated positively with FVC and FEV_1 [19]. Khedr et al. also reported that decreased hemi-diaphragmatic excursion was found in 41% of their patients, and it was associated with neurophysiological data of diaphragm, the degree of motor disability and respiratory dysfunction [20].

Table 3. Change in PFTs in stroke survivors undergoing a respiratory muscle training program.

Author/Date	Study Population/MEAN Age/Gender (M/F)	Type of Study	Type of Stroke	Follow-Up Time	Scale of Stroke Severity	Study Aim	Results
Aydogan A.S. et al. 2022 [35]	21 stroke patients: 11 in the treatment group (61.72 ± 10.77, 5/6) and 10 in the control group (66.10 ± 8.87, 2/8)	Single blinded randomised controlled trial	Any type	6 weeks	-	PFTs, stroke severity scores before and after a neurodevelopmental treatment program and IMT in the treatment arm	Significantly better PEF and MIP in the treatment group
Britto R.R. et al. 2011 [36]	18 patients with chronic stroke: 9 in the experimental group (56.66 ± 5.56, 5/4) and 9 in the control group (51.44 ± 15.98, 4/5)	Randomised controlled trial	Any type	8 weeks	-	Comparison of MIP, inspiratory muscular resistance before and after IMT	Significantly better values for MIP and inspiratory muscular resistance in the intervention group compared to baseline
Chen P.C. et al. 2016 [37]	21 patients with stroke and congestive heart failure: 11 in the IMT group (63.73 ± 14.64, 4/7) and 10 in the control group (67.50 ± 10.35, 4/6)	Randomised controlled trial	Any type	10 weeks	Barthel Index	Comparison of spirometry, MIP and MEP between IMT group and control	Significant better values of FEV_1, FVC, MIP and Barthel Index in the intervention group compared to baseline and in MIP compared to the control group
Cho J.E. et al. 2018 [38]	25 patients with stroke: 12 in the experimental group (47.58 ± 13.00, 7/5) and 13 in the control group (52.53 ± 9.06, 6/7)	Randomised controlled trial	Any type	6 weeks	-	Comparison of diaphragm thickness ratio, MIP and inspiratory muscle endurance between IMT group and control	Increased diaphragm thickness, MIP and inspiratory muscle endurance in the IMP group
Guillen-Sola A. et al. 2017 [39]	62 patients with dysphagia and stroke (69.0 ± 8.7, 38/24)	Randomised controlled trial	Ischemic stroke	3 months	NIHSS score on admission, mRS score, Barthel Index on admission at Rehabilitation	Comparison of dysphagia score, MIP and MEP after a 3-week rehabilitation program and in 3 months between standard shallow therapy group, standard shallow therapy with IEMS and standard shallow therapy and neuromuscular electric simulation	MIP and MEP significantly improved in the standard shallow therapy with IEMS, compared to the other groups

Table 3. *Cont.*

Author/Date	Study Population/MEAN Age/Gender (M/F)	Type of Study	Type of Stroke	Follow-Up Time	Scale of Stroke Severity	Study Aim	Results
Jung K.M. et al. 2017 [40]	12 patients with hemiparesis due to stroke: 6 in the experimental group (61.2 ± 4.2, 2/4) and 6 in the control group (62.2 ± 5.3, 3/3)	Randomised controlled trial	Any type	4 weeks	-	Comparison of PFTs and walking ability between IMT group vs. aerobic exercise group	Significant improvement of FEV_1, FVC in both groups, significantly better FEV_1, FVC in the IMT group
Kilicoglou M.S. et al. 2022 [41]	41 patients with stroke: 20 in the treatment group (64.6 ± 12.4, 10/10) and 21 in the control group (66.0 ± 10.3, 8/13)	Randomised-controlled trial	Any type	6 weeks	-	Effect of respiratory exercise program on PFTs and diaphragm ultrasound parameters	FVC, FEV_1, FEV_1/FVC and diaphragm ultrasound parameters were improved after treatment in the intervention group
Kim C.Y. et al. 2015 [42]	37 patients with post-stroke hemiplegia: 12 in the integrated training group (57.53 ± 7.73, 7/5), 13 in the respiratory muscle training group (59.20 ± 6.12, 6/7) and 12 in the control group (60.53 ± 0.38, 4/8)	Randomised controlled trial	Any type	6 weeks	-	Comparison of PFTs between controls, RMT and RMT plus abdominal drawing-in maneuver groups	Significantly better FEV_1, FVC and EMG diaphragm activation in the RMT and abdominal drawing-in maneuver group
Kim J. et al. 2014 [43]	20 stroke patients: 10 in the exercise group (54.10 ± 11.69) and 10 in the control group (53.90 ± 5.82)	Randomised-controlled trial	Any type	4 weeks	-	Effects of respiratory muscle and endurance training using an individualized training device for respiratory muscle training on PFTs and exercise capacity in stroke patients	FVC, FEV_1, PEF and 6MWT significantly better in the intervention group
Kulnik et al. 2015 [44]	82 patients with stroke within two weeks of stroke onset (64 ± 14, 49/33)	Single-blind randomized placebo-controlled trial	Any type	4 weeks	NIHSS score	Change in peak expiratory cough flow in patients with IMT, EMT and no respiratory muscle training	Significantly better values compared to baseline in all groups with no effect of training
Lee K. et al. 2019 [45]	25 chronic stroke patients, able to sit independently: 13 in the RMT group (58.62 ± 12.38, 7/6) and 12 in the TSE group (59.75 ± 13.38, 5/7)	Pilot randomised controlled trial	Any type	6 weeks	mRS score	Comparison of PFTs between patients with progressive RMT with and without trunk stabilisation exercise	The MEP, PEF, MIP and PIF were significantly increased in the RMT group than in the control group
Lee D.K et al. 2018 [46]	24 chronic stroke patients: 12 in the experimental group (61.7 ± 6.2, 6/6) and 12 the control group (59.2 ± 4.6, 6/6)	Randomised controlled trials	Any type	4 weeks	-	Comparison of PFTs, TIS and muscle activity of the trunk in patients who received neurodevelopmental treatment alone or with respiratory exercise	Significant better FVC, FEV_1, TIS, Rectus Abdominis, internal oblique and external oblique in the respiratory exercise group
Liaw M.Y. et al. 2020 [47]	21 patients with stroke within six months of unilateral stroke, dysphagia, dysarthria and respiratory muscle weakness (63.86 ± 11.16, 12/9)	Randomised controlled trial	Any type	6 weeks	mRS score, Barthel Index	Comparison of PFTs after IERMT and control group	FVC, FEV_1 and MIP were significantly better in the intervention group

Table 3. *Cont.*

Author/Date	Study Population/MEAN Age/Gender (M/F)	Type of Study	Type of Stroke	Follow-Up Time	Scale of Stroke Severity	Study Aim	Results
Messagi-Sartor M. et al. 2015 [48]	109 patients with subacute stroke (66.5 ± 11.2, 63/46)	Randomised controlled trial	Ischemic stroke	6 months	NIHSS score, Barthel Index, mRS score	Comparison of MIP and MEP in the IEMT and the control group	Improved respiratory muscle strength in the intervention and control group. In IEMT group significantly improved MIP and MEP. Respiratory complications at 6 months more often in the control group, risk reduction of 14%.
Oh D. et al. 2016 [49]	23 stroke patients: 11 in the experimental group (69.7 ± 6.8, 6/5) and 12 in the control group (71.6 ± 7.9, 7/5)	Randomised controlled trial	Any type	6 weeks	-	Comparison of abdominal muscle thickness and PFTs of the IMT group vs. conventional therapy group	FVC, FEV_1, deep abdominal muscle thickness and Berg Balance Scale scores significantly improved in the experimental group
Parreiras de Menezes K.K. et al. 2019 [50]	38 patients with stroke and respiratory muscle weakness: 19 in the experimental group (60 ± 14, 8/11), 19 in the control group (67 ± 11, 8/11)	Double-blind randomised trial	Any type	8 weeks	-	Comparison of MIP, MEP, respiratory complications in the RMT group with high-intensity home-based program vs. control group	Significant increase in MIP, MEP, endurance of respiratory muscles and reduction of dyspnea in intervention group
Ptaszkowska et al. 2019 [51]	60 stroke patients: 30 PNF-treated (64 ± 5, 20/10), 30 PNF-untreated (64 ± 7, 22/8)	Randomised controlled trial	Ischemic stroke	-	Barthel Index	Comparison of PFTs after respiratory stimulation through Proprioceptive Neuromuscular Facilitation (PNF) and controls	FEV_1/FVC% values in PNF-untreated group was substantially lower than in PNF-treated group
Rattes C. et al. 2018 [52]	10 stroke patients with right hemiparesis (60 ± 5.7, 8/2)	Randomised controlled trial	Any type	3 days	Barthel Index	Comparison of PFTs between respiratory stretching group and control	MIF, MEF and VT increased in respiratory stretching group compared to control group
Song G.B. et al. 2015 [53]	40 patients with stroke: 20 in the CRE group (55.50 ± 11.43, 12/8) and 20 in the CEE group (58.30 ± 11.10, 11/9)	Randomised controlled trial	Any type	8 weeks	-	Comparison of a chest resistance and a chest expansion intervention group regarding PFTs and TIS	Significantly better FVC, FEV_1 and TIS in both groups, TIS significantly better in the chest resistance intervention group
Sutbeyaz S.T. et al. 2010 [54]	45 patients with stroke, randomised in three groups: 15 in IMT (62.8 ± 7.2, 8.7); 15 in breathing retraining, diaphragmatic breathing and pursed-lips breathing (60.8 ± 6.8, 8/7); 15 control group (61.9 ± 6.15, 8/7).	Randomised controlled trial	Any type	6 weeks	Barthel Index	Effect of exercise--breathing retraining (BRT) and IMT--improve on cardiopulmonary functions	In IMT group significantly improved FEV_1, FVC, VC and $FEF_{25-75\%}$, compared with the BRT and control groups. PEF was increased significantly in the BTR group compared with the IMT and control groups. MIP and MEP increased in the BRT group and MIP in the IMT group compared with baseline and the control group.
Tovar-Alcaraz et al. 2021 [55]	16 stroke survivors in the subacute phase: 8 in the experimental group (58 ± 12.9, 6/2) and 8 in the control group (56 ± 9.2, 6/2)	Randomised controlled trial	Any type	8 weeks	Postural Scale for Stroke Patients (PASS), Berg scale	Comparison of MIP, PFTs, trunk and postural control in the IMT group vs. control	Significant increase in MIP compared to baseline in both groups, more significant in the IMT group

Table 3. *Cont.*

Author/Date	Study Population/MEAN Age/Gender (M/F)	Type of Study	Type of Stroke	Follow-Up Time	Scale of Stroke Severity	Study Aim	Results
Vaz L. et al. 2021 [56]	50 patients with stroke with inspiratory muscle weakness (53 ± 11, 21/29)	Randomised controlled trial	Any type	3 months	NIHSS score, Fugl-Meyer Assessment	Comparison of 6MWT, MIP, MEP in a group after IMT and without	Change in 6MWD in both groups but no difference in MIP, MEP after intervention
Yoo H.J et al. 2018 [57]	40 patients with stroke: 20 in the intervention group (14/6) and 20 in the control group (12/8)	Randomised controlled trial	Any type	3 weeks	NIHSS score, Modified Barthel Index, Berg Balance Scale, Fugl-Meyer Assessment	Comparison of PFTs and stroke severity scores in two groups assigned either to bedside IEMT or no intervention	PFTs significantly improved in the intervention group after 3 weeks of IEMT independent of the improvement in stroke-related disabilities
Zheng Y. et al. 2021 [58]	60 patients within two months post-stroke: 30 in the experimental group (63.50 ± 10.36, 24/6) and 30 in the control group (67.23 ± 9.15, 19/11)	Randomised controlled trial	Any type	3 weeks	Berg Balance Scale, Modified Barthel Index	Comparison of PFTs, stroke severity scores of the RMT group using Liuzijue Qigong vs. conventional respiratory training	Significant improvement in MIP, FVC and PEF in both groups, better MIP and MEP and TIS in the Liuzijue Qigong group

3.7. Change in PFTs after Respiratory Muscle Training in Stroke Survivors

We further reviewed randomised controlled studies examining the effect of respiratory muscle training on PFTs in stroke survivors (Table 3). The number of patients included varied from 10–109. Only one study included more than 100 patients.

3.8. Stroke Population

The population of stroke survivors included in the studies was heterogeneous. Patients had suffered a subacute stroke within one to three weeks [39,48] and within two weeks [44] in the minority of studies, before respiratory muscle training started. Study participants who had suffered a stroke within six months of the start of respiratory muscle training were included in the studies carried out by Jung K.M. et al. [40], Kim J. et al. [43] and Tovar-Alcaraz et al. [55]. Zheng et al. [58] included patients with a stroke within two months and Yoo H.L. et al. [57] included patients undergoing respiratory muscle training after having a stroke within three months. Most of the studies examining the effect of respiratory muscle training included patients having suffered a stroke more than three months prior [35,38,43,45,47,49–53]. A few studies included patients up to five years after having suffered a stroke [50,56].

Inclusion criteria for the various studies were also heterogeneous: in various studies hemiparesis was an inclusion criterion [36,39–42,46,48,51,54–56]. Dysphagia was an inclusion criterion in the studies carried out by Guillen-Sola et al. [39], Liaw M.Y. et al. [47] and heart failure in the study by Chen P.C. et al. [37]. Age span, stroke functional scores and MIP scores differed among the studies. Most of the studies excluded patients with known pulmonary disease, with the exception of the studies carried out by Guillen-Sola et al. [39], Lee D.K. et al. [46] and Oh D. et al. [49].

3.9. Type of Respiratory Muscle Training Intervention

There was a high variability in interventions regarding the type of respiratory muscle training applied, the number of sets per session, the repetitions of each set, the duration of the program and the frequency of the training. The number of sets per session varied between one and ten sets and the repetitions in each set ranged between five and thirty (Table 3). One-time, short-effect interventions were also reported [52].

Nine studies reported inspiratory and expiratory muscle training in the intervention group [39,41,42,45–48,50,57], while one study compared inspiratory with expiratory muscle training [44] and another compared chest resistance with chest expansion training [53]. Most of the studies included performed both inspiratory with expiratory muscle training. Standard respiratory muscle training was compared to the Liuzijue training protocol in one study [58]. The programs were performed at home, at bedside or at hospital. Control groups also varied from the conventional stroke rehabilitation program to placebo respiratory training (Table 3). There were also studies, in which the standard respiratory muscle training involved the control group [58] or standard neuromuscular electrical simulation in patients with dysphagia [39].

As shown in Table 3, different follow-up schedules were used. PFTs of stroke survivors were measured after a course of respiratory muscle training program. The maximum respiratory muscle training duration was ten weeks [37]. A median value of 6 weeks of observation was performed in most of the studies. In the majority of the studies, the follow-up after the stroke regarding change in the pulmonary function tests coincided with the duration in respiratory muscle training. Only in three studies, the reported follow-up exceeded the duration of respiratory muscle training: Guillen-Solla et al. and Vaz L. et al. reported their results after a follow-up of three months [39,56], while in the study by Messagi-Sartor et al., the follow-up lasted six months [48].

3.10. Outcomes Measured

Spirometry parameters including FEV_1 and FVC were measured in the majority of studies [35,37,40–43,45–47,49,51,53–55,57,58]. The cardinal parameters reported were MIP and MEP. Scores of functional assessment of stroke severity, assessed by Trunk Impairment Score (TIS), Berg Balance Scale, Functional Independence Measure and diaphragm function were also reported.

IMT seems to increase PFTs, in particular MIP. Aydogan et al. reported no difference between groups regarding FEV_1, FVC, FEV_1/FVC, PEF, MIP and MEP prior to IMT [35]. After IMT, they reported a statistically significant increase in the intervention group regarding FEV_1 (0.30 ± 0.22 L), PEF (1,34 ± 1.22 L), MIP (14.9 ± 16.41 cm H_2O and 15.76 ± 16.81% predicted) and MEP (13.54 ± 16.85 cm H_2O and 7.73 ± 8.45% predicted) [35]. They reported a ca. 220 mL increase in FVC, which did not reach statistical significance. In the control group, only FVC reached statistical significance (0.32 ± 0.43 L) [35]. When the post-intervention values in the IMT group and controls were compared, only PEF and MIP were significantly better in the IMT group [35]. Britto et al. showed that MIP and inspiratory muscular endurance were significantly better in the IMT group compared to the control and that MIP improved significantly only in the intervention group [36]: MIP prior to intervention was 67.8 ± 14.6 cm H_2O in the IMT group at baseline and rose to 102.2 ± 26.0 cm H_2O, while in the control group, MIP was 45.6 ± 13.8 cm H_2O and increased to 56.7 ± 8.7 cm H_2O [36]. In stroke patients with heart failure, a 10-week IMT program was associated with significantly better MIP, FEV_1, FEV_1% and FVC% compared to baseline [37]: IMT resulted in a 20.91 ± 19.73 cm H_2O increase in MIP, a 0.22 ± 0.28 L increase in FEV_1 and 12.47 ± 13.52% predicted of FEV_1% predicted and 4.95 ± 6.75% predicted of the FVC. While no statistically significant changes were seen in the control group, a comparison between the IMT group and controls showed a significant increase in MIP in the IMT group [37]. Compared to aerobic exercise, an IMT program for 4 weeks was associated with a greater improvement in FEV_1, FVC and 6MWT, although better FEV_1 and FVC were seen in both groups [40]. A 6-week IMT program was also associated with significant improvement in MIP, inspiratory muscle endurance and diaphragm thickness [38]: the IMT program resulted in an impressive increase in MIP from 50.05 ± 21.92 cm H_2O at baseline to 90.42 ± 30.91 cm H_2O post IMT, while the increase in the control group was only 10 cm H_2O [38].

Kilicoglu et al. reported a significant improvement in FEV_1 (0.15 ± 0.33 L or 7.30 ± 18.20% predicted) and FVC (0.31 ± 0.39 L or 2.65 ± 31.82% predicted) after respiratory training, which were associated with morphological parameters of diaphragm thickness, measured with ultrasound at baseline [41]. Similar results regarding FEV_1 and FVC were reported by Kim C.Y. et al. [42]. In the study by Oh et al., FEV_1 and FVC significantly increased by approximately 400 mL in the IMT group and the same applied for PEF (increase from 3.1 L/Min to 3.8 L/Min) [49]. Sutbeyaz et al. showed that IMT significantly improved spirometric values compared to a breathing retraining program or controls [54]: the mean increase in FEV_1 was 220 mL from 2.48 to 2.71 L compared to no difference in the other two groups and the mean increase in FVC was 230 mL, from 3.22 to 3.45 L [54]. Regarding pulmonary flows, no improvement was seen post-training in none of the groups while MIP and MEP increased in all three groups by approximately 2 to 7 cm H_2O [54].

A combined IMT-EMT intervention in patients with stroke and dysphagia showed the greater improvement in MIP and MEP, while all groups had no differences regarding respiratory complications [39]: MIP increased by 21.1 ± 13.1 cm H_2O at 3 weeks and by 18.3 ± 14.5 cm H_2O in the three months after the IMT-EMT intervention, an increase of approximately 21% and 18%, respectively, while MEP increased by 26.4 ± 16.9 cm H_2O in three weeks and 32.4 ± 21.2 cm H_2O in three months (increase of 26.4% and 19.4%, respectively), almost two to three times greater as in the other two groups [39]. MIP was significantly improved by 45.90 ± 29.31 cm H_2O after a combined respiratory muscle training program in patients with respiratory muscle weakness, dysphagia, and dysarthria compared to 5.45 ± 20.18 cm H_2O in the control group [47]. Significant changes between groups after the intervention were seen for FEV_1 and FVC [47]. FVC, FEV_1, PEF, 6MWT and Borg Dyspnea scores were significantly better after respiratory muscle training in the study by Kim J. et al. [43]. Improvements in FEV_1 and FVC after respiratory muscle training were related to better muscle activity of the trunk muscles [46,53]. In the study by Lee K. et al., the MIP, MEP, PEF, FEV_1, PIF and VC were significantly increased within the intervention and the control groups [45]. Regarding the between-group comparison, MIP, MEP, PEF and PIF were significantly increased in the RMT group compared with the control group [45]. MIP and MEP improvement after IEMT was also seen in patients with subacute stroke and was associated with a reduced rate of respiratory complications, including aspiration pneumonia [48]: MIP increased by 9.61 cm H_2O or 10.2% more in the IEMT group compared to the controls, while MEP increased by 10.2 cm H_20 or 7% more. Home-based respiratory muscle training programs have been shown to achieve similar results regarding MIP and MEP [50]. An IMT of increasing intensity from 15% to 60% of MIP was associated with a greater improvement in MIP compared to the control group with a fixed load 7 cm H_2O [55]; although there was no difference regarding FEV_1, FVC and PEF, MIP increased from 61.5 ± 31.5 to 80.5 ± 35.1 cm H_2O in the intervention group [55]. Zheng et al. reported better PFTs and stroke functional scores after an intervention program using Liuzijue respiratory training compared to standard respiratory training [58].

Other studies have failed to show an additive beneficial effect of IMT or EMT in patients with hemiplegia, as measured using peak expiratory cough flow [44] or MIP and MEP [56] and improvements in pulmonary function after intervention were not always associated with improvements in the functional status post-stroke [57]. A single session of respiratory stimulation through Proprioceptive Neuromuscular Facilitation (PNF) showed an increase in FEV_1/FVC but not of the other spirometric parameters, compared to the control group [51].

4. Discussion

Stroke represents the major cause of disability worldwide [59]. Stroke usually causes abnormalities in muscular tone, motor coordination and postural control [60]. Respiratory muscles can also be affected [61]. This leads to a decreased respiratory function due to respiratory muscle weakness. Besides the respiratory muscle insult, centrally induced changes in the respiratory efferent system can cause changes in the respiratory pattern and breathing frequency [61,62]. Moreover, since respiratory muscles do not function when isolated from the rest of the body, changes in the neck or trunk musculature due to spasticity and contracture in the hemiplegic site can lead to abnormal ventilatory patterns, with a dysfunction of the muscles of the affected side in cases of hemiparesis or hemiplegia and a compensatory hyperfunction of the muscles of the non-affected side, in order to ensure adequate minute ventilation [61]. These stroke-induced changes are mostly characterised by a restrictive pattern of lung function [15,61]. In extreme cases, the restrictive disturbance can be so severe that hypercapnia and hypoxemia may be seen, particularly in the acute and subacute phases [61,62].

Various studies have examined how respiratory muscle function, respiratory volumes, breathing rates, thoracic movement and cough efficacy are affected in stroke patients [16,33,63]. These changes are of particular interest in these patients since they can be used in order to assess the risk of a respiratory complication, such as aspiration or health-care-associated pneumonia or the risk of developing atelectasis [64].

Chronic dyspnea in patients after stroke is usually attributed either to the stroke per se, preexisting respiratory and cardiovascular conditions such as COPD or heart failure or to psychosocial reasons [65]. However, ventilatory changes are underdiagnosed [62]. This might lead to the initiation of treatments that are little effective (e.g., bronchodilators for patients with a restrictive pattern). Pulse oximetry and blood gas analysis are usually used to decide whether these patients need supplemental oxygen therapy.

PFTs are a simple, non-invasive, inexpensive method of assessing lung function and are widely used in respiratory medicine [6]. PFTs can also be performed at the bedside with simple equipment. Despite their obvious advantages, PFTs require the adequate cooperation of the examined subjects, in order to become valid results [6]. Due to aphasia, facial palsy, reduced consciousness, coordination disturbances or trunk muscle weakness, patients with stroke rarely undergo PFTs in order to assess their respiratory disorders [66,67]. More advanced techniques, such as cardiopulmonary exercise testing, are rarely used as diagnostic tools for the differential diagnosis of dyspnea in patients with stroke, since they are more time consuming and expensive and require excellent patient cooperation [15].

The aim of our review was to examine how PFTs change in post-stroke patients and whether respiratory muscle training can improve PFTs and thus respiratory function of these patients.

Results from the studies included in this review indicate that stroke patients have lower spirometric variables compared to healthy individuals and compared to normal reference values [12,17,19–21,23,25,31]. This involves mainly FEV_1 and FVC. The most common pattern observed was restrictive [12,17,19–21,23,25,31] while only one study reported an obstructive pattern, as suggested from the FEV_1/FVC ratio [12]. Other studies, however, have failed to show a reduction of forced dynamic lung volumes and only slow dynamic measurements such as VC were reduced in patients with stroke [11]. Interestingly, even if the above-mentioned values were normal, dynamic flow measurements such as PEF and $FEF_{25–75\%}$ were reduced, compared to healthy subjects [13,18]. PEF reduction is important, since it depicts a reduced cough reflex. More consistent results were seen when the inspiratory and expiratory pressures of MIP and MEP were examined, suggesting that both inspiration and expiration are affected [16,24,26,29,30].

Reductions in PFTs are associated with reduced functional status and reduced trunk control in patients with stroke in our review of the literature. FEV_1, FVC, MIP, MEP and PEF were associated with TIS, Barthel Index or Brunnstrom score [11,13,18,20,21,28–30]. These results are in accordance with a recently published meta-analysis of trunk control ability and respiratory function in stroke patients [68]. PFTs are also associated with diaphragm excursion in studies using ultrasonography [16,19,20]. However, there is not a single PFT value, which is consistently reduced in patients with stroke, therefore making standardisation of pulmonary functional impairment difficult. The lack of a single PFT value with a 100% sensitivity and specificity for the diagnosis of respiratory disease is not uncommon: even for extremely common pulmonary diseases, such as asthma and COPD, spirometry should be interpreted as a whole and many times repeated at different timepoints, in order to set a valid diagnosis [10,67,69]. The same principle seems to apply for patients with stroke.

It is important to notice that not only PFTs can be affected due to stroke but also that impaired lung function parameters are a risk factor for the development of cardiovascular complications, including stroke [10]. This is shown in many epidemiological studies, examining the predictive value of FEV_1 and FVC [10,69–71]. One could assume that reduced FEV_1 or FVC might represent underdiagnosed cases of COPD, which is known to affect the cardiovascular system, mainly due to the common detrimental effects of smoking in systemic inflammation [69]. However, this is not the only explanation, since restrictive PFTs (therefore exclusive of COPD) are also associated with increased cardiovascular risk [70]. In addition, COPD is clearly associated with an increased risk of hemorrhagic strokes, but it is unclear whether the increased frequency of ischemic stroke in COPD is the result of COPD per se or of a confounding effect [72]. Despite the presence or not of an etiologic correlation, the fact is that COPD patients who suffer a stroke have worse prognosis than patients with stroke without COPD [73].

On the other hand, the association of stroke severity indexes with reduced PFTs can increase our awareness of possible respiratory complications, thus enhancing a more proactive respiratory management of these patients. Therefore, we reviewed whether the respiratory function of patients with stroke can be improved, particularly with respiratory muscle training. PFTs showed a significant improvement after respiratory muscle training in the majority of the studies, particularly regarding MIP and MEP, although other PFT parameters were also increased [35–38,41,42,45,47,54]. Pai and Li reported that only MIP and MEP were significantly better after the respiratory muscle training and that the lack of improvement seen in FEV_1 and FVC might be the result of a too-short training period [68]. In their meta-analysis, Pozuelo-Carascosa et al. reported that respiratory muscle training improved FEV_1, FVC, PEF, MEP, MIP and walking ability assessed using the 6MWT [74]. The RMT interventions were associated with a 12.2% increase in FEV_1% predicted while FVC improved by 6.75% [74] PEF also increased by 46.97 L/sec and MEP and MIP improved their baseline values by 10.05 and 22.40 cm H_20, respectively [74]. Other meta-analyses have reported similar findings [2,75]. However, results are not consistent; Fabero-Garrido et al. reported in their meta-analysis that only MIP and PEF were affected from respiratory muscle training in the short term but not FEV_1 and MEP, although the difference in FEV_1 was close to statistical significance [15].

We did not examine changes in our review regarding walking capacity or exercise capacity using ergospirometry. Pozuelo-Carascosa et al. reported that the walking distance increased, although minimally [74]. However, small differences can have significant implications in patients' daily activities. Despite the PFTs improvement dyspnea, the Barthel Index and Berg Balance Scale were not improved [74]. Lack of improvement regarding quality of life and functional improvement has been reported in other studies [36,56].

The reasons for the lack of consistent changes in PFTs throughout the studies are not always clear. The small number of participants and the different protocols used make a results comparison difficult. However, it is important to keep in mind that the evaluation of pulmonary function directly after rehabilitation might not be showing all benefits: duration of follow-up lasted in most of the studies three to eight weeks and pulmonary function tests were reported directly at the end of the rehabilitation program. Only three studies in our review reported follow-up of three to six months after a rehabilitation program [39,48,56].

Both IMT and combined IMT and EMT showed positive results, although attempts have been undertaken to further improve the respiratory muscle training [58]. This is in accordance with the findings of Pozuelo-Carrascosa et al. [74]. The effects of respiratory muscle training were in addition to the physiotherapy program, which includes components known as physiotherapy, strength training and aerobic exercise.

The most important clinical consequence of the PFTs improvement after respiratory muscle training is the reduction of respiratory complications, as shown in the meta-analysis of Menezes et al. [76] analysing the results of the studies by Kulnik and Messagi-Sartor et al. [44,48]. Therefore, bedside and home-based care to improve trunk control and respiratory muscles can improve respiratory function.

5. Limitations

Several limitations apply when interpreting PFTs in stroke patients; apart from patient cooperation, studies recruited a small number of participants, in different stroke populations, with different intervention protocols and different measured outcomes. As mentioned in the discussion, measurements of PFTs were mostly carried out immediately at the end of the respiratory muscle training program, thus missing beneficial effects of training which could occur later in the course of recovery. Very few studies have reported consecutive PFTs in stroke patients. Thus, a baseline snapshot and a follow-up measurement in short time intervals cannot lead to safe conclusions regarding the long-term consequences of stroke regarding the respiratory function and the duration of improvement after a rehabilitation program. Moreover, to avoid bias, almost all studies have excluded patients with pulmonary comorbidities prior to stroke. However, this excludes the extrapolation of results in the majority of the patients in clinical practice who have concomitant cardiovascular and pulmonary risk factors.

6. Conclusions

Considering the results of the studies reviewed, it seems that PFTs are affected after stroke and this is seen in stroke survivors in the subacute and chronic phase. Although not performed routinely in stroke patients, PFTs can provide valuable information regarding the risk of further respiratory complications and correlate to stroke severity scores. Although, there is not a single PFT marker, which is reduced in all patients, MIP and MEP might be more sensitive to identify patients at risk. Respiratory muscle training can significantly improve PFTs, thus reducing respiratory complications. However, the small number of studies and study participants, with differences in the study protocols, underlines the need for further research in this field.

Author Contributions: F.D. and K.B. reviewed the literature, screened the abstracts of the reference list, deleted duplicates and citations not meeting the inclusion criteria and assessed the articles; P.S. and A.S. (Anastasia Sousanidou) solved any disagreements regarding screening or the selection process; F.D. and K.B. wrote the manuscript; A.G., P.S., F.C., C.K., A.S. (Aspasia Serdari) and D.T. reviewed the tables, the presentation of the data and the methodology; P.S., S.V., N.A. and K.V. carried out the review and editing; F.D., K.B. and C.-M.T. wrote the final version. All authors have read and agreed to the published version of the manuscript.

Funding: We acknowledge the support of this work by the project "Study of the interrelationships between neuroimaging, neurophysiological and biomechanical biomarkers in stroke rehabilitation (NEURO-BIO-MECH in stroke rehab)" (MIS 5047286), which was implemented under the action "Support for Regional Excellence" funded by the operational program "Competitiveness, Entrepreneurship and Innovation" (NSRFm2014-2020) and co-financed by Greece and the European Union (the European Regional Development Fund).

Institutional Review Board Statement: Not applicable.

Informed Consent Statement: Not applicable.

Data Availability Statement: All data discussed within this manuscript are available on PubMed.

Conflicts of Interest: The authors declare no conflicts of interest.

Abbreviations

BRT: exercise-breathing retraining, EMG: electromyography, EMT: expiratory muscle training, ERV: expiratory residual volume, $FEF_{25-75\%}$: forced expiratory flow 25–75, FEV_1: forced expiratory volume in 1 s, FEVC: forced expiratory vital capacity, FIVC: forced inspiratory vital capacity, FVC: forced vital capacity, IC: inspiratory capacity, IMT: inspiratory muscle training, MEF: maximal expiratory flow, MEP: maximal expiratory pressure, MIP: maximal inspiratory pressure, NIHSS: National Institutes of Health Stroke Scale, PEF: peak expiratory flow, PNF: Proprioceptive Neuromuscular Facilitation, PFTs: pulmonary function tests, RCT: randomised controlled trial, TIS: trunk impairment scale, VC: vital capacity

References

1. Gkantzios, A.; Tsiptsios, D.; Karatzetzou, S.; Kitmeridou, S.; Karapepera, V.; Giannakou, E.; Vlotinou, P.; Aggelousis, N.; Vadikolias, K. Stroke and Emerging Blood Biomarkers: A Clinical Prospective. *Neurol. Int.* **2022**, *14*, 784–803. [CrossRef] [PubMed]
2. Gomes-Neto, M.; Saquetto, M.B.; Silva, C.M.; Carvalho, V.O.; Ribeiro, N.; Conceição, C.S. Effects of Respiratory Muscle Training on Respiratory Function, Respiratory Muscle Strength, and Exercise Tolerance in Patients Poststroke: A Systematic Review With Meta-Analysis. *Arch. Phys. Med. Rehabil.* **2016**, *97*, 1994–2001. [CrossRef] [PubMed]
3. Javaheri, S.; Peker, Y.; Yaggi, H.K.; Bassetti, C.L.A. Obstructive sleep apnea and stroke: The mechanisms, the randomized trials, and the road ahead. *Sleep Med. Rev.* **2022**, *61*, 101568. [CrossRef]
4. Kishore, A.K.; Vail, A.; Chamorro, A.; Garau, J.; Hopkins, S.J.; Di Napoli, M.; Kalra, L.; Langhorne, P.; Montaner, J.; Roffe, C.; et al. How is pneumonia diagnosed in clinical stroke research? A systematic review and meta-analysis. *Stroke* **2015**, *46*, 1202–1209. [CrossRef]
5. Pelosi, P.; Severgnini, P.; Chiaranda, M. An integrated approach to prevent and treat respiratory failure in brain-injured patients. *Curr. Opin. Crit. Care* **2005**, *11*, 37–42. [CrossRef]

6. Miller, M.R.; Crapo, R.; Hankinson, J.; Brusasco, V.; Burgos, F.; Casaburi, R.; Coates, A.; Enright, P.; van der Grinten, C.P.M.; Gustafsson, P.; et al. General considerations for lung function testing. *Eur. Respir. J.* **2005**, *26*, 153–161. [CrossRef]
7. Hozawa, A.; Billings, J.L.; Shahar, E.; Ohira, T.; Rosamond, W.D.; Folsom, A.R. Lung function and ischemic stroke incidence: The Atherosclerosis Risk in Communities study. *Chest* **2006**, *130*, 1642–1649. [CrossRef] [PubMed]
8. Truelsen, T.; Prescott, E.; Lange, P.; Schnohr, P.; Boysen, G. Lung function and risk of fatal and non-fatal stroke. The Copenhagen City Heart Study. *Int. J. Epidemiol.* **2001**, *30*, 145–151. [CrossRef]
9. Wannamethee, S.G.; Shaper, A.G.; Ebrahim, S. Respiratory function and risk of stroke. *Stroke* **1995**, *26*, 2004–2010. [CrossRef]
10. Duong, M.; Islam, S.; Rangarajan, S.; Leong, D.; Kurmi, O.; Teo, K.; Killian, K.; Dagenais, G.; Lear, S.; Wielgosz, A.; et al. Mortality and cardiovascular and respiratory morbidity in individuals with impaired FEV_1 (PURE): An international, community-based cohort study. *Lancet Glob. Health* **2019**, *7*, e613–e623. [CrossRef]
11. Annoni, J.-M.; Ackermann, D.; Kesselring, J. Respiratory function in chronic hemiplegia. *Int. Disabil. Stud.* **1990**, *12*, 78–80. [CrossRef] [PubMed]
12. Ezeugwu, V.E.; Olaogun, M.; Mbada, C.E.; Adedoyin, R. Comparative lung function performance of stroke survivors and age-matched and sex-matched controls. *Physiother. Res. Int.* **2013**, *18*, 212–219. [CrossRef] [PubMed]
13. Jeong, Y.J.; Kim, G.S.; Jeong, Y.G.; Moon, H.I. Can Pulmonary Function Testing Predict the Functional Outcomes of Poststroke Patients?: An Observational Study. *Am. J. Phys. Med. Rehabil.* **2020**, *99*, 1145–1149. [CrossRef] [PubMed]
14. Van Criekinge, T.; Truijen, S.; Schröder, J.; Maebe, Z.; Blanckaert, K.; van der Waal, C.; Vink, M.; Saeys, W. The effectiveness of trunk training on trunk control, sitting and standing balance and mobility post-stroke: A systematic review and meta-analysis. *Clin. Rehabil.* **2019**, *33*, 992–1002. [CrossRef] [PubMed]
15. Fabero-Garrido, R.; del Corral, T.; Angulo-Díaz-Parreño, S.; Plaza-Manzano, G.; Martín-Casas, P.; Cleland, J.A.; Fernández-de-Las-Peñas, C.; López-de-Uralde-Villanueva, I. Respiratory muscle training improves exercise tolerance and respiratory muscle function/structure post-stroke at short term: A systematic review and meta-analysis. *Ann. Phys. Rehabil. Med.* **2022**, *65*, 101596. [CrossRef]
16. de Almeida, I.C.L.; Clementino, A.C.C.R.; Rocha, E.H.T.; Brandão, D.C.; de Andrade, A.D. Effects of hemiplegy on pulmonary function and diaphragmatic dome displacement. *Respir. Physiol. Neurobiol.* **2011**, *178*, 196–201. [CrossRef]
17. Fugl-Meyer, A.R.; Linderholm, H.; Wilson, A.F. Restrictive ventilatory dysfunction in stroke: Its relation to locomotor function. *Scand. J. Rehabil. Med. Suppl.* **1983**, *9*, 118–124.
18. Jandt, S.R.; Caballero, R.M.d.S.; Junior, L.A.F.; Dias, A.S. Correlation between trunk control, respiratory muscle strength and spirometry in patients with stroke: An observational study. *Physiother. Res. Int.* **2011**, *16*, 218–224. [CrossRef]
19. Jung, K.-J.; Park, J.-Y.; Hwang, D.-W.; Kim, J.-H.; Kim, J.-H. Ultrasonographic diaphragmatic motion analysis and its correlation with pulmonary function in hemiplegic stroke patients. *Ann. Rehabil. Med.* **2014**, *38*, 29–37. [CrossRef]
20. Khedr, E.M.; El Shinawy, O.; Khedr, T.; Abdel Aziz Ali, Y.; Awad, E.M. Assessment of corticodiaphragmatic pathway and pulmonary function in acute ischemic stroke patients. *Eur. J. Neurol.* **2000**, *7*, 509–516. [CrossRef]
21. Kimura, Y.; Takahashi, M.; Wada, F.; Hachisuka, K. Differences in the peak cough flow among stroke patients with and without dysphagia. *J. UOEH* **2013**, *35*, 9–16. [CrossRef] [PubMed]
22. Kulnik, S.T.; Birring, S.S.; Hodsoll, J.; Moxham, J.; Rafferty, G.F.; Kalra, L. Higher cough flow is associated with lower risk of pneumonia in acute stroke. *Thorax* **2016**, *71*, 474–475. [CrossRef]
23. Liaw, M.-Y.; Wang, L.-Y.; Pong, Y.-P.; Tsai, Y.-C.; Huang, Y.-C.; Yang, T.-H.; Lin, M.-C. Preliminary investigation of cardiopulmonary function in stroke patients with stable heart failure and exertional dyspnea. *Medicine* **2016**, *95*, e5071. [CrossRef] [PubMed]
24. Lista Paz, A.; Gonzalez Doniz, L.; Ortigueira Garcia, S.; Saleta Canosa, J.L.; Moreno Couto, C. Respiratory Muscle Strength in Chronic Stroke Survivors and Its Relation With the 6-Minute Walk Test. *Arch. Phys. Med. Rehabil.* **2016**, *97*, 266–272. [CrossRef]
25. Lista-Paz, A.; Kuisma, R.; Canosa, J.L.S.; Sebio García, R.; Gonzalez Doniz, L. Pulmonary function in patients with chronic stroke compared with a control group of healthy people matched by age and sex. *Physiother. Theory Pract.* **2023**, *39*, 918–926. [CrossRef] [PubMed]
26. Luvizutto, G.J.; Dos Santos, M.R.L.; Sartor, L.C.A.; da Silva Rodrigues, J.C.; da Costa, R.D.M.; Braga, G.P.; de Oliveira Antunes, L.C.; Souza, J.T.; de Carvalho Nunes, H.R.; Bazan, S.G.Z.; et al. Evaluation of Respiratory Muscle Strength in the Acute Phase of Stroke: The Role of Aging and Anthropometric Variables. *J. Stroke Cerebrovasc. Dis.* **2017**, *26*, 2300–2305. [CrossRef]
27. Min, S.W.; Oh, S.H.; Kim, G.C.; Sim, Y.J.; Kim, D.K.; Jeong, H.J. Clinical Importance of Peak Cough Flow in Dysphagia Evaluation of Patients Diagnosed With Ischemic Stroke. *Ann. Rehabil. Med.* **2018**, *42*, 798–803. [CrossRef]
28. Nunez Filha, M.C.; Mascarenhas, L.; Messias, D.; Furtado, C.; Dias, C.; Dantas, M.C.; Almeida, L.R.S.; Pinto, E.B. Stroke Severity and Maximum Inspiratory Pressure are Independently Associated with Functional Mobility in Individuals After Stroke. *J. Stroke Cerebrovasc. Dis.* **2020**, *29*, 105375. [CrossRef]
29. Pinheiro, M.B.; Polese, J.C.; Faria, C.D.; Machado, G.C.; Parreira, V.F.; Britto, R.R.; Teixeira-Salmela, L.F. Inspiratory muscular weakness is most evident in chronic stroke survivors with lower walking speeds. *Eur. J. Phys. Rehabil. Med.* **2014**, *50*, 301–307.
30. Santos, R.S.D.; Dall'alba, S.C.F.; Forgiarini, S.G.I.; Rossato, D.; Dias, A.S.; Forgiarini Junior, L.A. Relationship between pulmonary function, functional independence, and trunk control in patients with stroke. *Arq. Neuropsiquiatr.* **2019**, *77*, 387–392. [CrossRef]
31. Sezer, N.; Ordu, N.K.; Sutbeyaz, S.T.; Koseoglu, B.F. Cardiopulmonary and metabolic responses to maximum exercise and aerobic capacity in hemiplegic patients. *Funct. Neurol.* **2004**, *19*, 233–238. [PubMed]

32. Teixeira-Salmela, L.F.; Parreira, V.F.; Britto, R.R.; Brant, T.C.; Inácio, E.P.; Alcântara, T.O.; Carvalho, I.F. Respiratory pressures and thoracoabdominal motion in community-dwelling chronic stroke survivors. *Arch. Phys. Med. Rehabil.* **2005**, *86*, 1974–1978. [CrossRef] [PubMed]
33. Voyvoda, N.; Yücel, C.; Karataş, G.; Oğuzülgen, I.; Oktar, S. An evaluation of diaphragmatic movements in hemiplegic patients. *Br. J. Radiol.* **2012**, *85*, 411–414. [CrossRef] [PubMed]
34. Xiao, L.J.; Guo, Q.; Huang, F.Y.; Liao, M.X.; Zhang, L.L.; Yan, T.B. [Correlation between swallowing function and pulmonary ventilation function and respiratory muscles strength in patients with dysphagia after stroke]. *Zhonghua Yi Xue Za Zhi* **2020**, *100*, 504–508. [CrossRef]
35. Aydoğan Arslan, S.; Uğurlu, K.; Sakizli Erdal, E.; Keskin, E.D.; Demirgüç, A. Effects of Inspiratory Muscle Training on Respiratory Muscle Strength, Trunk Control, Balance and Functional Capacity in Stroke Patients: A single-blinded randomized controlled study. *Top. Stroke Rehabil.* **2022**, *29*, 40–48. [CrossRef]
36. Britto, R.R.; Rezende, N.R.; Marinho, K.C.; Torres, J.L.; Parreira, V.F.; Teixeira-Salmela, L.F. Inspiratory muscular training in chronic stroke survivors: A randomized controlled trial. *Arch. Phys. Med. Rehabil.* **2011**, *92*, 184–190. [CrossRef]
37. Chen, P.-C.; Liaw, M.-Y.; Wang, L.-Y.; Tsai, Y.-C.; Hsin, Y.-J.; Chen, Y.-C.; Chen, S.-M.; Lin, M.-C. Inspiratory muscle training in stroke patients with congestive heart failure: A CONSORT-compliant prospective randomized single-blind controlled trial. *Medicine* **2016**, *95*, e4856. [CrossRef]
38. Cho, J.-E.; Lee, H.-J.; Kim, M.-K.; Lee, W.-H. The improvement in respiratory function by inspiratory muscle training is due to structural muscle changes in patients with stroke: A randomized controlled pilot trial. *Top. Stroke Rehabil.* **2018**, *25*, 37–43. [CrossRef]
39. Guillén-Solà, A.; Messagi Sartor, M.; Bofill Soler, N.; Duarte, E.; Barrera, M.C.; Marco, E. Respiratory muscle strength training and neuromuscular electrical stimulation in subacute dysphagic stroke patients: A randomized controlled trial. *Clin. Rehabil.* **2017**, *31*, 761–771. [CrossRef]
40. Jung, K.-M.; Bang, D.-H. Effect of inspiratory muscle training on respiratory capacity and walking ability with subacute stroke patients: A randomized controlled pilot trial. *J. Phys. Ther. Sci.* **2017**, *29*, 336–339. [CrossRef]
41. Kılıçoğlu, M.S.; Yurdakul, O.V.; Çelik, Y.; Aydın, T. Investigating the correlation between pulmonary function tests and ultrasonographic diaphragm measurements and the effects of respiratory exercises on these parameters in hemiplegic patients. *Top. Stroke Rehabil.* **2022**, *29*, 218–229. [CrossRef] [PubMed]
42. Kim, C.-Y.; Lee, J.-S.; Kim, H.-D.; Kim, I.-S. Effects of the combination of respiratory muscle training and abdominal drawing-in maneuver on respiratory muscle activity in patients with post-stroke hemiplegia: A pilot randomized controlled trial. *Top. Stroke Rehabil.* **2015**, *22*, 262–270. [CrossRef]
43. Kim, J.; Park, J.H.; Yim, J. Effects of respiratory muscle and endurance training using an individualized training device on the pulmonary function and exercise capacity in stroke patients. *Med. Sci. Monit.* **2014**, *20*, 2543–2549. [CrossRef] [PubMed]
44. Kulnik, S.T.; Birring, S.S.; Moxham, J.; Rafferty, G.F.; Kalra, L. Does respiratory muscle training improve cough flow in acute stroke? Pilot randomized controlled trial. *Stroke* **2015**, *46*, 447–453. [CrossRef] [PubMed]
45. Lee, K.; Park, D.; Lee, G. Progressive Respiratory Muscle Training for Improving Trunk Stability in Chronic Stroke Survivors: A Pilot Randomized Controlled Trial. *J. Stroke Cerebrovasc. Dis.* **2019**, *28*, 1200–1211. [CrossRef] [PubMed]
46. Lee, D.-K.; Kim, S.-H. The effect of respiratory exercise on trunk control, pulmonary function, and trunk muscle activity in chronic stroke patients. *J. Phys. Ther. Sci.* **2018**, *30*, 700–703. [CrossRef] [PubMed]
47. Liaw, M.-Y.; Hsu, C.-H.; Leong, C.-P.; Liao, C.-Y.S.; Wang, L.-Y.; Lu, C.-H.; Lin, M.-C. Respiratory muscle training in stroke patients with respiratory muscle weakness, dysphagia, and dysarthria—A prospective randomized trial. *Medicine* **2020**, *99*, e19337. [CrossRef] [PubMed]
48. Messaggi-Sartor, M.; Guillen-Solà, A.; Depolo, M.; Duarte, E.; Rodríguez, D.A.; Barrera, M.-C.; Barreiro, E.; Escalada, F.; Orozco-Levi, M.; Marco, E. Inspiratory and expiratory muscle training in subacute stroke: A randomized clinical trial. *Neurology* **2015**, *85*, 564–572. [CrossRef]
49. Oh, D.; Kim, G.; Lee, W.; Shin, M.M.S. Effects of inspiratory muscle training on balance ability and abdominal muscle thickness in chronic stroke patients. *J. Phys. Ther. Sci.* **2016**, *28*, 107–111. [CrossRef]
50. Parreiras de Menezes, K.K.; Nascimento, L.R.; Ada, L.; Avelino, P.R.; Polese, J.C.; Mota Alvarenga, M.T.; Barbosa, M.H.; Teixeira-Salmela, L.F. High-Intensity Respiratory Muscle Training Improves Strength and Dyspnea Poststroke: A Double-Blind Randomized Trial. *Arch. Phys. Med. Rehabil.* **2019**, *100*, 205–212. [CrossRef]
51. Ptaszkowska, L.; Ptaszkowski, K.; Halski, T.; Taradaj, J.; Dymarek, R.; Paprocka-Borowicz, M. Immediate effects of the respiratory stimulation on ventilation parameters in ischemic stroke survivors: A randomized interventional study (CONSORT). *Medicine* **2019**, *98*, e17128. [CrossRef] [PubMed]
52. Rattes, C.; Campos, S.L.; Morais, C.; Gonçalves, T.; Sayão, L.B.; Galindo-Filho, V.C.; Parreira, V.; Aliverti, A.; de Andrade, A.D. Respiratory muscles stretching acutely increases expansion in hemiparetic chest wall. *Respir. Physiol. Neurobiol.* **2018**, *254*, 16–22. [CrossRef] [PubMed]
53. Song, G.B.; Park, E.C. Effects of chest resistance exercise and chest expansion exercise on stroke patients' respiratory function and trunk control ability. *J. Phys. Ther. Sci.* **2015**, *27*, 1655–1658. [CrossRef] [PubMed]
54. Sutbeyaz, S.T.; Koseoglu, F.; Inan, L.; Coskun, O. Respiratory muscle training improves cardiopulmonary function and exercise tolerance in subjects with subacute stroke: A randomized controlled trial. *Clin. Rehabil.* **2010**, *24*, 240–250. [CrossRef] [PubMed]

55. Tovar-Alcaraz, A.; de Oliveira-Sousa, S.L.; Leon-Garzon, M.C.; Gonzalez-Carrillo, M.J. Effects of inspiratory muscle training on respiratory function and balance in stroke survivors: A randomized controlled trial. *Rev. Neurol.* **2021**, *72*, 112–120. [CrossRef] [PubMed]
56. Vaz, L.d.O.; Almeida, J.d.C.; Froes, K.S.d.S.O.; Dias, C.; Pinto, E.B.; Oliveira-Filho, J. Effects of inspiratory muscle training on walking capacity of individuals after stroke: A double-blind randomized trial. *Clin. Rehabil.* **2021**, *35*, 1247–1256. [CrossRef] [PubMed]
57. Yoo, H.-J.; Pyun, S.-B. Efficacy of Bedside Respiratory Muscle Training in Patients With Stroke: A Randomized Controlled Trial. *Am. J. Phys. Med. Rehabil.* **2018**, *97*, 691–697. [CrossRef]
58. Zheng, Y.; Zhang, Y.; Li, H.; Qiao, L.; Fu, W.; Yu, L.; Li, G.; Yang, J.; Ni, W.; Yong, Z.; et al. Comparative Effect of Liuzijue Qigong and Conventional Respiratory Training on Trunk Control Ability and Respiratory Muscle Function in Patients at an Early Recovery Stage From Stroke: A Randomized Controlled Trial. *Arch. Phys. Med. Rehabil.* **2021**, *102*, 423–430. [CrossRef]
59. Mozaffarian, D.; Benjamin, E.J.; Go, A.S.; Arnett, D.K.; Blaha, M.J.; Cushman, M.; de Ferranti, S.; Després, J.-P.; Fullerton, H.J.; Howard, V.J.; et al. Heart disease and stroke statistics—2015 update: A report from the American Heart Association. *Circulation* **2015**, *131*, e29–e322. [CrossRef]
60. Canning, C.G.; Ada, L.; Adams, R.; O'Dwyer, N.J. Loss of strength contributes more to physical disability after stroke than loss of dexterity. *Clin. Rehabil.* **2004**, *18*, 300–308. [CrossRef]
61. Lanini, B.; Bianchi, R.; Romagnoli, I.; Coli, C.; Binazzi, B.; Gigliotti, F.; Pizzi, A.; Grippo, A.; Scano, G. Chest wall kinematics in patients with hemiplegia. *Am. J. Respir. Crit. Care Med.* **2003**, *168*, 109–113. [CrossRef] [PubMed]
62. Lanini, B.; Gigliotti, F.; Coli, C.; Bianchi, R.; Pizzi, A.; Romagnoli, I.; Grazzini, M.; Stendardi, L.; Scano, G. Dissociation between respiratory effort and dyspnoea in a subset of patients with stroke. *Clin. Sci.* **2002**, *103*, 467–473. [CrossRef] [PubMed]
63. Cohen, E.; Mier, A.; Heywood, P.; Murphy, K.; Boultbee, J.; Guz, A. Diaphragmatic movement in hemiplegic patients measured by ultrasonography. *Thorax* **1994**, *49*, 890–895. [CrossRef] [PubMed]
64. Martino, R.; Foley, N.; Bhogal, S.; Diamant, N.; Speechley, M.; Teasell, R. Dysphagia after stroke: Incidence, diagnosis, and pulmonary complications. *Stroke* **2005**, *36*, 2756–2763. [CrossRef] [PubMed]
65. Ramsburg, H.; Moriarty, H.J.; MacKenzie Greenle, M. End-of-Life Symptoms in Adult Patients With Stroke in the Last Two Years of Life: An Integrative Review. *Am. J. Hosp. Palliat. Care* **2023**. [CrossRef]
66. Narain, S.; Puckree, T. Pulmonary function in hemiplegia. *Int. J. Rehabil. Res.* **2002**, *25*, 57–59. [CrossRef]
67. Stanojevic, S.; Kaminsky, D.A.; Miller, M.R.; Thompson, B.; Aliverti, A.; Barjaktarevic, I.; Cooper, B.G.; Culver, B.; Derom, E.; Hall, G.L.; et al. ERS/ATS technical standard on interpretive strategies for routine lung function tests. *Eur. Respir. J.* **2022**, *60*, 2101499. [CrossRef]
68. Pai, H.-C.; Li, C.-C. Relationship Between Trunk Control Ability and Respiratory Function in Stroke Patients: A Scoping Review and Meta-Analysis. *Asian Nurs. Res. (Korean Soc. Nurs. Sci.)* **2023**, *17*, 61–69. [CrossRef]
69. Young, R.P.; Hopkins, R.; Eaton, T.E. Forced expiratory volume in one second: Not just a lung function test but a marker of premature death from all causes. *Eur. Respir. J.* **2007**, *30*, 616–622. [CrossRef]
70. Kulbacka-Ortiz, K.; Triest, F.J.J.; Franssen, F.M.E.; Wouters, E.F.M.; Studnicka, M.; Vollmer, W.M.; Lamprecht, B.; Burney, P.G.J.; Amaral, A.F.S.; Vanfleteren, L.E.G.W. Restricted spirometry and cardiometabolic comorbidities: Results from the international population based BOLD study. *Respir. Res.* **2022**, *23*, 34. [CrossRef]
71. Silvestre, O.M.; Nadruz, W., Jr.; Querejeta Roca, G.; Claggett, B.; Solomon, S.D.; Mirabelli, M.C.; London, S.J.; Loehr, L.R.; Shah, A.M. Declining Lung Function and Cardiovascular Risk: The ARIC Study. *J. Am. Coll. Cardiol.* **2018**, *72*, 1109–1122. [CrossRef] [PubMed]
72. Corlateanu, A.; Covantev, S.; Mathioudakis, A.G.; Botnaru, V.; Cazzola, M.; Siafakas, N. Chronic Obstructive Pulmonary Disease and Stroke. *COPD* **2018**, *15*, 405–413. [CrossRef] [PubMed]
73. Szylińska, A.; Kotfis, K.; Bott-Olejnik, M.; Wańkowicz, P.; Rotter, I. Post-Stroke Outcomes of Patients with Chronic Obstructive Pulmonary Disease. *Brain Sci.* **2022**, *12*, 106. [CrossRef] [PubMed]
74. Pozuelo-Carrascosa, D.P.; Carmona-Torres, J.M.; Laredo-Aguilera, J.A.; Latorre-Román, P.; Párraga-Montilla, J.A.; Cobo-Cuenca, A.I. Effectiveness of Respiratory Muscle Training for Pulmonary Function and Walking Ability in Patients with Stroke: A Systematic Review with Meta-Analysis. *Int. J. Environ. Res. Public Health* **2020**, *17*, 5356. [CrossRef]
75. Menezes, K.K.; Nascimento, L.R.; Ada, L.; Polese, J.C.; Avelino, P.R.; Teixeira-Salmela, L.F. Respiratory muscle training increases respiratory muscle strength and reduces respiratory complications after stroke: A systematic review. *J. Physiother.* **2016**, *62*, 138–144. [CrossRef]
76. Menezes, K.K.; Nascimento, L.R.; Avelino, P.R.; Alvarenga, M.T.M.; Teixeira-Salmela, L.F. Efficacy of Interventions to Improve Respiratory Function After Stroke. *Respir. Care* **2018**, *63*, 920–933. [CrossRef]

neurology international

MDPI

Review

A Comprehensive Review on the Role of Resting-State Functional Magnetic Resonance Imaging in Predicting Post-Stroke Motor and Sensory Outcomes

Foteini Christidi [1,†], Ilias Orgianelis [1,†], Ermis Merkouris [1], Christos Koutsokostas [1], Dimitrios Tsiptsios [1,*], Efstratios Karavasilis [2], Evlampia A. Psatha [2], Anna Tsiakiri [1], Aspasia Serdari [3], Nikolaos Aggelousis [4] and Konstantinos Vadikolias [1]

1 Neurology Department, School of Medicine, Democritus University of Thrace, 68100 Alexandroupolis, Greece; christidi.f.a@gmail.com (F.C.); iliaorgi@med.duth.gr (I.O.); ermimerk@med.duth.gr (E.M.); chriskoutsokostas2001@gmail.com (C.K.); anniw_3@hotmail.com (A.T.); vadikosm@yahoo.com (K.V.)
2 Department of Radiology, School of Medicine, Democritus University of Thrace, 68100 Alexandroupolis, Greece; ekaravas@med.duth.gr (E.K.); eviepsatha@yahoo.gr (E.A.P.)
3 Department of Child and Adolescent Psychiatry, School of Medicine, Democritus University of Thrace, 68100 Alexandroupolis, Greece; aserntar@med.duth.gr
4 Department of Physical Education and Sport Science, Democritus University of Thrace, 69100 Komotini, Greece; nagelous@phyed.duth.gr
* Correspondence: tsiptsios.dimitrios@yahoo.gr
† These authors contributed equally to this work.

Abstract: Stroke is a major leading cause of chronic disability, often affecting patients' motor and sensory functions. Functional magnetic resonance imaging (fMRI) is the most commonly used method of functional neuroimaging, and it allows for the non-invasive study of brain activity. The time-dependent coactivation of different brain regions at rest is described as resting-state activation. As a non-invasive task-independent functional neuroimaging approach, resting-state fMRI (rs-fMRI) may provide therapeutically useful information on both the focal vascular lesion and the connectivity-based reorganization and subsequent functional recovery in stroke patients. Considering the role of a prompt and accurate prognosis in stroke survivors along with the potential of rs-fMRI in identifying patterns of neuroplasticity in different post-stroke phases, this review provides a comprehensive overview of the latest literature regarding the role of rs-fMRI in stroke prognosis in terms of motor and sensory outcomes. Our comprehensive review suggests that with the advancement of MRI acquisition and data analysis methods, rs-fMRI emerges as a promising tool to study the motor and sensory outcomes in stroke patients and evaluate the effects of different interventions.

Keywords: functional magnetic resonance imaging; resting-state functional connectivity; stroke; motor outcomes; sensory outcomes; biomarkers

1. Introduction

Stroke is one of the leading causes of chronic disability due to its impact on the motor, sensory, and cognitive functions of patients. The presence of these deficits is one of the factors that hinder the full recovery of patients [1]. During the post-stroke period, the brain inherently tries to restore its functional status that has been affected [2]. There can be recovery, where the affected function is fully restored, or compensation, where mechanisms emerge through which other brain regions participate in achieving a previous level of functionality [3]. However, compensatory mechanisms may sometimes lead to incomplete recovery, exacerbating the existing deficit through maladaptive plasticity [4]. The range of these mechanisms partially explains the range of varying degrees of effectiveness of intervention programs [5]. It also highlights the need to study additional factors or markers that could contribute to designing better rehabilitation programs, predicting motor and sensory outcomes, and better assessing intervention effectiveness.

Functional magnetic resonance imaging (fMRI) is the most commonly used method of functional neuroimaging, allowing the non-invasive study of brain activity [6]. It is based on the investigation of the differences between the paramagnetic properties of oxygenated and non-oxygenated hemoglobin, examined through the blood oxygen level dependent (BOLD) signal in different brain regions either during a task (task fMRI) or at rest (resting state fMRI; rs-fMRI) within the MRI scanner. During rs-fMRI, the subject remains still and calm inside the MRI scanner with eyes usually closed, without receiving external stimuli or thinking about anything. The time-dependent coactivation of different brain regions at rest is described as resting-state (rs) activation. Distinct brain areas that are co-activated in response to stimuli (e.g., during task-fMRI) also show specific or functional organization at rest and constitute the so-called rs functional networks (Figure 1). Thus, rs-fMRI is a task-independent approach that is based on intrinsic low-frequency fluctuations (typically <0.1 Hz) in the BOLD signal. The latter can be used to compute the functional connectivity between spatially remote areas based on the temporal correlations of the BOLD signals between these areas [7]. These functionally connected networks are named after the function they are related to when activated [8]. Each rs functional network is composed of individual brain regions (structures/nodes), whose functional connectivity (FC) is based on the quantitative correlation of the perfusion rate of individual structures. The magnitude of the correlations of the BOLD signal time series between nodes constituting a functional structure and between individual structures constituting a network is considered an indication of the adequacy of FC [9]. Two of the main ways of detecting FC between individual structures/nodes are: (a) correlation of the BOLD signal time series of individual structures, (b) independent component analysis. As a non-invasive task-independent functional neuroimaging approach, rs-fMRI may therefore provide therapeutically useful information on both the focal vascular lesion and the connectivity-based reorganization and subsequent functional recovery in stroke patients [10].

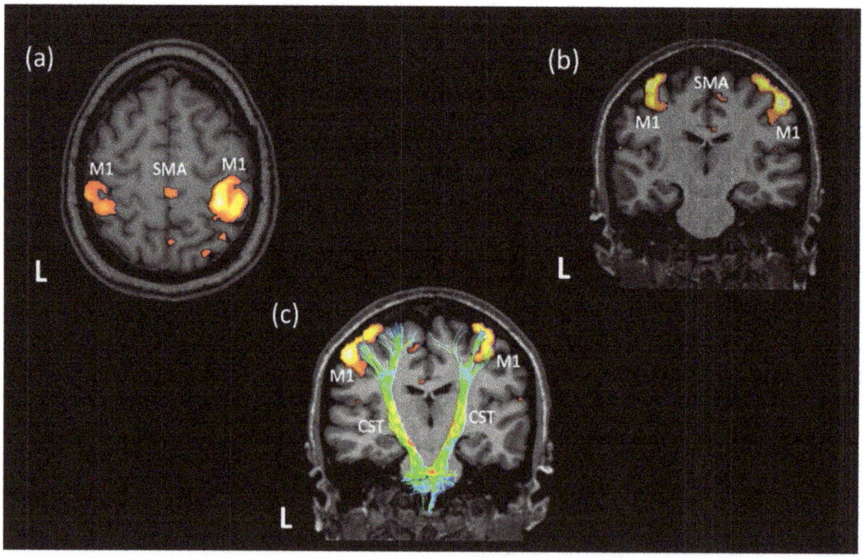

Figure 1. The characteristic patterns of functional connectivity at rest between the regions of the motor network of a healthy individual. The nodes of the motor network (left and right M1, SMA) are overlaid on a 3D-T1 sequence and presented in the axial (**a**) and coronal (**b**) planes. The locations of the left and right M1 nodes and their association with the distribution of the major motor white-matter pathway (CST) is presented in (**c**). Note that the CST had been reconstructed using tractography on

available DTI data for the same subject and is colored based on the distribution of the fractional anisotropy values along the tract. The analysis of the DTI data, as well as the rs-fMRI data and the identification of the motor network at rest, have been performed on a healthy subject (who provided written informed consent for the data acquisition, analysis, and presentation) using the Brainance MD platform (Advantis Medical Imaging). rs-fMRI = resting-state functional magnetic resonance imaging; DTI = diffusion tensor imaging; L = left hemisphere; M1 = primary motor cortex; SMA = supplementary motor area; CST = corticospinal tract.

Considering the role of prompt and accurate prognosis in stroke survivors along with the potential of rs-fMRI in identifying patterns of neuroplasticity in different post-stroke phases, this review provides a comprehensive overview of the latest literature regarding the role of rs-fMRI in stroke prognosis in terms of motor and sensory outcomes.

2. State-of-the-Art of Current Literature on rs-fMRI in Motor and Sensory Outcomes after Stroke

To address the main aim of this comprehensive review, we reviewed most representative human studies (cross-sectional or longitudinal) during the last decade which used rs-fMRI and focused on stroke motor and sensory outcomes (Supplementary Material: Supplementary Tables S1 and S2, Supplementary Figure S1).

2.1. Prediction of Motor Outcomes Using rs-fMRI

2.1.1. Changes in FC Based on Traditional and Relatively New rs-fMRI Indices

The topological characteristics of the cortical motor-related network in patients with subcortical stroke were examined in a study by Yin et al. (2013) in which all areas of interest were distal to the initial lesion [11]. The results indicated decreased FC between the ipsilesional M1 and the contralesional middle frontal gyrus, as well as between the ipsilesional postcentral gyrus and the contralesional postcentral gyrus. Accordingly, disruption of interhemispheric FC in the somatomotor network was found to be strongly associated with post-stroke upper limb dysfunction, whereas it was not associated with performance in the same network. Additionally, the study identified diverse patterns of functional remodeling in both cortical connectivity and localized homogeneity, suggesting varying results in hand functionality. Specifically, decreased betweenness centrality (that assesses the amount of influence a node has over the flow of information in a graph) of the ipsilesional dorsolateral premotor cortex indicated poor outcomes in hand functionality.

According to Li et al. (2014), patients with dysphagia showed greater rs functional architecture irregularities than patients who experienced no difficulty swallowing [12]. These irregularities are largely related to alterations in the coherent intrinsic neuronal activity of BOLD fluctuations detected using rs-fMRI. The results imply that effective recovery is linked to brain activation that is relevant to cortical swallowing representation in the compensatory hemisphere or other recruited regions of the unaffected hemisphere.

Lam et al. (2018), with respect to FC as a stroke biomarker, investigated left (L) and right (R) motor cortex rs-FC (L M1-R M1), which they considered to be a sign of functional integrity [13]. Stroke patients with higher L M1-R M1 rs-connectivity were less disabled than those with lower connectivity, consistent with the fact that L M1-R M1 patterns at rest are comparable to those that arise during motor task performance and contribute to the variability of motor behaviors. Contrary to previous studies on this topic, in which both a positive and a negative relationship between the two were found, possibly because of different stroke parameters, there was no significant interaction between corticospinal tract (CST) injury and L M1-R M1 rs-connectivity in the regression models obtained in this study. According to the authors, this underscores the need for larger-scale studies that assess stroke patients at all stages of chronicity and severity. The authors speculate that CST injury and L M1-R M1 rs connectivity may serve as different indices of motor outcomes considering the above.

It is well documented that after rehabilitation, ipsilesional intrahemispheric FC decreases and interhemispheric FC increases. Furthermore, the decrease in FC between bilateral motor cortices correspond more strongly with neurological function than the decrease in FC inside ipsilesional motor cortices. Chi et al. (2018) studied patients with mild acute ischemic stroke (AIS) stroke and found that although the interhemispheric FC of the motor network was lower in patients than in healthy subjects, there were differences between patients with favorable and poor outcomes only in the M1 and in the contralesional dorsal premotor area [14]. The authors concluded that in functional recovery after AIS, interhemispheric FC may be more important than ipsilesional intrahemispheric FC. Additionally, FC may represent a potentially therapeutic target in stroke patients as well as a reliable and prompt imaging biomarker for AIS using neurostimulation techniques.

Zou et al. (2018) revealed that the frontoparietal system is essential for higher-order motor–cognitive processing, including motor imagery and the regulation of goal-directed movements, whereas the sensorimotor system is primarily active during motor execution and somatosensory information processing. Both systems have been found to be involved in cerebral remodeling, characterized by aberrant task-evoked brain activation as well as disturbed rs-FC [15]. The random reorganization hypothesis is supported by the fact that when a stroke affects the brain, connectivity reorganization occurs in two distinct phases: first, existing links are broken and rewired, and second, new links are created due to connections being established randomly across all possible brain regions. The authors also concluded that there was a correlation between the degree of motor deficits and lower interhemispheric FC between cortical motor regions, even in individuals who had made an excellent clinical recovery.

In the study conducted by Hong et al. (2019), only the completely paralyzed patients showed reduced FC in perceptual areas, although both patients with completely and partially paralyzed hands showed reduced within-network FC in the contralesional superior parietal cortex and ipsilesional supplementary motor area (SMA) [16]. Because continuous dominant stimulation of the contralesional sensorimotor cortex has been shown to interfere with normal function of the paretic hand, the authors suggested that this region may be involved in functional outcomes of the ipsilateral hand. They discovered that patients whose hands were completely paralyzed had increased network FC in the contralesional sensorimotor cortex compared with patients who were only partially paralyzed, suggesting that the severe hand impairments were followed by excessive mobilization of contralesional sensorimotor assets. In addition, there was a negative correlation between the Fugl–Meyer Assessment (FMA) scores and the mean FC in the contralesional sensorimotor cortex, indicating that contralesional hemisphere involvement is not a desirable indicator in stroke patients in the chronic stage.

More sophisticated approaches to rs-fMRI data have also been applied during the past few years. Liu et al. (2015) revealed that the amplitude of low-frequency fluctuation (ALFF), which reflects regional spontaneous neuronal activity, was increased in the bilateral M1 area during the initial phase of subcortical infarction [17]. The authors hypothesized that this is due to several factors, including altered vasomotor activity and neurovascular coupling, although the exact neurophysiological basis remains unknown. The findings indicate that increased spontaneous neuronal activity, determined via rs-fMRI, might possibly contribute to the functional reorganization of physically impaired brain regions during post-stroke motor recovery.

Moreover, in a study by Zhao et al. (2018), patients with completely paralyzed hands showed decreased regional homogeneity (ReHo) in the bilateral cerebellar posterior lobes and increased ReHo in the contralesional SMA compared with patients with less severe paralysis in the typical frequency band [18]. Similar findings were noted in the slow and subfrequency bands, supporting the authors' theory that the frequency of low-frequency oscillations and stroke severity influence spontaneous brain activity after stroke. In addition, there was a strong association between the mean ReHo values in these regions and the

FMA-HW scores in all patients, suggesting that frequency-specific changes in ReHo may be related to hand function recovery in stroke patients.

Brain entropy (BEN), an fMRI-based approach, was utilized by Liang et al. (2020) to map the temporal complexity of the entire brain and its capacity to handle incoming or outgoing data [19]. Lower BEN values often signify decreased neural activity and erratic information processing capabilities. It has been noted that lower BEN values in the contralesional precentral gyrus and ipsilesional M1 may be indicative of a reduced aptitude of information exploration in the contralesional sensorimotor system. Furthermore, lower BEN values were observed in the bilateral SMA, suggesting that the operational framework of motor execution and planning was disturbed. It is possible that decreased brain complexity in this area is connected to the impaired integrity of the cortical structure. Additionally, the BEN values in the ipsilesional SMA showed a positive correlation with the FMA scores, suggesting that the absence of brain complexity in this region could contribute to poor motor outcomes in stroke patients. This occurs because the SMA is essential for the progression of motor rehabilitation.

Additionally, Bonkhoff et al. (2020), in a study of patients with mild to severe stroke, demonstrated for the first time that moderately impaired patients spent significantly more time in state 2, a less connected state, while patients with severe motor symptoms significantly preferred state 1, a state with strong intradominal connectivity [20]. Unexpectedly, they discovered that moderately impaired patients had more pronounced changes in FC (between the ipsi- and contralesional sensorimotor cortices and the same sensorimotor network and bilateral paracentral cortex) in both static and dynamic analyses, despite milder clinical deficits. However, the researchers found impaired interhemispheric connectivity between sensorimotor cortical components in severe strokes, with motor symptoms only in dynamic analysis. Bonkhoff et al. (2021) showed that in the first three months after stroke, the observed recovery in stroke severity was strongly connected with—or showed strong trends in—dynamic connectivity between three pairs of networks [21]. The change in stroke severity and the dynamic connectivity of default mode networks in the bilateral intraparietal lobule and left angular gyrus were significantly inversely correlated in state 1. Dynamic connectivity of the bilateral putamen and superior temporal gyrus and first and bilateral anterior insula also showed positive associations with recovery of stroke severity at stage 3. Recovery of motor symptoms, which make up a large proportion of the NIHSS score, may be related to the subcortical putamen, revealing its importance in stroke outcomes. It is known that the default mode and cognitive control networks control basic brain activities, which explains why individuals who have lesions in these regions tend to have worse stroke outcomes. In addition, the strong negative connectivity between networks suggests that there is less communication between different functional areas. However, the researchers found no significant association between the degree of segregation and changes in stroke severity in the first three months after stroke.

2.1.2. Changes in FC and Their Association with the Application of Specific Rehabilitation Methods

In a study by Fan et al. (2015), who used robot-assisted bilateral arm therapy (RBAT) for stroke patients, changes in M1–M1 connectivity before and after treatment were associated with better motor and functional progress, and these differences were thought to serve as a significant mediator of the relationship between changes in some disability scores [22]. As a result, rs-FC can demonstrate this flexibility, and bilateral arm training can alter functional connections between sensorimotor brain regions and restore hemispheric imbalances in stroke recovery.

Zhang et al. (2016) examined the functional connectivity (FC) alterations between the ipsilesional primary motor cortex (M1) and the entire brain of stroke patients compared to normal controls, as well as in stroke patients preceding and following traditional rehabilitation and motor imagery therapy [23]. Relative to the controls, the FC in the patient group was substantially increased between the ipsilesional M1 and other areas of the brain.

This may be a form of compensation when interhemispheric FC balance is lost due to the stroke. Following the treatment, the FC between the ipsilesional and contralesional M1 increased, yet the FC between the ipsilesional M1 and other regions decreased. Specifically, the decreased FC between the ipsilesional M1 and left paracentral lobules—areas involved in the motion control and the sensation of the limbs—hinted at ipsilateral hemispheric FC recovery. The study revealed positive correlations between the FC change and the motor function recovery of stroke patients with hemiplegia, indicating that the FC could be used as a biomarker of motor function recovery. Moreover, Li et al. (2016) used repetitive transcranial magnetic stimulation (rTMS) over the ipsilesional M1 to increase CST activity, which aided in the recovery of motor function in stroke patients [24]. The authors attributed this to enhanced FC between the bilateral M1 and the contralesional supplementary motor area (SMA), which serves as an anatomical basis for motor recovery once M1 output is disrupted. Increased FC between the ipsilesional M1 and bilateral thalamus, which is a critical component of the extrapyramidal system and strongly related to motor coordination, was also noted, and this may be considered a possible compensatory mechanism for recovery of motor function after stroke.

On the same basis, Lefebvre et al. (2017) showed that FC increased exclusively in the somatomotor network after combined dual transcranial direct current stimulation (dual-tDCS)/motor skill learning relative to the period before intervention, which altered the spontaneous activity of the somatomotor network [25]. Consistent with previous studies, they supported the possibility that stroke could cause a net deleterious inhibition from the normal hemisphere to the damaged side. However, they also suggested another mechanism, namely, that stroke patients rely more heavily on their nondamaged upper limb in daily activities, which could strengthen the cortical FC of this limb.

Furthermore, Li et al. (2017), in their study of FC in subcortical stroke patients using conventional Western medical treatment and acupuncture, found a statistically significant association between pretreatment FC values between bilateral M1 and the percent changes in neurological deficit scores [26]. Given that ischemic stroke damage causes anatomical and functional changes in perilesional and distal brain regions, communication and connectivity between the two hemispheres were also influenced. Intra-hemispheric FC in regions of interests in the right contralateral hemisphere showed a considerable rise in post-stroke patients. This finding can be interpreted based on the mechanism of compensation for decreased FC of brain regions in the contralateral hemisphere to maintain motor function.

On the other hand, Guo et al. (2021) studied the motor networks of stroke patients with motor deficits after treatment with rTMS. Motor recovery could be due to the different functional recovery and reorganization that both high-frequency rTMS and low-frequency rTMS induced in the motor network [27]. Of note, after high-frequency rTMS and low-frequency rTMS, alterations were observed in the ipsilesional and contralateral hemispheres, respectively. This could be explained by the different processes of the various rTMS modes. High-frequency rTMS could promote cortical excitability, whereas low-frequency rTMS might reduce the abnormally excessive inhibition of lesioned M1 and support regrowth of the damaged hemisphere. As a result, the compensation of unequal interhemispheric excitability and connections may be related to motor recovery after stroke. A similar outcome was found by Chen et al. (2022), who investigated how coupled inhibitory and facilitative rTMS treatment affects early reorganization of motor networks in cerebral infarction patients [28]. They discovered that rTMS can induce functional remodeling of cortical motor networks by altering the intensity of intra- and interhemispheric FC. In addition, they note that FC changes were associated with a regaining of motor function and may serve as a future focus of neurorehabilitation interventions for early cerebral stroke patients.

2.1.3. Changes in FC and Structural Connectivity

Chen et al. (2013) demonstrated a correlation between the upper extremities FMA (UE-FMA) score and the left and right M1 rs-FC, as well as the white matter integrity in the transcallosal M1–M1 tract [29]. However, they did not find any discernible association between WM integrity of the transcallosal M1–M1 tract and the rs-interhemispheric M1 FC. According to their theories, which are consistent with previous findings, the ipsilesional M1 may no longer have the same inhibitory effect on the contralesional M1 as before, and/or the contralesional M1 may have an excessively imbalanced inhibitory effect on the ipsilesional M1. It is therefore plausible that patients who recovered more quickly from their motor impairments had greater interhemispheric M1 connectivity and a more normal inhibitory effect from ipsilesional to contralesional M1.

The investigation by Kalinosky et al. (2017) successfully combined structural and functional connectivity techniques. While rs-FC alone can identify abnormalities in stroke patients that are associated with function, structuro-functional connectivity (SFC) detects the maximum FC to a voxel over various distances and may provide insight into network-specific processes underlying pathology and recovery [30]. The cerebellum, midbrain, and thalamus—integrative brain regions that are crucial for motor function—were found to have decreased SFC. Intrinsic SFC in the cerebellum was substantially linked with hand motor performance, serving as a marker of sensorimotor function. Additionally, stroke patients were found to have decreased intrinsic SFC between regions connected by longer fiber pathways, especially in areas associated with integrative cortical network nodes, including the cerebellum and prefrontal cortex: the center of the default mode network. This supports the possibility that alterations in these structuro-functional networks reflect compensatory or maladaptive mechanisms of reorganization and that the stroke-affected sensorimotor and cerebellar networks shift towards creating shorter-distance connections to the default mode and prefrontal network nodes.

Lin et al. (2018) discovered that structural and functional features may have good predictive value for initial motor impairment [31]. FC between motor areas improved in the long-term phase, but mainly in the first three months, which was substantially associated with motor function throughout the same period, but not at 12 months post-stroke. Therefore, the integrity of CST was found to influence interhemispheric FC, and recovery developed mainly within the first three months after the stroke, never reaching the optimal levels.

Research by Lee et al. (2019) shows changes in structural and functional connectivity after stroke measured using diffusion tensor imaging (DTI) and rs-fMRI [32]. Patients with mild stroke experienced restructuring of brain networks in motor-related areas, whereas the severe stroke group showed no significant alterations in FC, possibly indicating significant structural damage in the affected brain regions. The severity of motor impairment is influenced by alterations in myelination and axonal density, which may be due to differences in structural connectivity between the two groups. During poststroke reorganization, the severe group cannot successfully recover motor function through remyelination processes. Like other research, spared projections may be able to control paretic muscles post-stroke through perilateral reconfiguration of the motor network, which would lead to an increase in FC between specific motor-related brain regions such as M1 and the SMA.

With respect to rs-fMRI as a prognostic tool, Lu et al. (2019) support that FC of neuronal networks in the brain fluctuates dynamically because of synchronization of intrinsic neuronal interactions [33]. Compared to the baseline, the interhemispheric FC between ROIs was decreased at day 7, then gradually increased from day 7 to 90 and returned to normal at day 90. Of note, from day 7 to day 90 post-stroke, FA in the bilateral (ipsilesional, contralesional) CST increased, NHISS scores decreased, and FMA and BI progressively increased. The early decline in interhemispheric FC after stroke may point to the disruption of ischemic stroke-damaged networks. The elimination of transient hemispheric diaschisis and the formation of new axons forming new connections and projections might be associated with the restoration of interhemispheric FC. Therefore, the above results suggest that the response of the contralesional hemisphere to ipsilesional brain activity changes when the two hemispheres cannot communicate effectively at rest. Proliferative reorganization in the ipsilesional and contralesional CST and functional reorganization in the SMN may support and promote neurological functional recovery after internal capsule infarction.

Similarly, the study by Xia et al. (2021) involving participants with mild to moderate stroke showed longitudinal improvements in their interhemispheric FC [34]. Notably, interhemispheric FC recovery largely took place within 4 weeks of stroke, making this the ideal time window for patients to experience interhemispheric FC reorganization. Considering these, it was found that increased interhemispheric FC and recovery of corticospinal pathways were positively correlated during the first 4 weeks after stroke but not between weeks 4 and 12.

2.2. Prediction of Sensory Outcome Using rs-fMRI

Bannister et al. (2015) found increased interhemispheric FC of S1 at 6 months in stroke patients with a lower degree of touch impairment that was not present at 1 month and lower interhemispheric connectivity in stroke patients with impaired touch compared with healthy controls [35]. The stroke group also showed FC at 6 months that was absent in the visual occipital regions and frontoparietal attention regions at 1 month. These results suggest a pattern of disrupted FC between somatosensory and other cortical regions that eventually exhibited some signs of recovery. Moreover, the improvements in the TDT score over time and the changes in FC between the 1- and 6-month periods were all located in the contralesional hemisphere. These emphasize the importance of an intact contralesional hemispheric FC, particularly in the somatosensory cortex. Moreover, Goodin et al. (2018) examined the rs-FC in stroke patients with both right and left hemisphere lesions and discovered that the right lesion group displayed greater FC in comparison with the left hemisphere group based on the connection between the contralesional left primary somatosensory cortex (S1) seed to the left superior parietal and mid-occipital regions [36]. This may indicate that connections between the somatosensory and visual networks are disrupted when only the left hemisphere, and not the right, is damaged. In addition, disruption of interhemispheric FC from both hemispheres was found regardless of the lesion location. Consequently, injury to the somatosensory system in one hemisphere not only affects interhemispheric inhibition but also exerts a negative effect on recovery in general.

3. Discussion

The current comprehensive analysis of the latest literature within the field of the application of rs-fMRI in stroke patients suggests that rs-fMRI can be a promising biomarker for the prediction of motor and sensory outcomes.

While task-fMRI focuses on mapping brain activity during specific tasks, rs-fMRI explores intrinsic connectivity patterns in the absence of explicit tasks. Both approaches contribute valuable information to the understanding of brain function and are often used together to provide a more comprehensive view of brain function. The combination allows for the identification of task-specific activations and the exploration of the underlying intrinsic connectivity patterns. Functional networks identified using rs-fMRI strongly overlap with the brain regions activated by task-fMRI [37], and spontaneous activity at rest correlates with trial-to-trail fluctuations in task-evoked responses [38] and behavior [39]. In contrast to task fMRI, resting-state fMRI is a non-invasive imaging technique, allowing for the examination of functional connectivity without the need of any external stimuli or tasks. The fact that the participants can simply rest with their eyes closed makes it suitable for a wide range of populations, including stroke patients. In addition, compared to task-fMRI, where participants perform specific cognitive tasks during MRI scanning, resting-state fMRI is relatively simple to acquire. The non-invasive character, the absence of specific experiment and participants' engagement during scanning, and the ease of acquisition facilitates the integration of resting-state fMRI into experimental protocols and clinical settings. For instance, a growing literature in brain tumor patients demonstrates that the results obtained from resting-state fMRI during pre-surgical language mapping can be comparable to and, in some respects, superior to task-fMRI [40,41]. Thus, resting-state fMRI offers a great opportunity to study and understand the functional network changes caused by stroke, better assisting in patients' stratification in clinical trials and enabling rehabilitation interventions to be tailored to improve recovery.

Lesion volume or lesion location, age, stroke severity, and comorbid factors may influence stroke outcomes [42–48], but even mild stroke severity is not enough to provide an accurate prediction of favorable outcomes [49]. Thus, novel prognostic biomarkers are crucial for the development of personalized recovery treatment strategies [50] and the amelioration of patients' and caregivers' quality of life [51]. Brain plasticity remains a crucial issue in stroke rehabilitation, and advances in MRI techniques have the potential to improve the clinical and research opportunities and knowledge in that field. Clinical recovery, whether motor or sensory, is supported through reorganization of surviving brain tissue, alterations in interhemispheric lateralization, involvement of associative cortices linked to injured areas, and organization of cortical representation maps [52]. The variability of motor and sensory impairment after stroke has been associated with both structural [53] and functional alterations [52] in networks proximal and/or distal to the brain lesions. In contrast to task-related fMRI, rs-fMRI minimizes the subject's involvement and the interindividual differences in task performance, thus enabling the integration of rs-fMRI into clinical MRI and the monitoring of functional alterations during different post-stroke phases (e.g., subacute vs. chronic) and therapeutic approaches (e.g., physiotherapy, motor imagery, neuromodulation, etc.).

Some critical comments regarding methodological aspects of the available literature are necessary before reaching a conclusion. We found that some studies focused on motor outcomes using classic and more sophisticated approaches for the rs-fMRI analysis (Supplementary Material: Supplementary Table S2, section A). Other studies used rs-fMRI and focused on motor outcomes following specific rehabilitation interventions (Supplementary Material: Supplementary Table S2, section B), while other studies examined motor outcome in association with functional connectivity and structural connectivity (Supplementary Material: Supplementary Table S2, section C). However, few studies focused on sensory outcomes (Supplementary Material: Supplementary Table S2, section D). Most studies included patients with either ischemic stroke or hemorrhagic stroke. The total number of stroke patients included in all the studies varied significantly, ranging from 10 to 67 patients. Of note, the stroke patients were contrasted to demographically matched healthy individuals in most but not all studies, and none of the studies included a disease control group other than stroke patients. We also identified significant variability in the time of MRI acquisition post-stroke, ranging from a few days (i.e., 2–3 days) post-stroke to 60.0 +/− 69.6 months post-stroke. Most studies used 3T, while the rs-fMRI acquisition time (when available) ranged from 5 min to 10.25 min. Several approaches were applied to analyze the raw data, including independent component analysis (ICA), seed-to-whole brain FC analysis, region-of-interest (ROI) to ROI FC analysis, regional homogeneity (ReHO) analysis, amplitude of low-frequency fluctuations (ALFF) analysis, intrinsic structural–functional connectivity (iSFC) analysis, and static and dynamic functional network connectivity analyses. Some studies also included other MRI modalities and reported lesion analyses (location and extent), white matter hyperintensity analyses, structural integrity of white matter using diffusion tensor imaging (DTI), or cortical thickness of rs-fMRI ROIs. The most commonly used clinical measures were the National Institutes of Health Stroke Scale (NIHSS), the Fugl–Meyer Assessment (FMA), and the Barthel Index (BI).

Despite the available research on the application of rs-fMRI in post-stroke motor and sensory outcome so far, several questions remain unanswered, leaving room for further investigations. The studies included in the current comprehensive review are characterized by a high heterogeneity in terms of clinical characteristics. Thus, an attempt to organize this review based on clinical variables, including the type of stroke or lesion classification, might ultimately be less informative for the readers. A more in-depth analysis of existing literature through a meta-analysis may further address this issue, and if possible, integrate information about lesion location and its impact at the connectome level (i.e., patterns of functional connectivity). The sample size of most studies was small, and together with the sample heterogeneity (e.g., stroke severity, post-stroke phase), the generalization of the current findings may be limited. Of note, different post-stroke phases are associated with different static and dynamic functional and structural features; thus, the inclusion of patients under different post-stroke phases may limit the clinical significance of these findings. In addition, the interventional studies included interventions of different durations, while the follow-up period differed among the studies. As a result, a dose–effect of interventions on FC patterns could not be easily found, whereas the durability of this effect requires further investigation in longitudinal studies with longer follow-up periods. Furthermore, even though the majority of the studies used the FMA for the evaluation of motor functional disability, gold-standard clinical measures (e.g., NIHSS) were not uniformly used in all the studies. From a methodological point of view, detailed descriptions of rs-fMRI protocols were not available for all the studies (e.g., duration of rs-fMRI sequence); thus, replication of the findings may not be easily obtained for all the studies. Finally, different and more sophisticated methodological approaches to rs-fMRI raw data may definitely yield informative details regarding the usefulness of rs-fMRI in research settings, yet the application in every-day clinical practice and prediction of single-subject prognosis has yet to be established. Thus, longitudinal studies with larger sample sizes and well-characterized groups in terms of clinical features are warranted to enable the identification of possible rs-fMRI biomarkers for the prognosis of stroke motor and sensory outcomes not only at group level but also—and more importantly—at the single-case level. In line with existing literature in pre-surgical planning [40,41], further studies contrasting the usefulness, sensitivity, and specificity of task-fMRI and resting-state fMRI in predicting stroke outcomes are also warranted.

4. Conclusions

With the advancement of MRI acquisition and data analysis methods, rs-fMRI emerges as a promising tool to study motor and sensory outcome in stroke patients and to evaluate the effects of different interventions, including recently used rTMS and tDCS. The implementation of rs-fMRI in stroke may further establish the predictive value of rs-fMRI in stroke clinical practice and provide valuable information for more accurate patient stratification into clinical trials and evidence-based clinical decision making.

Supplementary Materials: The following supporting information can be downloaded at: https://www.mdpi.com/article/10.3390/neurolint16010012/s1, Table S1: Search strategy and selection criteria for the most representative rs-fMRI studies in association to motor and sensory outcome after stroke which were included in the comprehensive review; Figure S1: Study flow chart of the process of selection the most representative rs-fMRI studies in association to motor and sensory outcome after stroke; Table S2: Representative rs-fMRI studies on motor and sensory outcome after stroke.

Author Contributions: Conceptualization, F.C. and C.K.; methodology, I.O.; formal analysis, D.T. and A.S.; writing—original draft preparation, E.M., E.K. and E.A.P.; writing—review and editing, A.T., F.C. and N.A.; supervision, K.V. All authors have read and agreed to the published version of the manuscript.

Funding: This research was funded by the project "Study of the interrelationships between neuroimaging, neurophysiological and biomechanical biomarkers in stroke rehabilitation (NEURO-BIO-MECH in stroke rehab)" (MIS 5047286), which was implemented under the action "Support for Regional Excellence" funded by the operational program "Competitiveness, Entrepreneurship and Innovation" (NSRFm2014-2020) and co-financed by Greece and the European Union (the European Regional Development Fund).

Institutional Review Board Statement: Not applicable.

Informed Consent Statement: The MRI (rs-fMRI and DTI) data (Figure 1) were acquired from a healthy control subject who provided written informed consent for the acquisition, analysis, and presentation of the data.

Data Availability Statement: No new data were created or analyzed in this study. Data sharing is not applicable to this article.

Acknowledgments: We thank the healthy control subject for participating in the acqusition of the MRI data. We also thank Advantis Medical Imaging for providing access to the Brainance MD platform to analyze the rs-fMRI and DTI data.

Conflicts of Interest: The authors declare no conflicts of interest.

References

1. Sachdev, P.S.; Brodaty, H.; Valenzuela, M.J.; Lorentz, L.; Looi JC, L.; Wen, W.; Zagami, A.S. The neuropsychological profile of vascular cognitive impairment in stroke and TIA patients. *Neurology* **2004**, *62*, 912–919. [CrossRef] [PubMed]
2. Loubinoux, I.; Brihmat, N.; Castel-Lacanal, E.; Marque, P. Cerebral imaging of post-stroke plasticity and tissue repair. *Rev. Neurol.* **2017**, *173*, 577–583. [CrossRef]
3. Hylin, M.J.; Kerr, A.L.; Holden, R. Understanding the Mechanisms of Recovery and/or Compensation following Injury. *Neural Plast.* **2017**, *2017*, 7125057. [CrossRef]
4. Hommel, M.; Detante, O.; Favre, I.; Touze, E.; Jaillard, A. How to Measure Recovery? Revisiting Concepts and Methods for Stroke Studies. *Transl. Stroke Res.* **2016**, *7*, 388–394. [CrossRef] [PubMed]
5. Sarraj, A.; Grotta, J.C. Stroke: New horizons in treatment. *Lancet Neurol.* **2014**, *13*, 2–3. [CrossRef]
6. Chen, J.E.; Glover, G.H. Functional Magnetic Resonance Imaging Methods. *Neuropsychol. Rev.* **2015**, *25*, 289–313. [CrossRef] [PubMed]
7. Friston, K.J.; Frith, C.D.; Liddle, P.F.; Frackowiak, R.S. Functional Connectivity: The Principal-Component Analysis of Large (PET) Data Sets. *J. Cereb. Blood Flow. Metab.* **1993**, *13*, 5–14. [CrossRef]
8. Papanicolaou, A.C. *The Oxford Handbook of Functional Brain Imaging in Neuropsychology and Cognitive Neurosciences*; Oxford University Press: New York, NY, USA, 2017.
9. Stamatakis, E.A.; Orfanidou, E.; Papanicolaou, A.C. Functional Magnetic Resonance Imaging. In *The Oxford Handbook of Functional Brain Imaging in Neuropsychology and Cognitive Neurosciences*; Oxford University Press: New York, NY, USA, 2017; pp. 43–60.
10. Ovadia-Caro, S.; Margulies, D.S.; Villringer, A. The Value of Resting-State Functional Magnetic Resonance Imaging in Stroke. *Stroke* **2014**, *45*, 2818–2824. [CrossRef]
11. Yin, D.; Song, F.; Xu, D.; Sun, L.; Men, W.; Zang, L.; Yan, X.; Fan, M. Altered topological properties of the cortical motor-related network in patients with subcortical stroke revealed by graph theoretical analysis. *Hum. Brain Mapp.* **2013**, *35*, 3343–3359. [CrossRef]
12. Li, S.; Zhou, M.; Yu, B.; Ma, Z.; Chen, S.; Gong, Q.; He, L.; Huang, X.; Lui, S.; Wang, X.; et al. Altered default mode and affective network connectivity in stroke patients with and without dysphagia. *J. Rehabil. Med.* **2014**, *46*, 126–131. [CrossRef]
13. Lam, T.K.; Binns, M.A.; Honjo, K.; Dawson, D.R.; Ross, B.; Stuss, D.T.; Black, S.E.; Chen, J.J.; Fujioka, T.; Chen, J.L. Variability in stroke motor outcome is explained by structural and functional integrity of the motor system. *Sci. Rep.* **2018**, *8*, 9480. [CrossRef]
14. Chi, N.-F.; Ku, H.-L.; Chen, D.Y.-T.; Chen, D.Y.T.; Tseng, Y.C.; Chen, C.J.; Lin, Y.C.; Hsieh, Y.C.; Chan, L.; Chiou, H.Y.; et al. Cerebral Motor Functional Connectivity at the Acute Stage: An Outcome Predictor of Ischemic Stroke. *Sci. Rep.* **2018**, *8*, 16803. [CrossRef] [PubMed]
15. Zou, Y.; Zhao, Z.; Yin, D.; Fan, M.; Small, M.; Liu, Z.; Hilgetag, C.C.; Kurths, J. Brain anomaly networks uncover heterogeneous functional reorganization patterns after stroke. *NeuroImage Clin.* **2018**, *20*, 523–530. [CrossRef] [PubMed]
16. Hong, W.; Lin, Q.; Cui, Z.; Liu, F.; Xu, R.; Tang, C. Diverse functional connectivity patterns of resting-state brain networks associated with good and poor hand outcomes following stroke. *Neuroimage Clin.* **2019**, *24*, 102065. [CrossRef] [PubMed]
17. Liu, G.; Dang, C.; Peng, K.; Xie, C.; Chen, H.; Xing, S.; Chen, X.; Zeng, J. Increased spontaneous neuronal activity in structurally damaged cortex is correlated with early motor recovery in patients with subcortical infarction. *Eur. J. Neurol.* **2015**, *22*, 1540–1547. [CrossRef]
18. Zhao, Z.; Tang, C.; Yin, D.; Wu, J.; Gong, J.; Sun, L.; Jia, J.; Xu, D.; Fan, M. Frequency-specific alterations of regional homogeneity in subcortical stroke patients with different outcomes in hand function. *Hum. Brain Mapp.* **2018**, *39*, 4373–4384. [CrossRef] [PubMed]
19. Liang, L.; Hu, R.; Luo, X.; Feng, B.; Long, W.; Song, R. Reduced Complexity in Stroke with Motor Deficits: A Resting-State fMRI Study. *Neuroscience* **2020**, *434*, 35–43. [CrossRef] [PubMed]
20. Bonkhoff, A.K.; Espinoza, F.A.; Gazula, H.; Vergara, V.M.; Hensel, L.; Michely, J.; Paul, T.; Rehme, A.K.; Volz, L.J.; Fink, G.R.; et al. Acute ischaemic stroke alters the brain's preference for distinct dynamic connectivity states. *Brain* **2020**, *143*, 1525–1540. [CrossRef]

21. Bonkhoff, A.K.; Schirmer, M.D.; Bretzner, M.; Etherton, M.; Donahue, K.; Tuozzo, C.; Nardin, M.; Giese, A.K.; Wu, O.; Calhoun, V.D.; et al. Abnormal dynamic functional connectivity is linked to recovery after acute ischemic stroke. *Hum. Brain Mapp.* **2021**, *42*, 2278–2291. [CrossRef]
22. Fan, Y.; Wu, C.; Liu, H.; Lin, K.C.; Wai, Y.Y.; Chen, Y.I. Neuroplastic changes in resting-state functional connectivity after stroke rehabilitation. *Front. Hum. Neurosci.* **2015**, *9*, 546. [CrossRef]
23. Zhang, Y.; Liu, H.; Wang, L.; Yang, J.; Yan, R.; Zhang, J.; Sang, L.; Li, P.; Wang, J.; Qiu, M. Relationship between functional connectivity and motor function assessment in stroke patients with hemiplegia: A resting-state functional MRI study. *Neuroradiology* **2016**, *58*, 503–511. [CrossRef] [PubMed]
24. Li, J.; Zhang, X.-W.; Zuo, Z.-T.; Lu, J.; Meng, C.L.; Fang, H.Y.; Xue, R.; Fan, Y.; Guan, Y.Z.; Zhang, W.H. Cerebral Functional Reorganization in Ischemic Stroke after Repetitive Transcranial Magnetic Stimulation: An fMRI Study. *CNS Neurosci. Ther.* **2016**, *22*, 952–960. [CrossRef] [PubMed]
25. Lefebvre, S.; Dricot, L.; Laloux, P.; Desfontaines, P.; Evrard, F.; Peeters, A.; Jamart, J.; Vandermeeren, Y. Increased functional connectivity one week after motor learning and tDCS in stroke patients. *Neuroscience* **2017**, *340*, 424–435. [CrossRef]
26. Li, Y.; Wang, Y.; Liao, C.; Huang, W.; Wu, P. Longitudinal Brain Functional Connectivity Changes of the Cortical Motor-Related Network in Subcortical Stroke Patients with Acupuncture Treatment. *Neural Plast.* **2017**, *2017*, 5816263. [CrossRef] [PubMed]
27. Guo, Z.; Jin, Y.; Bai, X.; Jiang, B.; He, L.; McClure, M.A.; Mu, Q. Distinction of High- and Low-Frequency Repetitive Transcranial Magnetic Stimulation on the Functional Reorganization of the Motor Network in Stroke Patients. *Neural Plast.* **2021**, *2021*, 8873221. [CrossRef] [PubMed]
28. Chen, X.; Liu, F.; Lin, S.; Yu, L.; Lin, R. Effects of Virtual Reality Rehabilitation Training on Cognitive Function and Activities of Daily Living of Patients with Poststroke Cognitive Impairment: A Systematic Review and Meta-Analysis. *Arch. Phys. Med. Rehabil.* **2022**, *103*, 1422–1435. [CrossRef]
29. Chen, J.L.; Schlaug, G. Resting state interhemispheric motor connectivity and white matter integrity correlate with motor impairment in chronic stroke. *Front. Neurol.* **2013**, *4*, 178. [CrossRef]
30. Kalinosky, B.T.; Berrios Barillas, R.; Schmit, B.D. Structurofunctional resting-state networks correlate with motor function in chronic stroke. *Neuroimage Clin.* **2017**, *16*, 610–623. [CrossRef]
31. Lin, L.Y.; Ramsey, L.; Metcalf, N.V.; Rengachary, J.; Shulman, G.L.; Shimony, J.S.; Corbetta, M. Stronger prediction of motor recovery and outcome post-stroke by cortico-spinal tract integrity than functional connectivity. *PLoS ONE* **2018**, *13*, e0202504. [CrossRef]
32. Lee, J.H.; Kyeong, S.; Kang, H.; Kim, D.H. Structural and functional connectivity correlates with motor impairment in chronic supratentorial stroke: A multimodal magnetic resonance imaging study. *NeuroReport* **2019**, *30*, 526–531. [CrossRef]
33. Lu, Q.; Huang, G.; Chen, L.; Li, W.; Liang, Z. Structural and functional reorganization following unilateral internal capsule infarction contribute to neurological function recovery. *Neuroradiology* **2019**, *61*, 1181–1190. [CrossRef] [PubMed]
34. Xia, Y.; Huang, G.; Quan, X.; Qin, Q.; Li, H.; Xu, C.; Liang, Z. Dynamic Structural and Functional Reorganizations Following Motor Stroke. *Med. Sci. Monit.* **2021**, *27*, e929092. [CrossRef] [PubMed]
35. Bannister, L.C.; Crewther, S.G.; Gavrilescu, M.; Carey, L.M. Improvement in Touch Sensation after Stroke is Associated with Resting Functional Connectivity Changes. *Front. Neurol.* **2015**, *6*, 165. [CrossRef] [PubMed]
36. Goodin, P.; Lamp, G.; Vidyasagar, R.; McArdle, D.; Seitz, R.J.; Carey, L.M. Altered functional connectivity differs in stroke survivors with impaired touch sensation following left and right hemisphere lesions. *NeuroImage Clin.* **2018**, *18*, 342–355. [CrossRef] [PubMed]
37. Fox, M.D.; Raichle, M.E. Spontaneous fluctuations in brain activity observed with functional magnetic resonance imaging. *Nat. Rev. Neurosci.* **2007**, *8*, 700–711. [CrossRef] [PubMed]
38. Fox, M.D.; Snyder, A.Z.; Vincent, J.L.; Raichle, M.E. Intrinsic fluctuations within cortical systems account for intertrial variability in human behavior. *Neuron* **2007**, *56*, 171–184. [CrossRef] [PubMed]
39. Fox, M.D.; Snyder, A.Z.; Zacks, J.M.; Raichle, M.E. Coherent spontaneous activity accounts for trial-to-trial variability in human evoked brain responses. *Nat. Neurosci.* **2006**, *9*, 23–25. [CrossRef] [PubMed]
40. Park, K.Y.; Lee, J.J.; Dierker, D.; Marple, L.M.; Hacker, C.D.; Roland, J.L.; Marcus, D.S.; Milchenko, M.; Miller-Thomas, M.M.; Benzinger, T.L.; et al. Mapping language function with task-based vs. resting-state functional MRI. *PLoS ONE* **2020**, *15*, e0236423. [CrossRef]
41. Lemée, J.-M.; Berro, D.H.; Bernard, F.; Chinier, E.; Leiber, L.M.; Menei, P.; Minassian, A.T. Resting-state functional magnetic resonance imaging versus task-based activity for language mapping and correlation with perioperative cortical mapping. *Brain Behav.* **2019**, *9*, e01362. [CrossRef]
42. Christidi, F.; Tsiptsios, D.; Sousanidou, A.; Karamanidis, S.; Kitmeridou, S.; Karatzetzou, S.; Aitsidou, S.; Tsamakis, K.; Psatha, E.A.; Karavasilis, E.; et al. The Clinical Utility of Leukoaraiosis as a Prognostic Indicator in Ischemic Stroke Patients. *Neurol. Int.* **2022**, *14*, 76. [CrossRef]
43. Gkantzios, A.; Karapepera, V.; Tsiptsios, D.; Liaptsi, E.; Christidi, F.; Gkartzonika, E.; Karatzetzou, S.; Kokkotis, C.; Kyrtsopoulos, M.; Tsiakiri, A.; et al. Investigating the Predictive Value of Thyroid Hormone Levels for Stroke Prognosis. *Neurol. Int.* **2023**, *15*, 926–953. [CrossRef] [PubMed]

44. Karatzetzou, S.; Tsiptsios, D.; Sousanidou, A.; Fotiadou, S.; Christidi, F.; Kokkotis, C.; Gkantzios, A.; Stefas, E.; Vlotinou, P.; Kaltsatou, A.; et al. Elucidating the Role of Baseline Leukoaraiosis on Forecasting Clinical Outcome of Acute Ischemic Stroke Patients Undergoing Reperfusion Therapy. *Neurol. Int.* **2022**, *14*, 923–942. [CrossRef] [PubMed]
45. Karatzetzou, S.; Tsiptsios, D.; Sousanidou, A.; Fotiadou, S.; Christidi, F.; Kokkotis, C.; Gkantzios, A.; Stefas, E.; Sousanidou, A.; Kaltsatou, A.; et al. Copeptin Implementation on Stroke Prognosis. *Neurol. Int.* **2023**, *15*, 83–99. [CrossRef]
46. Liaptsi, E.; Merkouris, E.; Polatidou, E.; Tsiptsios, D.; Gkantzios, A.; Kokkotis, C.; Petridis, F.; Christidi, F.; Karatzetzou, S.; Karaoglanis, C.; et al. Targeting Neutrophil Extracellular Traps for Stroke Prognosis: A Promising Path. *Neurol. Int.* **2023**, *15*, 1212–1226. [CrossRef] [PubMed]
47. Orgianelis, I.; Merkouris, E.; Kitmeridou, S.; Tsiptsio, D.; Karatzetzou, S.; Sousanidou, A.; Gkantzios, A.; Christidi, F.; Polatidou, E.; Beliani, A.; et al. Exploring the Utility of Autonomic Nervous System Evaluation for Stroke Prognosis. *Neurol. Int.* **2023**, *15*, 661–696. [CrossRef]
48. Tziaka, E.; Christidi, F.; Tsiptsios, D.; Sousanidou, A.; Karatzetzou, S.; Tsiakiri, A.; Doskas, T.K.; Tsamakis, K.; Retzepis, N.; Konstantinidis, C.; et al. Leukoaraiosis as a Predictor of Depression and Cognitive Impairment among Stroke Survivors: A Systematic Review. *Neurol. Int.* **2023**, *15*, 238–272. [CrossRef] [PubMed]
49. Khatri, P.; Conaway, M.R.; Johnston, K.C. Ninety-day outcome rates of a prospective cohort of consecutive patients with mild ischemic stroke. *Stroke* **2012**, *43*, 560–562. [CrossRef]
50. Chang, W.H. Personalized Approaches to Stroke: One Step Forward for Functional Recovery of Stroke Patients. *J. Pers. Med.* **2022**, *12*, 822. [CrossRef]
51. Tsiakiri, A.; Vlotinou, P.; Paschalidou, A.; Konstantinidis, C.; Christidi, F.; Tsiptsios, D.; Detsaridou, G.; Petridou, A.; Gkantzios, A.; Karatzetzou, S.; et al. A Scoping Review on Coping Strategies and Quality of Life of Stroke Caregivers: Often Underestimated Variables in Stroke Recovery Process? *BioMed* **2023**, *3*, 349–368. [CrossRef]
52. Cramer, S.C. Repairing the human brain after stroke: I. Mechanisms of spontaneous recovery. *Ann. Neurol.* **2008**, *63*, 272–287. [CrossRef]
53. Christidi, F.; Tsiptsios, D.; Fotiadou, A.; Kitmeridou, S.; Karatzetzou, S.; Tsamakis, K.; Sousanidou, A.; Psatha, E.A.; Karavasilis, E.; Seimenis, I.; et al. Diffusion Tensor Imaging as a Prognostic Tool for Recovery in Acute and Hyperacute Stroke. *Neurol. Int.* **2022**, *14*, 841–874. [CrossRef]

neurology international

MDPI

Article

Safety and Efficacy of Low-Dose Eptifibatide for Tandem Occlusions in Acute Ischemic Stroke

Paweł Latacz [1], Tadeusz Popiela [2], Paweł Brzegowy [2], Bartłomiej Lasocha [3], Krzysztof Kwiecień [1] and Marian Simka [4,*]

1 Department of Vascular Surgery and Angiology, Brothers of Mercy St. John of God Hospital, 31-061 Krakow, Poland; pawlat@me.com (P.L.); krzyskwiecien@yahoo.com (K.K.)
2 Chair of Radiology, Jagiellonian University Medical College, 31-008 Krakow, Poland; msjpopie@cyf-kr.edu.pl (T.P.); pawelbrzegowy007@gmail.com (P.B.)
3 Diagnostic Imaging Unit, University Hospital, 31-501 Krakow, Poland; blasocha@su.krakow.pl
4 Institute of Medical Sciences, University of Opole, 45-040 Opole, Poland
* Correspondence: msimka@uni.opole.pl

Abstract: Objectives: The optimal treatment strategy for ischemic stroke in patients presenting with tandem occlusions of the internal carotid artery remains controversial. Several studies have demonstrated better clinical outcomes after eptifibatide, which is a short half-life antiplatelet agent. This retrospective analysis focused on the safety and efficacy of low-dose eptifibatide administration in stroke patients with tandem lesions. Methods: We evaluated the results of endovascular treatment in 148 stroke patients with tandem lesions. Patients in whom balloon angioplasty alone resulted in satisfactory cerebral flow did not receive eptifibatide (33 patients); others received this drug together with stent implantation (115 patients). Eptifibatide was given as an intravenous bolus of 180 μg/kg and then in a modified low dose of 1 μg/kg/min for 24 hours. Results: There were no statistically significant differences between both groups regarding 30-day mortality, frequency of thrombotic events, or hemorrhagic complications. An analysis of clinical status at 30-day follow-up revealed that the administration of eptifibatide was associated with a statistically significant better outcome: a higher rate of either no neurological symptoms or only mild symptoms (4 NIHSS points maximally). Conclusions: The administration of eptifibatide in stroke patients presenting with tandem lesions is relatively safe. Moreover, treatment with this drug can improve clinical outcomes in these challenging patients.

Keywords: ischemic stroke; tandem lesion; epifibatide; stent

1. Introduction

The optimal treatment strategy for acute ischemic stroke in patients presenting with so-called tandem lesions—i.e., comprising both occlusion or high-grade stenosis of the extracranial segment of the internal carotid artery (ICA) coexisting with occlusion of the intracranial cerebral arteries (intracranial segments of the ICA, the middle, or the anterior cerebral artery)—still remains controversial [1–3]. Data that come from clinical trials are scarce, since such patients typically represent only 10–20% of the cohorts studied. Quite often, patients presenting with tandem lesions are excluded from randomized controlled trials [4–7]. However, it is known that intravenous trombolysis in these patients is able to recanalyze occlusions in only about 5% of the cases, while the risk of hemorrhagic complications is high. Moreover, good clinical outcomes after standard intravenous trombolysis in stroke patients with tandem lesions are at the level of 17% with a mortality rate at the level of 50% [8,9]. On the other hand, data coming from observational studies suggest that reperfusion rates and clinical outcomes are better if extracranial lesions in these patients are managed with angioplasty and stenting [9–12].

Antiplatelet therapy is another controversial issue associated with the treatment of strokes in patients presenting with tandem lesions [11–15]. Recommended pharmacological treatments accompanying elective carotid stent placement include dual-antiplatelet therapy and intraprocedural anticoagulation. Yet, such a strategy is controversial in the setting of acute stroke because of the risk of intracranial hemorrhage, which is especially high if heparin is used. Nonetheless, current guidelines also recommend antiplatelet agents in these patients, particularly intravenous aspirin, although good-quality data regarding safety profile and clinical efficacy are needed [16].

This risk of intracranial hemorrhage in stroke patients particularly concerns tandem occlusions. An increased risk of intracranial bleeding is also seen when antiplatelet agents are used in these patients. Therefore, usually, these pharmacological agents are not used at the beginning of treatment. Moreover, bleeding associated with long half-life antiplatelet agents, such as aspirin or clopidogrel, cannot be easily controlled and is almost always associated with neurologic deterioration and poor clinical outcomes [10,11,14,15,17–19].

This problem could be overcome if a short half-life drug was used. Theoretically, the early administration of such an antiplatelet agent in patients managed with angioplasty and stenting should decrease the risk of intraprocedural cerebral embolic events as well as the risk of stent thrombosis. For this purpose, GP IIB/IIIA receptor antagonists have been used with the most common agent being eptifibatide.

Epifibatide is cleared by the kidneys, and its pharmacokinetic plasma half-life is about 4 h. Because of its short half-life, in patients without renal insufficiency, a cessation of epifibatide infusion relatively quickly restores normal hemostatic platelet function [20]. Its biological half-life is estimated to be at the level of 2.5 h, while the "real-life" stop of a prolonged bleeding, demonstrated in many clinical trials and in every-day hospital practice, is within 30 min. It should be noted that eptifibatide is contraindicated in patients undergoing hemodialysis. It is also contraindicated in patients with thrombocytopenia with a platelet count of less than 100,000/μL. This drug is primarily used in patients receiving coronary angioplasty and presenting with unstable angina or non-ST elevation myocardial infarction. Another current clinical application of epifibatide is the management of acute ischemic stroke especially in patients presenting with tandem lesions. In these patients, there is a high risk of thrombotic reocclusion, while the risk of hemorrhagic transformation of ischemic necrosis is also high. Thus, there is a need for a short half-life and a reversible antiplatelet drug.

Yet, data regarding the safety and efficacy of early administration of eptifibatide in this particular group of stroke patients are scarce and inconclusive [18,19,21–23]. Moreover, according to the producer, this pharmacological agent is contraindicated in stroke patients; thus, all published reports on eptifibatide in stroke patients concern its off-label administration. A meta-analysis by Liu et al. revealed that the treatment of stroke patients with eptifibatide neither had an effect on favorable neurological outcomes nor did it increase mortality [24]. On the other hand, several studies have demonstrated that treatment with eptifibatide in stroke patients can be relatively safe and can improve stroke outcomes. It is possible that these dissimilar observations of this agent resulted from its different dosing. Accompanying coronary stent implantation, eptifibatide is typically administered as an intravenous infusion at a rate of 2 μg/kg/min. It has already been suggested that perhaps such dosing in stroke patients is too high, and indeed, in several trials, it has been lowered to 0.75 μg/kg/min or 1 μg/kg/min.

The aim of this single-center retrospective analysis was to assess the safety and efficacy of the administration of the modified low dosage of eptifibatide (Integrelin, GlaxoSmithKline, Brentford, UK) accompanying endovascular angioplasty and stenting of the ICA in patients presenting with tandem lesions in the acute phase of ischemic stroke. This drug was administered in the standard initial intravenous bolus of 180 μg/kg, but instead of the usually recommended infusion at the rate of 2 μg/kg/min, a dose of 1 μg/kg/min was used. This modification was primarily aimed at reducing potential hemorrhagic complications.

2. Materials and Methods

This study is a post hoc analysis, which focused on the safety of the early administration of eptifibatide in stroke patients with tandem lesions. We retrospectively evaluated the results of the endovascular treatment in 148 consecutive patients (114 men and 34 women) presenting with acute ischemic stroke due to tandem lesions, who were managed from 2016 to 2023. All of them were treated following our standard hospital protocol for the management of ischemic stroke caused by tandem lesions (Figure 1). This survey includes both patients with strokes of atherosclerotic etiology and those presenting with stroke due to artery dissection.

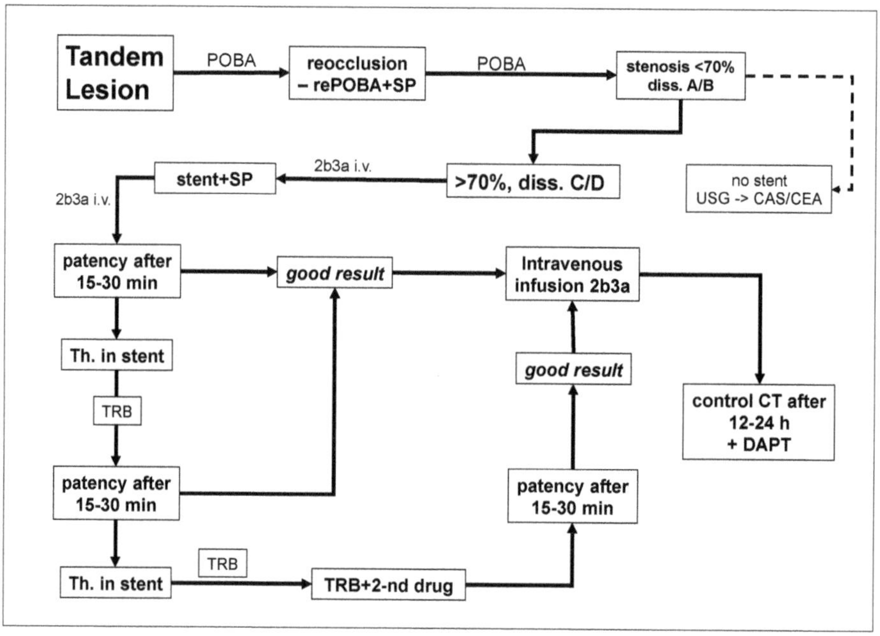

Figure 1. Algorithm of the management of stroke patients with tandem lesions. POBA—balloon angioplasty; SP—protection system; diss. (A/B or C/D)—dissection type A–D); 2b3a i.v.—eptifibatide intravenous infusion; patency after 15–30 min.—control angiography after 15–30 min in cathlab; Th. in stent—thrombosis in stent; TRB—thrombectomy; TRB+2-nd drug—thrombectomy + second antiplatelet drug; DAPT—dual antiplatelet treatment; control CT—CT after 12–24 h.

Patients were aged 23–98 years (mean: 65 ± 12.8). Overall, 111 patients (75%) presented with total occlusion of the ICA, while the remaining 37 patients (25%) had 80–99% stenosis of this artery. All of them had also occlusions in the intracranial arteries supplying the brain—details are given in (Table 1).

The primary endpoint of this study was the proportion of patients who had 30-day mortality of any cause, any intracranial bleeding or any periprocedural complications comprising target artery dissection, perforation and/or new embolic cerebral infarction. Symptomatic intracranial hemorrhage, according to the European Cooperative Acute Stroke Study II, was defined as any intraparenchymal, subarachnoid or intraventricular bleeding, which was associated with clinical worsening of at least 4 points in the National Institutes of Health Stroke Scale (NIHSS). Stent thrombosis and/or hemorrhagic complications were diagnosed either intraprocedurally, or during routine CT or MR imaging, which was performed 24 h after the procedure, or during diagnostics that was started immediately in a case of clinical worsening or recurrence of neurologic symptoms. For the latter purpose, Doppler sonography together with CT or MR angiography were utilized. The secondary endpoint of this study was neurological status measured with the NIHSS at 30-day follow-up.

Table 1. Clinical characteristics of the patients, lesion characteristics and endovascular techniques used. NS—difference statistically insignificant.

	Patients Receiving Eptifibatide (n = 115)	**Patients Not Receiving Eptifibatide (n = 33)**	**p Value (Fisher Exact Test)**
Age (years), mean (SD)	65.0 (12.7)	63.5 (14.1)	NS
Males/females—n (%)	86/29 (75/25)	28/5 (85/15)	NS
	Medical history		
hypertension—n (%)	67 (58)	18 (55)	NS
diabetes mellitus—n (%)	42 (37)	6 (18)	NS
coronary disease—n (%)	30 (26)	6 (18)	NS
history of myocardial infarction—n (%)	19 (17)	5 (15)	NS
NIHSS at admission—mean (SD)	14 (5.3)	16 (4.7)	NS
ASPECT at admission—mean (SD)	8 (1.5)	7 (1.4)	NS
	Extracranial carotid lesion		
Stenosis $\geq$ 80%—n (%)	28 (24)	9 (27)	NS
Occlusion—n (%)	87 (76)	24 (73)	NS
	Intracranial occlusion location		
Middle cerebral artery or anterior cerebral artery—n (%)	78 (68)	25 (76)	NS
Intracranial part of ICA—n (%)	37 (32)	8 (24)	NS
	Intracranial thrombectomy technique		
stent retriever—n (%)	33 (29)	13 (39)	NS
aspiration alone—n (%)	82 (71)	20 (61)	NS
Extracranial carotid procedure			
Angioplasty alone	4	33	
Stenting	111	0	

For the purpose of this survey, all patients were divided into two groups: (i) receiving a short half-life GP IIB/IIIA receptor antagonist (eptifibatide) intra- and postprocedurally, and (ii) patients in whom antiplatelet treatment was postponed until the second postprocedural day, when—after exclusion of intracranial bleeding—all patients were allowed to receive standard antiplatelet therapy if such a therapy was clinically justified. No patients received heparin during endovascular procedure.

Decision upon the administration of eptifibatide was primarily associated with early stent implantation. If endovascular angioplasty of extracranial occlusion/stenosis of the ICA, using a long inflation of the balloon, resulted in satisfactory dilatation of the artery and good inflow (there should be less than 70% residual stenosis of the extracranial lesion after primary angioplasty (Figure 1) and good inflow to the cerebral arteries), the decision upon stent implantation was postponed until 2–7 days postprocedure. Otherwise, if angioplasty failed or there was an immediate reocclusion, stent was implanted and eptifibatide was administered. This agent was given in an initial intravenous bolus of 180 μg/kg (total dose of such a bolus in our patients: 11–16 mg) and then as intravenous infusion 1 μg/kg/min for 24 h. Eptifibatide was also administered in the same doses if there were the above-described indications for stent implantation, yet the stent was not implanted due to technical problems; in this case series, there were two such patients. Given these indications, patients receiving eptifibatide presented with an unfavorable morphology of tandem lesions, potentially with a high risk of thrombotic reocclusion, which justified antiplatelet treatment. Of note, the reduced dose of eptifibatide—1 μg/kg/min instead of standard 2 μg/kg/min—was primarily aimed at the reduction in hemorrhagic complications.

Out of the total of 148 patients assessed in this survey, 115 individuals (78%) received eptifibatide. In this group, 111 patients had stent implantation, including 7 patients in whom two stents were implanted. The remaining 33 patients (22%) did not receive early antiplatelet treatment and had no stents implanted.

The clinical and radiologic characteristics of strokes in both groups (i.e., receiving or not receiving early antiplatelet therapy) were similar. Median scores of the NIHSS at the baseline were 15 and 16, respectively, and this difference was statistically insignificant (see Table 1). The average ASPECT score (the Alberta stroke program early CT score) in patients receiving eptifibatide was 7.7, while in those without antiplatelet treatment, it was 7.3. Similarly, postprocedural recanalization rates of the middle cerebral artery in both groups were comparable. The Thrombolysis in Cerebral Infarction (TICI) score in patients receiving eptifibatide was TICI2b in 20% of them, TICI2c in 8%, and TICI3 in 72%. In patients without antiplatelet treatment, these scores were TICI2b in 27% of the patients, TICI2c in 12%, and TICI3 in 61%. Mean volumes of the necrosis and penumbra in patients receiving eptifibatide were 16.2 mL and 114.3 mL, respectively, while in those without antiplatelet treatment, the mean volumes were 28.6 mL and 135.8 mL, respectively. The difference between the groups regarding necrosis volume was statistically significant ($p = 0.01$; assessed using the two-sample t-test), while the difference in terms of penumbra volume was statistically insignificant. The preprocedural clinical characteristics of the patients are given in Table 1.

The algorithm of patients' management is presented in Figure 1. Extracranial occlusions or stenoses of the ICA were managed with balloon angioplasty with or without stent implantation. In addition, all patients received standard intravenous thrombolysis with alteplase. Decision upon stent implantation, as well as upon the administration of antiplatelet agent was left to the discretion of doctor performing the endovascular procedure. For all patients, the team tried to perform thrombectomy of the intracranial occlusion first and to address extracranial lesion thereafter [25]. If introduction of the reperfusion catheter into the distal segments of the ICA was not possible, the proximal critical lesion was managed firstly either with angioplasty alone or with angioplasty and stenting.

Protection systems were used in 44 patients. The proximal protection devices used comprised the Mo.Ma Ultra 8F (Medtronic, Minneapolis, MN, USA) and the Cello (Covidien, ev3 Endovascular, Inc., Plymouth, MN, USA); the SpiderFX™ Embolic Protection Device (Medtronic, Minneapolis, MN, USA) was used as a distal protection system. Decision upon distal or proximal protection depended on the morphology of occluding lesions as well as on the angioarchitecture of the carotid arteries. Stents were implanted if balloon angioplasty alone failed to reopen the extracranial occlusion or if the flow improvement in this segment lasted only a few minutes and control intraprocedural angiography revealed re-occlusion of the artery. We used several types of stents: the Carotid WallstentTM (Boston Scientific, Natick, MA, USA), the RoadSaver™ stent (Terumo, Tokyo, Japan), and the XACT stent (Abbott Vascular, Abbott Park, IL, USA). The type, length and diameter of the stents were tailored to the angioarchitecture of the target lesion.

In order to reveal potential reocclusion and/or bleeding control, intraprocedural angiography was performed in each patient 5, 10 and 15 min after the stent implantation. The angiographic success of the procedure was evaluated using the modified Treatment in Cerebral Infarction score (mTICI) with the assessment of the previously occluded part of cerebral circulation at the end of endovascular repair.

In all patients, in order to assess the efficacy of the treatment, patency of the stent and/or target artery, and to reveal potential early intracranial bleeding, CT or MR of the head was performed 24 h after the procedure. Then, if there were no hemorrhagic complications, patients were recommended to take dual antiplatelet therapy (aspirin + clopidogrel or ticagrelor) for 1–3 months.

Statistical Analysis

In order to compare the preprocedural clinical characteristics of the patients, as well as the rates of severe complications in both groups, such as bleeding, thrombosis or death, Fisher's exact test was used. To compare parametric pre-interventional neurologic status (NIHSS scores, ASPECT score, TICI score, necrosis volume) and the duration of hospital stay, we used the two-sample t-test; in the latter case, patients who died were excluded. To compare neurologic status at 30-day follow-up, we categorized the patients into six ordinal groups: without neurologic symptoms (NIHSS score: 0), with mild stroke symptoms (NIHSS score: 1–4), moderate stroke symptoms (NIHSS score: 5–14), severe stroke symptoms (NIHSS score: 15–24), very severe stroke symptoms (NIHSS score: 25 and more), and those who died for any reason. Significance of the differences regarding neurologic status at the 30-day follow-up was calculated using the Mann–Whitney U test. In order to assess the correlation between the pre-procedural necrosis volume and the NIHSS score at the 30-day follow-up, Pearson's r coefficient was calculated. The significance of the p values of all statistical tests used was set at $p < 0.05$. Statistical analysis was performed using the PAST data analysis package (version 3.0; University of Oslo, Oslo, Norway).

3. Results

Clinical assessment at 30-day follow-up was possible in all patients in this series except for those who died. There were 12 (8.1%) fatalities: 9 (7.8%) in the group of patients receiving eptifibatide and 3 death (9.1%) in the group of patients who were not managed with this drug. This difference was statistically insignificant. Of note, the frequencies of thrombotic and hemorrhagic complications were higher in patients not receiving eptifibatide, yet these differences were not statistically significant. The duration of hospital stay was similar in both groups of patients. Baseline neurological status measured with the NIHSS was almost the same in both groups (15 vs. 16 points). Yet, at the 30-day follow-up, the clinical status of the patients who received eptifibatide was better (median NIHSS: 3 vs. 7 points). Details are given in Table 2.

Table 2. Complication rates and clinical status at the 30-day follow-up. NS—difference statistically non-significant; * patients who have died during 30-day follow-up were excluded; details regarding statistical test used are given in statistical analysis.

	Patients Receiving Eptifibatide (n = 115)	Patients Not Receiving Eptifibatide (n = 33)	Statistical Significance
Stent and/or ICA thrombosis	6 (5.2%)	5 (15.2%)	NS
Intracranial bleeding	11 (9.5%)	5 (15.2%)	NS
Intracranial bleeding or thrombosis	16 (13.9%)	9 (27.3%)	NS
30-day mortality of any cause	9 (7.8%)	3 (9.1%)	NS
Mortality due to stroke	4 (3.5%)	2 (6.1%)	NS
Median duration of hospital stay duration (days) *	9 (3–38)	10 (4–57)	NS
Median score of the National Institutes of Health Stroke Scale at baseline	15 (2–23)	16 (5–24)	NS
Median score of the National Institutes of Health Stroke Scale at 30-day follow-up *	3 (0–36)	7 (0–21)	NS

Detailed analysis of neurological status at the 30-day follow-up has revealed that the early administration of this short half-life antiplatelet agent in a modified lower dose was associated with a higher rate of either no neurological symptoms or symptoms of a mild stroke (maximally 4 NIHSS points). This beneficial effect of the modified early administration of eptifibatide was statistically significant, while the rates of severe adverse events, including fatalities, were not significantly different from those in the group of patients not receiving such a medication. Details of this analysis are given in Table 3 and also shown in Figure 2. Since there was a statistically significant difference between the groups regarding baseline necrosis volume, we have checked whether this parameter was correlated with the clinical outcome. Still, Pearson's r coefficient was 0.21, revealing a weak correlation only.

Table 3. Neurological status of the patients at 30-day follow-up. p value calculated with the Mann–Whitney U-test was 0.04.

	Patients Receiving Eptifibatide (n = 115)	Patients Not Receiving Eptifibatide (n = 33)
Patients without neurologic symptoms (NIHSS score: 0)	26 (22.6%)	3 (9.1%)
Patients with mild stroke symptoms (NIHSS score: 1–4)	41 (35.7%)	9 (27.3%)
Patients with moderate stroke symptoms (NIHSS score: 5–14)	31 (27.0%)	15 (45.5%)
Patients with severe stroke symptoms (NIHSS score: 15–24)	7 (6.1%)	3 (9.1%)
Patients with very severe stroke symptoms (NIHSS score: ≥25)	1 (0.9%)	0
Death before 30-day follow-up	9 (7.8%)	3 (9.1%)

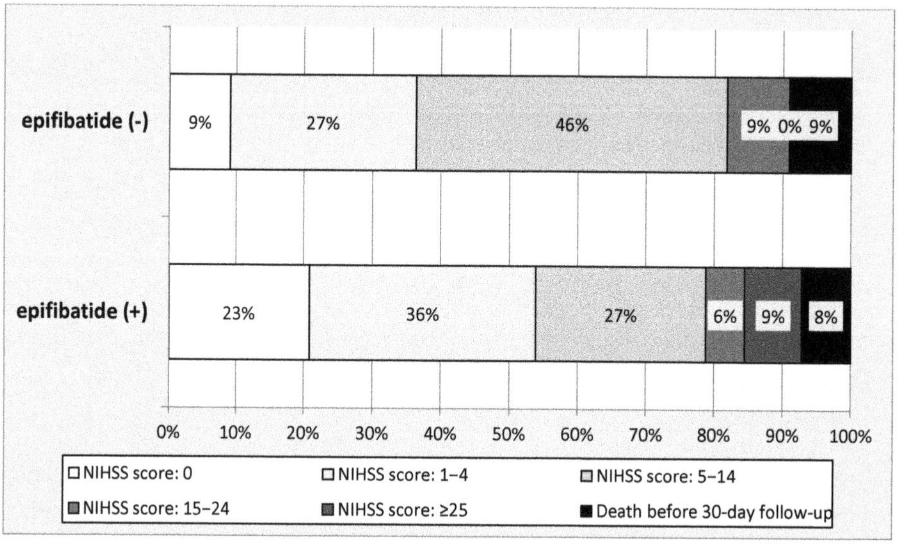

Figure 2. Neurological status measured by National Institutes of Health Stroke Scale at 30-day follow-up; upper graph—patients not receiving eptifibatide, lower graph—patients receiving eptifibatide. Differences regarding clinical outcome were statistically significant ($p = 0.04$; Mann–Whitney U-test) in favor of patients managed with eptifibatide.

4. Discussion

In this retrospective analysis, we have found that the administration of eptifibatide in stroke patients presenting with tandem lesions was relatively safe. The rates of thromboses of the target artery, as well as the rates of intracranial hemorrhages, were not statistically different from those in the group of patients not receiving this antiplatelet agent. The mortality rate was not significantly higher after administration of eptifibatide, either. The frequencies of these adverse events were similar to those reported by other studies [10,12,17–19,21].

A quite unexpected finding from our survey was significantly better neurological outcome at the 30-day follow-up in patients who were managed with eptifibatide. Theoretically, since this antiplatelet agent was given only to patients with unsuccessful first balloon angioplasty and thus generally with an unfavorable status of the target artery, a worse clinical outcome should be expected. Perhaps, these better clinical outcomes in the eptifibatide group resulted from lower rates of thrombotic and hemorrhagic complications in comparison with patients not receiving this drug (even if these differences in our material were not statistically significant). Theoretically, these differences could be associated with fewer thrombotic occlusions in the cerebral microvasculature [16,22–24]. Of note, reocclusion of the ICA or suboptimal thrombectomy are independent predictors of the intracranial bleeding in the settings of ischemic stroke [16,26]. In the study by Renu et al., it was found that the highest rate of bleeding complications occurred in the group in which stent implantation was used in combination with antiplatelet drug, but there was either a suboptimal intracranial recanalization (TICI $\leq$ 2a) or there was stent thrombosis [23]. Probably, incomplete recanalization and damage to the blood–brain barrier, in combination with the antiplatelet drugs, increased the risk of intracranial bleeding [26].

Considering the retrospective nature of this study, our results should be interpreted with caution; perhaps all stroke patients with tandem lesions should be managed with a low dose of eptifibatide. Indeed, such an approach was reported by Jost et al. They were giving this antiplatelet agent in all tandem lesion patients with stroke irrespective of whether they were treated with stent placement or angioplasty [21]. These authors also recommend the dose of 1 μg/kg/min of eptifibatide, since they have found that the dosing of 0.5 μg/kg/min was associated with high reocclusion rates, while other studies suggested that 2 μg/kg/min can result in an increased risk of potentially fatal intracranial bleeding.

Our study showed that the stenting of extracranial ICA lesions and intracranial thrombectomy augmented with low-dose eptifibatide can be a good strategy for the management of difficult to recanalyze tandem lesions. Perhaps such a medication should be given to all and not only selected tandem lesion patients. A similar beneficial effect of the low dose of eptifibatide in tandem lesion patients has recently been demonstrated in the retrospective analysis by Waters et al. [27]. Still, prospective trials should unequivocally demonstrate safety and efficacy of eptifibatide or other short half-life antiplatelet agents in these challenging stroke patients. Such trials should also establish the optimal dose of this antiplatelet agent. For the time being, no such data are available. Of note, other antiplatelet agents, such as aspirin, clopidogrel, prasugrel or tirofiban, have also been proposed for the management in tandem lesion stroke patients. Still, evidence coming from high-quality clinical trials is scarce, and there is no general consensus regarding the use of these drugs in this particular subgroup of stroke patients [16,28]. We acknowledge that there are important limitations of our survey, comprising its retrospective character, relatively small number of patients assessed and the high heterogeneity of the cohort studied.

5. Conclusions

The administration of a modified low dose of eptifibatide in stroke patients presenting with tandem lesions is relatively safe. Moreover, treatment with this drug can possibly improve clinical outcomes in these patients. Yet, prospective randomized clinical trials are needed to unequivocally evaluate the clinical value of this antiplatelet agent.

Author Contributions: Conceptualization, P.L. and M.S.; methodology, P.L. and T.P.; validation, M.S. and P.L; formal analysis, M.S.; investigation, P.L., T.P., P.B., B.L. and K.K.; data curation, M.S.; writing—original draft preparation, P.L.; writing—review and editing, M.S.; supervision, P.L.; project administration, P.L. and T.P. All authors have read and agreed to the published version of the manuscript.

Funding: This research received no external funding.

Institutional Review Board Statement: This was a retrospective analysis of anonymized data associated with standard treatment—local ethics committee approval was not needed for such a study.

Informed Consent Statement: Informed consent was obtained from all patients involved in this survey.

Data Availability Statement: Anonymized data presented in this study are available on request from the corresponding author. Preprint version of this paper is available at https://doi.org/10.20944/preprints202312.2305.v1.

Conflicts of Interest: The authors declare no conflicts of interest.

References

1. Linfante, I.; Llinas, R.H.; Selim, M.; Chaves, C.; Kumar, S.; Parker, R.A.; Caplan, L.R.; Schlaug, G. Clinical and vascular outcome in internal carotid artery versus middle cerebral artery occlusions after intravenous tissue plasminogen activator. *Stroke* **2002**, *33*, 2066–2071. [CrossRef] [PubMed]
2. Pechlaner, R.; Knoflach, M.; Matosevic, B.; Ruecker, M.; Schmidauer, C.; Kiechl, S.; Willeit, J. Recanalization of extracranial internal carotid artery occlusion after i.v. thrombolysis for acute ischemic stroke. *PLoS ONE* **2013**, *8*, e55318. [CrossRef] [PubMed]
3. Rubiera, M.; Ribo, M.; Delgado-Mederos, R.; Santamarina, E.; Delgado, P.; Montaner, J.; Alvarez-Sabín, J.; Molina, C.A. Tandem internal carotid artery/middle cerebral artery occlusion: An independent predictor of poor outcome after systemic thrombolysis. *Stroke* **2006**, *37*, 2301–2305. [CrossRef] [PubMed]
4. Berkhemer, O.A.; Fransen, P.S.; Beumer, D.; Berg, L.A.V.D.; Lingsma, H.F.; Yoo, A.J.; Schonewille, W.J.; Vos, J.A.; Nederkoorn, P.J.; Wermer, M.J.H.; et al. A randomized trial of intraarterial treatment for acute ischemic stroke. *N. Engl. J. Med.* **2015**, *372*, 11–20. [CrossRef] [PubMed]
5. Campbell, B.C.; Mitchell, P.J.; Kleinig, T.J.; Dewey, H.M.; Churilov, L.; Yassi, N.; Yan, B.; Dowling, R.J.; Parsons, M.W.; Oxley, T.J.; et al. Endovascular therapy for ischemic stroke with perfusion-imaging selection. *N. Engl. J. Med.* **2015**, *372*, 1009–1018. [CrossRef] [PubMed]
6. Goyal, M.; Demchuk, A.M.; Menon, B.K.; Eesa, M.; Rempel, J.L.; Thornton, J.; Roy, D.; Jovin, T.G.; Willinsky, R.A.; Sapkota, B.L.; et al. Randomized assessment of rapid endovascular treatment of ischemic stroke. *N. Engl. J. Med.* **2015**, *372*, 1019–1030. [CrossRef]
7. Veunac, L.; Saliou, G.; Knebel, J.F.; Bartolini, B.; Puccinetlli, F.; Michel, P.; Haljdu, S.D. Revascularization of carotid artery occlusion using stenting versus non stenting in endovascular management of tandem occlusion stroke. *J. Clin. Neurosci.* **2022**, *98*, 15–20. [CrossRef]
8. Jovin, T.G.; Chamorro, A.; Cobo, E.; De Miquel, M.A.; Molina, C.A.; Rovira, A.; Román, L.S.; Serena, J.; Abilleira, S.; Ribo, M.; et al. Thrombectomy within 8 hours after symptom onset in ischemic stroke. *N. Engl. J. Med.* **2015**, *372*, 2296–2306. [CrossRef]
9. Jadhav, A.P.; Zaidat, O.O.; Liebeskind, D.S.; Yavagal, D.R.; Haussen, D.C.; Hellinger, F.R., Jr.; Jahan, R.; Jumaa, M.A.; Szeder, V.; Nogueira, R.G.; et al. Emergent management of tandem lesions in acute ischemic stroke: Analysis of the STRATIS registry. *Stroke* **2019**, *50*, 428–433. [CrossRef] [PubMed]
10. Papanagiotou, P.; Haussen, D.C.; Turjman, F.; Labreuche, J.; Piotin, M.; Kastrup, A.; Steglich-Arnholm, H.; Holtmannspötter, M.; Taschner, C.; Eiden, S.; et al. (TITAN Investigators), Carotid stenting with antithrombotic agents and intracranial thrombectomy leads to the highest recanalization rate in patients with acute stroke with tandem lesions. *ACC Cardiovasc. Interv.* **2018**, *11*, 1290–1299. [CrossRef] [PubMed]
11. Wallocha, M.; Chapot, R.; Nordmeyer, H.; Fiehler, J.; Weber, R.; Stracke, C.P. Treatment methods and early neurologic improvement after endovascular treatment of tandem occlusions in acute ischemic stroke. *Front. Neurol.* **2019**, *10*, 127. [CrossRef]
12. Gory, B.; Haussen, D.C.; Piotin, M.; Steglich-Arnholm, H.; Holtmannspötter, M.; Labreuche, J.; Kyheng, M.; Taschner, C.; Eiden, S.; Nogueira, R.G.; et al. Impact of intravenous thrombolysis and emergent carotid stenting on reperfusion and clinical outcomes in patients with acute stroke with tandem lesion treated with thrombectomy: A collaborative pooled analysis. *Eur. J. Neurol.* **2018**, *25*, 1115–1120. [CrossRef]
13. Neuberger, U.; Moteva, K.; Vollherbst, D.F.; Schönenberger, S.; Reiff, T.; Ringleb, P.A.; Bendszus, M.; Pfaff, J.A.R.; Möhlenbruch, M. Tandem occlusions in acute ischemic stroke: Impact of antithrombotic medication and complementary heparin on clinical outcome and stent patency. *J. Neurointerv Surg.* **2020**, *12*, 1088–1093. [CrossRef]
14. Heck, D.; Brown, M. Carotid stenting and intracranial thrombectomy for treatment of acute stroke due to tandem occlusions with aggressive antiplatelet therapy may be associated with a high incidence of intracranial hemorrhage. *J. Neurointerv Surg.* **2015**, *7*, 170–175. [CrossRef]

15. Pop, R.; Zinchenko, I.; Quenardelle, V.; Mihoc, D.; Manisor, M.; Richter, J.; Severac, F.; Simu, M.; Chibbaro, S.; Rouyer, O. Predictors and clinical impact of delayed stent thrombosis after thrombectomy for acute stroke with tandem lesions. *Am. J. Neuroradiol.* **2019**, *40*, 533–539. [CrossRef]
16. Goyal, M.; Orlov, K.; Jensen, M.E.; Taylor, A.; Majoie, C.; Jayaraman, M.; Liu, J.; Milot, G.; Brouwer, P.; Yoshimura, S.; et al. A DELPHI consensus statement on antiplatelet management for intracranial stenting due to underlying atherosclerosis in the setting of mechanical thrombectomy. *Neuroradiology* **2021**, *63*, 627–632; Erratum in *Neuroradiology* **2021**, *63*, 1391–1392. [CrossRef]
17. Poppe, A.Y.; Jacquin, G.; Roy, D.; Stapf, C.; Derex, L. Tandem carotid lesions in acute ischemic stroke: Mechanisms, therapeutic challenges, and future directions. *AJNR Am. J. Neuroradiol.* **2020**, *41*, 1142–1148. [CrossRef] [PubMed]
18. Goyal, M.; Yoshimura, S.; Milot, G.; Fiehler, J.; Jayaraman, M.; Dorn, F.; Taylor, A.; Liu, J.; Albuquerque, F.; Jensen, M.; et al. Considerations for antiplatelet management of carotid stenting in the setting of mechanical thrombectomy: A Delphi consensus statement. *AJNR Am. J. Neuroradiol.* **2020**, *41*, 2274–2279. [CrossRef] [PubMed]
19. Osteraas, N.D.; Crowley, R.W.; Panos, N.; Dafer, R.M. Eptifibatide use following emergent carotid stenting in acute anterior circulation ischemic stroke with tandem occlusion. *J. Stroke Cerebrovasc. Dis.* **2020**, *29*, 105021. [CrossRef] [PubMed]
20. Pasala, T.; Sattayaprasert, P.; Bhat, P.K.; Athappan, G.; Gandhi, S. Clinical and economic studies of eptifibatide in coronary stenting. *Ther. Clin. Risk Manag.* **2014**, *10*, 603–614. [CrossRef] [PubMed]
21. Jost, A.; Roels, C.; Brown, M.; Janjua, R.; Heck, D. Low-dose eptifibatide for tandem occlusion in stroke: Safety and carotid artery patency. *AJNR Am. J. Neuroradiol.* **2021**, *42*, 738–742. [CrossRef]
22. Gaoting, M.; Sun, X.; Cheng, H.; Burgin, W.S.; Luo, W.; Jia, W.; Liu, Y.; He, W.; Geng, X.; Zhu, L.; et al. Combined approach to eptifibatide and thrombectomy in acute ischemic stroke because of large vessel occlusion: A matched-control analysis. *Stroke* **2022**, *53*, 1580–1588.
23. Luo, L.; Lin, J.; Deng, Y.; Li, Z.; Yuan, Y.; Zhang, W. Treatment of progressive ischemic stroke with low dose eptifibatide: A retrospective case control study. *Exp. Ther. Med.* **2023**, *25*, 22. [CrossRef]
24. Liu, J.; Yang, Y.; Liu, H. Efficacy outcomes and safety measures of intravenous tirofiban or eptifibatide for patients with acute ischemic stroke: A systematic review and meta-analysis of prospective studies. *J. Thromb. Thrombolysis* **2022**, *53*, 898–910. [CrossRef]
25. Marnat, G.; Mourand, I.; Eker, O.; Machi, P.; Arquizan, C.; Riquelme, C.; Ayrignac, X.; Bonafe, A.; Costalat, V. Endovascular management of tandem occlusion stroke related to internal carotid artery dissection using a distal to proximal approach: Insight from the RECOST Study. *AJNR Am. J. Neuroradiol.* **2016**, *37*, 1281–1288. [CrossRef] [PubMed]
26. Renu, A.; Blasco, J.; Laredo, C.; Llull, L.; Urra, X.; Obach, V.; López-Rueda, A.; Rudilosso, S.; Zarco, F.; González, E.; et al. Carotid stent occlusion after emergent stenting in acute ischemic stroke: Incidence, predictors and clinical relevance. *Atherosclerosis* **2020**, *313*, 8–13. [CrossRef] [PubMed]
27. Waters, M.J.; Vargas, J.; Turk, A.; Chaudry, I.; Turner, R.D. Safety and efficacy of eptifibatide in acute ischemic stroke requiring extracranial carotid artery stenting. *Interv. Neuroradiol.* **2023**. [CrossRef] [PubMed]
28. Sepp, D.; Berndt, M.; Mönch, S.; Ikenberg, B.; Wunderlich, S.; Maegerlein, C.; Zimmer, C.; Boeckh-Behrens, T.; Friedrich, B. Outcome and risk of hemorrhage in patients with tandem lesions after endovascular treatment: A propensity score-matched case-control study. *Heliyon* **2023**, *9*, e14508. [CrossRef] [PubMed]

neurology international

MDPI

Article

The Usefulness of Factor XIII Concentration Assessment in Patients in the Acute Phase of Ischaemic Stroke Treated with Thrombolysis

Małgorzata Wiszniewska [1,2,*], Urszula Włodarczyk [2], Magdalena Sury [2], Artur Słomka [3], Natalia Piekuś-Słomka [4], Anna Żdanowicz [5] and Ewa Żekanowska [3]

1 Emergency Medical Services, University of Applied Sciences, 64-920 Piła, Poland
2 Neurological Department with Stroke Unit, Specialist Hospital, 64-920 Piła, Poland
3 Department of Pathophysiology, Nicolaus Copernicus University in Torun, Ludwik Rydygier Collegium Medicum, Faculty of Pharmacy, 85-067 Bydgoszcz, Poland; artur.slomka@cm.umk.pl (A.S.)
4 Department of Inorganic and Analytical Chemistry, Nicolaus Copernicus University in Toruń, Ludwik Rydygier Collegium Medicum, 85-067 Bydgoszcz, Poland; natalia.piekus@cm.umk.pl
5 Department of Nursing, Stanislaw Staszic State University of Applied Science, 64-920 Piła, Poland
* Correspondence: mpwisz@gmail.com or mwiszniewska@ans.pila.pl

Abstract: Background and Aims: In recent years, there has been a growing interest in factor XIII in ischaemic stroke. The study's main aim was to assess the usefulness of factor XIII concentration determination in patients with acute ischaemic stroke (AIS) treated with thrombolysis with recombinant tissue plasminogen activator (t-PA). Methods: The study was conducted in two groups of 84 patients with AIS: group I—with thrombolytic therapy and group II—without thrombolysis. A physical examination, neurological status (using the National Institutes of Health Stroke Scale, NIHSS), daily patients' activities measured with the Barthel Index and Modified Rankin Scale (mRS), and blood parameters were conducted on day 1 and day 7. The following parameters were assessed: highly sensitive C-reaction protein (CRP), fibrinogen, D-dimers (DD), neutrophil–lymphocyte ratio (NLR index), and the concentration of factor XIII-A. Results: In group I, the concentration of XIII-A decreased significantly between day 1 and 7 ($p < 0.001$). In group I, the concentration of XIII-A on day 7 in Total Anterior Circulation Infarct (TACI) was significantly lower than in non-TACI stroke. XIII-A concentration in group I was significantly lower in patients < 31 points with Acute Stroke Registry and Analysis of Lausanne (ASTRAL). A greater decrease in XIII-A between the first sampling on day 1 and the second sampling on day 7 was associated with a worse patient neurological state in group I. Conclusions: In patients with AIS treated with t-PA, factor XIII concentrations decrease in the acute phase of stroke, and the largest decrease occurs in the TACI stroke. Determination of factor XIII concentration in patients with AIS can be used in clinical practice as an additional parameter supporting the assessment of stroke severity and may play a role in the prognosis; lower factor XIII-A activity may be a predictor of a worse prognosis.

Keywords: factor XIII; acute ischaemic stroke; thrombolysis

1. Introduction

Stroke is a clinical syndrome characterised by sudden focal or generalised brain dysfunction with symptoms lasting longer than 24 h or leading to death that have no other cause than vascular [1]. According to the new definition, stroke can be diagnosed when symptoms persist for a period shorter than indicated above, for example, following thrombolytic treatment or when neuroimaging examination shows unambiguous evidence of an ischaemic focus [2]. Intravenous thrombolytic treatment with recombinant tissue plasminogen activator (t-PA) continues to be the gold standard for the treatment of the acute phase of ischaemic stroke (IS) [3].

Factor XIII, called fibrinase or fibrin stabilising factor, consisting of two potentially active A subunits and two carrier B subunits, plays an important yet not fully understood role in IS [4–6]. It participates in the development of atherosclerosis through the active binding of FXIII-A to the receptor for angiotensin-1 on monocytes [4]. It plays a significant role in the final phase of the coagulation cascade, in which fibrinogen molecules are cleaved into fibrin monomers and fibrinopeptides A and B [7]. Fibrin monomers form a fibrin network that is stabilised by active factor XIII. Factor XIII is converted by thrombin into FXIII-A2B2, which, in turn, is broken down into two fractions under the influence of Ca^{2+} ions: active FXIII-A2 (FXIII-A) and inactive FXIII-B. Factor XIII-A acts as a catalyst for the synthesis of isopeptide bonds between the free amino groups of one fibrin monomer with the amino groups of the glutamine of the other fibrin monomer, creating a fibrin network [8]. On the other hand, factor FXIII-A2 also exhibits anti-fibrinolytic activity by binding itself to proteins of the fibrinolytic system: α2-antiplasmin, thrombin activatable fibrinolysis inhibitor (TAFI), plasminogen activation inhibitor-1 (PAI-1), inhibiting the activity of the fibrinolytic system [9].

The FXIII-A2 active subunit can be used to detect the acute phase of IS, as the presence of this peptide in plasma indicates an active coagulation process [10].

Aim of the Study

The aim of the study was to compare the effects of the t-PA thrombolytic treatment and conservative treatment on factor XIII levels in AIS.

Evaluate the clinical utility of factor XIII levels determination in patients with AIS treated with t-PA.

Evaluate the relationship between the risk factors of IS and factor XIII levels, as well as selected inflammatory and coagulation protein parameters.

Evaluate the dynamics of changes in the following parameters: factor XIII, CRP, DD, and fibrinogen levels in relation to the neurological state of the patient in the acute phase of IS.

2. Materials and Methods

The study group consisted of 84 patients with AIS treated in the Neurology Department with Stroke Unit in the Stanislaw Staszic Specialist Hospital in Piła in the period from 2016 to 2019.

IS was diagnosed based on a CT scan performed within 20 min of arrival at the hospital.

The patients were divided into two groups. Group I consisted of 52 patients who received thrombolytic treatment with t-PA within a time window of up to 4.5 h from the onset of the stroke. The inclusion criterion for thrombolytic treatment was based on the guidelines applicable at the time [11]. Group II consisted of 32 patients with AIS who did not meet the criteria for t-PA treatment, and all of them exceeded the time window.

Each patient underwent a clinical assessment performed by doctors working in the Neurology Department with Stroke Unit, who conducted a physical examination on admission to the unit and on days 1 and 7. In each patient, the stroke risk factors were analysed together with age, weight, waist–hip ratio (WHR), and blood pressure (BP). The clinical stroke syndrome was assessed according to the Oxford Community Stroke Project classification (OSCP), distinguishing total anterior circulation infarct (TACI), partial anterior circulation infarct (PACI), lacunar infarct (LACI), and posterior circulation infarct (POCI) [12]. Stroke aetiology was assessed by applying the TOAST classification [13]. Each patient underwent ultrasound examination consisting of a Doppler ultrasound of the cephalic arteries, electrocardiogram, Holter electrocardiogram, and echocardiography.

Before starting the t-PA treatment, each patient had routine laboratory tests performed, including neutrophil–lymphocyte ratio (NLR), coagulation parameters, and analysis of lipids. Factor XIII-A2 concentration was determined for each patient on admission using an enzyme-linked immunosorbent assay (ELISA). Factor XIII-A concentrations were measured with an assay (Zymutest Factor XIII-A, Hyphen BioMed, SAS, Neuille-sur-Oise, France) and expressed as a percentage. On day 7, the following parameters were measured again: factor XIII-A2 concentration, haemoglobin, haematocrit, erythrocytes, leukocytes, platelets, activated partial thromboplastin time (APTT), and international normalized ratio (INR). Each patient's motor skills were assessed on the seventh day using the modified Rankin scale (mRs).

The computer programme STATISTICA v.13.3 from TIBCO (Palo Alto, CA, USA) was used for statistical analysis. Non-parametric tests were used to analyse the differences between the parameters studied in each group: Mann–Whitney U test for comparison between the two groups and Kruskal–Wallis rank ANOVA (with Dunn's post hoc test) for quantitative variables. A chi-square test was used to compare the percentage distributions of the categorised variables, and Spearman's correlation analysis was used to assess the monotonic statistical relationship between the variables. Differences were considered statistically significant for $p < 0.05$.

3. Results

Demographic and Clinical Characteristics of the Study Participants

A total of 52 patients treated with t-PA (group I) and 32 patients not treated with t-PA (group II) were included in the analysis. The gender distribution was similar in both groups: in group I, women accounted for 42.3% of the members, and 41.5% in group II (a statistically insignificant difference). Table 1 shows the demographic and clinical characteristics of the patients.

Table 1. Demographic and clinical characteristics of patients with ischaemic stroke treated with t-PA (group I) and not treated with t-PA (group II).

Parameters	Group I (Treated with t-PA)	Group II (Not Treated with t-PA)	*p* (Group I vs. Group II)
Age, years (Me)	39–88 (70)	39–88 (70)	0.72
Weight, kg (Me)	55–130 (80)	56–136 (80)	0.58
WHR (Me)	0.79–1.38 (1.02)	0.72–1.38 (1.03)	0.53
Height, m (Me)	1.45–1.94 (1.69)	1.42–1.89 (1.69)	0.93
Stroke classification	**Group I N (%)**	**Group II N (%)**	***p* (group I vs. group II)**
	According to OSCP		
PACI	19 (36.54)	9 (28.13)	0.51
POCI	8 (15.38)	9 (28.13)	
LACI	20 (38.46)	12 (37.50)	
TACI	5 (9.62)	1 (6.24)	
	According to TOAST		
Cardiovascular thromboembolic	12 (23.08)	12 (37.50)	0.34
Atherosclerosis of large blood vessels	10 (29.23)	5 (15.62)	
Unknown aetiology	22 (44.23)	8 (28.13)	
Disease of small blood vessels	7 (13.46)	18 (18.75)	

Analysis of stroke based on stroke location according to the Oxford Community Stroke Project—OSCP and stroke aetiology according to Trial of Org 10,172 in Acute Stroke Treatment—TOAST. TACI—Total Anterior Circulation Infarct, PACI—Partial Anterior Circulation Infarct, LACI—Lacunar Infarct, POCI—Posterior Circulation Infarct, Me—median, WHR—waist–hip ratio.

Lacunar infarct was predominant in groups I and II (38.5% and 37.5% of patients, respectively), while TACI stroke was the least common (9.6% and 6%, respectively). In group l, cardioembolic stroke occurred in 23% of the patients, while in group II, it occurred in 37% of the patients (Table 1). Tables 2 and 3 show the correlations between the factor XIII levels and CRP, DD, and fibrinogen levels on admission on day 1 and on day 7. In the first sampling on day 1, CRP, DD, and fibrinogen values were similar in both groups. On day 7, the CRP value increased but only in group I (Me was 4.6 mg/L on day 7 vs. 1.65 mg/L on day 1), which was considered as a statistically significant increase ($p < 0.01$) (Tables 2 and 3).

Table 2. Correlations between the factor XIII-A levels in groups I and II on admission (day 1) and on day 7.

XIII-A Concentration (%)	Group I N = 52	Group II N = 32	*p*
1—day Me (Q1–Q3)	119.54 (109.71–131.39)	109.10 (97.92–122.20)	0.011
7—day Me (Q1–Q3)	108.61 (97.84–122.80)	111.20 (99.77–119.25)	0.744
Me Difference (Q1–Q3) between day 1 and day 7	−11.26 (−25.88–3.6)	1.20 (−8.15–10.09)	0.006
p for difference	<0.001	0.751	

Me—median, Q1—first quartile, Q3—third quartile.

Table 3. Correlations between the distributions of CRP, D-dimers, and fibrinogen concentrations in groups I and II on admission (day 1) and on day 7.

Variables	Group I N = 52	Group II N = 32	*p*
CRP day 1 [mg/L] Me (Q1–Q3)	1.65 (1.00–3.85)	2.05 (1.35–7.55)	0.06
CRP day 7 [mg/L] Me (Q1–Q3)	4.60 (1.40–12.80)	4.35 (2.00–8.75)	0.98
CRP difference [mg/L] Me (Q1–Q3)	3.30 (−0.30–10.10)	0.20 (−2.70–6.80)	0.16
p for difference between day 1 and day 7	<0.01	0.76	
DD day 1 [ngFEU/mL] Me (Q1–Q3)	914.5 (441.5–1248.5)	765.5 (411.0–1430.5)	0.77
DD day 7 [ngFEU/mL] Me (Q1–Q3)	664.5 (437.0–1051.0)	736.00 (295.5–1311.5)	0.91
DD difference [ng FEU/mL] Me (Q1–Q3)	−105.0 (−347.0–128.0)	−14.5 (−201.0–258.5)	0.30
p for difference between day 1 and day 7	0.19	0.98	
Fibrinogen day 1 [mg/dL] Me (Q1–Q3)	332.0 (302.5–417.0)	356.0 (306.0–464.0)	0.31
Fibrinogen day 7 [mg/dL] Me (Q1–Q3)	360.0 (285.0–453.5)	377.5 (305.5–431.5)	0.85
Fibrinogen difference [mg/dL] Me (Q1–Q3)	23.5 (−57.0–96.0)	−3.0 (−62.0–83.0)	0.61
p for difference between day 1 and day 7	0.17	0.88	

Me—median, Q1—first quartile, Q3—third quartile, CRP—acute phase protein, DD—D dimers.

The concentration of factor XIII-A in citrated plasma on the day of the onset of the stroke in group I was significantly higher than that in group II (Me was 119.54% vs. 109.90%, respectively; $p = 0.011$) (Table 2). In group I, factor XIII-A levels decreased significantly between day 1 and day 7 ($p < 0.001$), whereas this was not observed in group II (Table 2). The decrease in XIII-A levels in group I did not show any relation with either gender or age of the patients. Factor XIII-A concentrations varied over time in diverse types of strokes (Table 4). In group II, on day 7, factor XIII-A levels were significantly higher in patients with a cardioembolic aetiology of stroke compared to patients with stroke of undetermined aetiology (Me was 117.3% vs. 89.5%, respectively; $p = 0.03$). In stroke caused by small vessel disease, factor XIII-A concentrations were lower on day 1 compared to day 7; however, these differences were not statistically significant. In the case of large artery pathology, patients in group I treated with t-PA had lower factor XIII-A concentrations on day 7 than in group I, while no such decrease was noted in group II (Table 4).

Table 4. Comparison of factor XIII concentrations in relation to stroke aetiology as defined by the TOAST scale.

Group	Aetiology	N	XIII-A [%] Me (Q1–Q3) Day 1	XIII-A [%] Me (Q1–Q3) Day 7
I	Cardiovascular thromboembolic	12	114.22 (105.61–27.58)	103.51 (89.20–108.71)
	Unknown aetiology	23	127.89 (114.62–133.85)	109.32 (99.60–124.51)
	Changes in large blood vessels	10	117.20 (103.26–119.73)	107.00 (96.22–119.72)
	Disease of small blood vessels	7	110.68 (109.20–131.53)	116.79 (111.10–124.16)
	p		0.18	0.25
II	Cardiovascular thromboembolic	12	119.90 (102.05–131.85)	117.30 (110.15–125.80)
	Unknown aetiology	9	98.86 (83.46–117.90)	89.48 (84.78–110.30)
	Changes in large blood vessels	5	110.00 (106.40–116.90)	112.60 (112.10–117.00)
	Disease of small blood vessels	6	105.40 (95.06–124.00)	107.00 (95.34–117.30)
	p		0.27	0.03

Me—median, Q1—first quartile, Q3—third quartile.

When comparing factor XIII-A levels in patients with cardiogenic stroke to patients with non-cardiogenic stroke, statistically significantly higher factor XIII-A levels were observed in patients with cardiogenic stroke in group II (Me = 117.3% vs. Me = 109.9%; $p = 0.03$) in the sampling on day 7. In cardiogenic stroke in group I, factor XIII-A concentrations were lower on day 7 compared to group II (Table 5).

Table 5. Comparison of factor XIII-A levels in relation to stroke aetiology.

Group	Aetiology	N	XIII-A [%] Me (Q1–Q3) Day 1	XIII-A [%] Me (Q1–Q3) Day 7	XIII-A [%] Me (Q1–Q3) Difference
I	Cardiogenic	12	114.22 (105.61–27.58)	103.51 (89.20–108.71)	−15/07 (−35.76–0.18)
	Non-cardiogenic	40	120.08 (110.20–131.39)	110.89 (98.74–124.08)	−9.87 (−23.41–5.45)
	p		0.37	0.08	0.45
II	Cardiogenic	12	119.90 (102.05–131.85)	117.30 (110.15–125.80)	2.20 (−7.80–13.25)
	Non-cardiogenic	20	104.80 (96.02–118.05)	109.85 (91.60–114.80)	0.61 (−8.80–9.07)
	p		0.15	0.03	0.71

Me—median, Q1—first quartile, Q3—third quartile.

Correlation analysis showed a slight negative correlation between the difference in factor XIII concentrations in sampling on day 7 and on day 1 and the mRS value, which was only observed in group I treated with t-PA. This means that, in group I treated with t-PA, but not in group II, a more significant reduction in factor XIII was associated with higher mRS values. If there were no differences or when factor XIII concentration increased, the mRS values were lower (Table 6).

Table 6. Correlations between XIII-A and mRS values.

Pair of Variables	Pair of Variables	Group I		Group II	
		R	*p*	R	*p*
XIII-A I sample on day 1	mRS I sampling	0.15	0.28	0.00	1.00
	mRS II sampling	0.10	0.48	−0.11	0.55
	mRs difference II-I	−0.01	0.95	−0.14	0.46
XIII-A II sample on day 7	mRS I sampling	0.09	0.54	−0.06	0.76
	mRS II sampling	0.01	0.97	−0.10	0.60
	mRs difference II-I	−0.16	0.27	−0.15	0.41
XIII-A difference (II-I)	mRS I sampling	0.03	0.85	0.05	0.77
	mRS II sampling	−0.18	0.20	0.08	0.68
	mRs difference II-I	−0.29	0.03	−0.05	0.77

R—difference, mRS—modified Rankin scale.

4. Discussion

In recent years, factor XIII in thromboembolic conditions has been receiving more attention in the field of research on haemostasis in the context of cardiovascular diseases, including stroke. Factor XIII, as one of the last elements of the coagulation cascade, is involved in the formation of the fibrin clot. Cross-linked fibrin shows greater resistance to mechanical deformation and less susceptibility to fibrinolysis. In contrast, t-Pa (Actilyse) administered to treat the acute phase of IS mimics endogenous t-Pa, activating inactive plasminogen to plasmin, which acts on fibrin to dissolve it, removing clots from the vessels. The products of digestion of stabilised fibrin include fragments double-D, so-called D-dimers, in which the two molecules are linked by cross-links. In the clotting process, factor XIII is used to form a fibrin clot, which causes its levels to drop in the blood. In the available literature, there is little information on the effect of thrombolysis on coagulation parameters, their correlation with each other, and the possible predictive value of any of the haemostatic parameters in relation to the patient's neurological state. This study confirmed previous findings, according to which the IS patients treated with thrombolysis experience a decrease in factor XIII-A levels [14,15]. In our patients treated with Actylise up to 4.5 h after stroke onset, the level of factor XIII-A values on day 7 after administrating the treatment was significantly decreased ($p < 0.001$). In a study that included 132 patients with AIS treated with thrombolysis in the timespan of 4.5 h, Szekely et al. [14] showed that factor XIII levels decreased immediately after the Actylise infusion, as well as 24 h after the end of treatment (factor XIII value before thrombolysis equalled 126.6% ± 36.1%, 24 h after the end of treatment—116% ± 35.0%; $p = 0.034$). The authors tried to explain this phenomenon by the mechanism of consumption and the degradation of factor XIII. The decrease in factor XIII concentration after stroke could be due to the consumption of this factor into the thrombus that is formed and a decrease in its activity through the action of proteolytic enzymes such as plasmin. However, the reduction mechanism of factor XIII levels is not fully understood. Schroeder et al., after studying patients with IS, found a similar decrease in factor XIII-A levels on day 2 after the stroke in both patients treated with thrombolysis and untreated patients [16]. Hence, they put forward the hypothesis that t-Pa treatment alone does not affect the decrease in factor XIII. However, in our study, we showed a significant decrease in the treatment group. On the other hand, Hur et al., in an in vitro study similar to ours, found that the concentration of factor XIII-A decreases significantly after an action of t-PA on the thrombus [15]. They explained this phenomenon by the proteolytic action of plasmin on factor XIII-A found in fibrin. A decrease in factor XIII levels was also observed by Gemnati et al. in other acute thromboembolic diseases, such as pulmonary embolism and acute myocardial infarction, as the disease progressed, and this decrease was greatest in patients with cardiac heart failure and in those in whom death occurred [17,18].

Like other authors, we did not observe statistically significant correlations between factor XIII-A, D-dimers, and fibrinogen in our study [19,20].

In our study, a group of IS patients, who underwent thrombolytic treatment, experienced a statistically significant decrease in factor XIII-A levels, accompanied by a statistically significant increase in CRP levels between days 1 and 7. However, there was no statistically significant correlation between factor XIII-A and CRP. There are several papers showing a link between factor XIII-A and inflammation, as well as a decrease in factor XIII-A during inflammation [21–23]. This may be due, among other factors, to coagulation in the micro-circulation due to inflammation and the consumption of factor XIII-A. In addition, in inflammation, there is a reduction in the number of platelets that contain factor XIII-A. However, thromboembolic stroke is not a simple inflammatory condition, and it should be noted that most papers aimed at investigating the correlation between factor XIII-A and CRP in thromboembolic conditions have not shown such a correlation [14,24,25]. The issue certainly requires further research.

In the study conducted in group I (treatment group with Actilyse), the level of factor XIII-A decreased significantly on day 7 after the onset of the stroke compared to the baseline value on day 1. In contrast, no such decrease was observed in patients who did not receive thrombolytic treatment. However, Schroeder et al. showed no greater decrease in factor XIII levels in patients treated with thrombolytic therapy compared to patients who did not receive this treatment [16]. This indicates that factor XIII levels in the acute phase of stroke and in other thrombotic incidents require further detailed studies, both experimental and observational.

Our own study showed that there is a correlation between factor XIII-A levels and stroke aetiology, as assessed by the TOAST classification. In group II, which did not receive t-PA, patients with stroke caused by a cardiogenic embolism had significantly higher factor XIII-A levels 7 days after stroke onset (the second sampling) compared to patients with stroke of a non-cardiogenic cause (Me was 117.30% in cardiogenic stroke vs. 109.85% in non-cardiogenic stroke, respectively; $p = 0.03$). The results that were obtained are in line with those of Schroeder et al., who also showed that in stroke caused by a cardiogenic embolism, factor XIII-A levels were higher than in IS due to small vessel disease or large vessel atherosclerosis of undetermined aetiology, but there was a difference that did not reach statistical significance [16]. The authors explained this phenomenon as follows: the thrombus in cardioembolic stroke forms relatively earlier, before occlusion with thrombotic material of the cerebral artery occurs, unlike in strokes of non-cardiogenic aetiology. Thus, factor XIII-A is consumed earlier, and at the time of the laboratory measurement, its level rises again at the second measurement on day 7, whereas in non-cardiogenic aetiology, it is consumed during thrombus formation. In our study, we found that among patients treated with t-PA, the lowest factor XIII-A levels were found in the second sampling on day 7 in the most severe type of TACI stroke according to the OCSP scale, and there was a statistically significant difference ($p = 0.04$) compared to non-TACI stroke, which is in line with the results of Kohler et al. [26]. It is, thus, possible to draw an indirect conclusion that lower factor XIII-A levels indicate greater consumption of the factor during thrombus formation and are associated with a more severe stroke course and higher mortality.

We observed that in the treatment group (group I), the NIHSS score on day 7 was statistically significantly lower than in the group that did not receive thrombolytic treatment. This is evidence of the effectiveness of thrombolytic treatment in patients, which is consistent with many studies [26–28]. Factor XIII-A may play a significant role as a prognostic element in IS. In our study, we observed that a decrease in XIII-A levels between days 1 and 7 was associated with higher mRS scores and that this implies a worse functional state of the patient. Our results are consistent with the observations of Szekely et al., who concluded that factor XIII levels were an independent predictor of higher short-term (14-day) mortality in patients receiving thrombolytic therapy [14]. Both our results and the results of Szekely et al. suggest that patients treated with t-PA with a greater decrease in factor XIII-A concentration would require more surveillance, which could improve their prognosis. In addition, it seems that this group of patients would need to be analysed more closely to determine additional factors that could influence this decrease in factor XIII levels. In the study conducted by Sun et al., factor XIII levels did not correlate with the patient's neurological state at 3-month follow-up [29].

It is worth mentioning that we also noted a significant statistical relationship between factor XIII-A levels and ASTRAL scores on day 1, where a score of more than 31 points means that the prognosis will be poor for more than 50% of the population, but there are no similar reports in other studies. However, it can be hypothesised that factor XIII-A may have a prognostic value in patients treated with thrombolytic therapy. Warach et al. elucidated the pharmacological advantages of tenecteplase in the treatment of acute ischemic stroke, particularly its enhanced fibrin specificity and extended half-life, which facilitate a more efficient administration relative to alteplase. These pharmacodynamic properties suggest that factor XIII could be explored as a prognostic biomarker during tenecteplase therapy, potentially offering insights into patient outcomes based on its modulation during treatment [30].

Limitations of the above study may include the relatively small size of the study group, which consisted of 52 people, and the control group of 32 people. Another limitation of the study is the selection of the control group. Patients in the study group who received thrombolytic treatment represented a highly selected patient population with strict inclusion criteria for thrombolytic treatment, while the control group consisted of patients with varying stroke onset (over 4.5 h), making this group very heterogeneous. Another limitation appears to be the wide disparity in terms of the age of the patients studied, and this makes for a very heterogeneous group, especially in terms of co-morbidities that may affect the coagulation system, including factor XIII. In addition, there was no long-term follow-up of patients in the study, for example, 30- or 90-day follow-up, which could provide an answer as to whether factor XIII-A levels on admission are a predictor of patients' health status in the long-term perspective. It seems that more homogeneous groups of patients in terms of age, for example, from 50 to 70 years, could also influence the results obtained for factor XIII-A. Each patient underwent lipid profiling; however, there were no statistically significant results confirming a relationship between lipid levels and factor XIII. Therefore, these data were not included in this article. The majority of patients were not on statin therapy prior to their stroke. Furthermore, the prognostic significance of lipid profiles post-stroke, as detailed in the study by Vitturi and Gagliardi, underscores the potential influence of lipid levels on stroke outcomes, suggesting that lipid management could play a crucial role in secondary stroke prevention. Therefore, the present study suggests the possibility of further research on factor XIII and lipid profiles in ischemic stroke [31].

5. Conclusions

In patients with AIS, XIII-A concentration decreases after t-PA treatment in the acute phase of stroke, and the largest decrease occurs in the TACI stroke.

Determination of factor XIII concentration in patients with AIS can be used in clinical practice as an additional parameter to support the assessment of stroke severity and may play a role in the prognosis; lower factor XIII-A activity may be a predictor of a worse prognosis.

Author Contributions: Conceptualization, M.W., U.W., A.S., N.P.-S. and E.Ż.; Methodology, M.W., U.W., A.S. and N.P.-S.; Software, N.P.-S.; Validation, N.P.-S.; Formal analysis, M.W., A.S. and N.P.-S.; Investigation, N.P.-S.; Resources, M.W. and N.P.-S.; Data curation, M.W.; Writing—original draft, M.W., U.W., M.S., A.S., N.P.-S., A.Ż. and E.Ż.; Writing—review & editing, M.W., A.S. and N.P.-S.; Visualization, N.P.-S.; Supervision, M.W. and N.P.-S.; Project administration, N.P.-S. and E.Ż. All authors have read and agreed to the published version of the manuscript.

Funding: This research received no external funding.

Institutional Review Board Statement: The study was conducted in accordance with the Declaration of Helsinki, and approved by the the Ethics Committee of the Bioethics Committee at the Nicolaus Copernicus University in Toruń, Collegium Medicum in Bydgoszcz (project identification code: KB 694/2016).

Informed Consent Statement: Patient consent was waived due to all patients are anonymous and there are no any sensitive information is used.

Data Availability Statement: Data is contained within the article.

Conflicts of Interest: The authors declare no conflict of interest.

References

1. WHO Force. Report of the WHO Task Force on Stroke and Other Cerebrovascular Disease. *Stroke* **1989**, *20*, 1407–1431.
2. Powers, W.J.; Rabinstein, A.A.; Ackerson, T.; Adeoye, O.M.; Bambakidis, N.C.; Becker, K.; Biller, J.; Brown, M.; Demaerschalk, B.M.; Hoh, B.; et al. Guidelines for the early management of patients with acute ischemic stroke: 2019 update to the 2018 Guidelines for the early management of acute ischemic stroke: A guideline for Healthcare professionals from the American Heart Association/American Stroke Associations. *Stroke* **2019**, *50*, e344–e418. [CrossRef]
3. Berge, E.; Whiteley, W.; Audebert, H.; De Marchis, G.M.; Fonseca, A.C.; Padiglioni, C.; de la Ossa, N.P.; Strbian, D.; Tsivgoulis, G.; Turc, G. European Stroke Organisation (ESO) guidelines on intravenous thrombolysis for acute ischaemic stroke. *Eur. Stroke J.* **2021**, *6*, I–LXII. [CrossRef]
4. Shroeder, V.; Kohler, H.P. New developments in the area of factor XIII. *J. Thromb. Haemost.* **2013**, *11*, 234–244. [CrossRef]
5. Board, P.G.; Webb, G.C.; JMcKee Ichinose, A. Localization of the coagulation factor XIII. A subunit gene(F13A) to chromosome bands 6 p 24–p 25. *Cytogenet. Cell Genet.* **1988**, *48*, 25–27. [CrossRef]
6. Pelcovits, A.; Schiffman, F.; Niroula, R. Factor XIII deficiency: A review of clinical presentation and management. *Hematol. Oncol. Clin. N. Am.* **2021**, *35*, 1171–1180. [CrossRef]
7. Hoffman, M.; Monroe, D.M. Coagulation 2006. A modern view of hemostasis. *Hematol. Oncol. Clin. N. Am.* **2007**, *21*, 1–11. [CrossRef] [PubMed]
8. Rijken, D.C.; de Willige, S.U. Inhibitor of Fibrinolysis by Coagulation Factor XIII. *BioMed Res. Int.* **2017**, *2017*, 1209676. [CrossRef]
9. Byrnes, J.R.; Wolberg, A.S. Newly-recognized roles of factor XIII in thrombosis. *Semin. Thromb. Hemost.* **2016**, *42*, 445–454. [CrossRef]
10. Ortner, E.; Schroeder, V.; Walser, R.; Zerbe, O.; Kohler, H.P. Sensitive and selective detection of free FXIII activation peptide: A potential marker of acute thrombotic events. *Blood* **2010**, *115*, 5089–5096. [CrossRef]
11. Członkowska, A.; Niewada, M. Stroke management guidelines. *Pol. Prz. Neurol.* **2019**, *15* (Suppl. A), A1–A156.
12. Bamford, J.; Sandercock, P.; Dennis, M.; Burn, J.; Warlow, C. Classification and natural history of clinically identifiable subtypes of cerebral infarction. *Lancet* **1991**, *337*, 1521–1526. [CrossRef]
13. Adams, H.P., Jr.; Bendixen, B.H.; Kappelle, L.J.; Biller, J.; Love, B.B.; Gordon, D.L.; Marsh, E.E., 3rd. Classification of acute ischemic stroke. Definitions for use in a multicenter clinical trial. TOAS. Trial of Org 10172 in Acute Stroke Treatment. *Stroke* **1993**, *24*, 35–41. [CrossRef]
14. Székely, E.G.; Czuriga-Kovács, K.R.; Bereczky, Z.; Katona, É.; Mezei, Z.A.; Nagy, A.; Tóth, N.K.; Berényi, E.; Muszbek, L.; Csiba, L.; et al. Low factor XIII levels after intravenous thrombolysis predict short-term mortality in ischemic stroke patients. *Sci. Rep.* **2018**, *8*, 7662. [CrossRef]
15. Hur, W.S.; Mazinani, N.; Lu, X.D.; Britton, H.M.; Byrnes, J.R.; Wolberg, A.S.; Kastrup, C.J. Coagulation factor XIIIa is inactivated by plasmin. *Blood* **2015**, *126*, 2329–2337. [CrossRef]
16. Schroeder, V.; Ortner, E.; Mono, M.L.; Galimanis, A.; Meier, N.; Findling, O.; Fischer, U.; Brekenfeld, C.; Arnold, M.; Mattle, H.P. Coagulation factor XIII activation peptide and subunit levels in patients with acute ischaemic stroke: A pilot study. *Thromb. Res.* **2010**, *126*, 122–127. [CrossRef]
17. Gemmati, D.; Zeri, G.; Orioli, E.; Mari, R.; Moratelli, S.; Vigliano, M.; Marchesini, J.; Grossi, M.E.; Pecoraro, A.; Cuneo, A.; et al. Factor XIII-A dynamics in acute myocardial infarction: A novel prognostic biomarker? *Thromb. Haemost.* **2015**, *114*, 123–132.
18. Alkjaersig, N.; Fletcher, A.P.; Lewis, M.; Ittyerah, R. Reduction of coagulation factor XIII concentration in patients with myocardial infarction, cerebral infarction, and other thromboembolic disorders. *Thromb. Haemost.* **1977**, *38*, 863–873. [CrossRef]

19. Roffe, C.; Nevatte, T.; Sim, J.; Bishop, J.; Ives, N.; Ferdinand, P.; Gray, R.; Stroke Oxygen Study Investigators and the Stroke Oxygen Study Collaborative Group. Effect of routine low-dose oxygen supplementation on death and disability in adults with acute stroke: The Stroke Oxygen Study Randomized Clinical Trial. *JAMA* **2017**, *318*, 1125–1135.
20. Jadhav, A.P.; Desai, S.M.; Jovin, T.G. Indications for mechanical throbectomy for acute ischemic stroke: Current guidelines and beyond. *Neurology* **2021**, *16*, S126–S136. [CrossRef]
21. Loof, T.G.; Mörgelin, M.; Johansson, L.; Oehmcke, S.; Olin, A.I.; Dickneite, G.; Norrby-Teglund, A.; Theopold, U.; Herwald, H. Coagulation, an Ancestral Serine Protease Cascade, Exerts a Novel Function in Early Immune Defense. *Blood* **2011**, *118*, 2589–2598. [CrossRef] [PubMed]
22. Zeerleder, S.; Schroeder, V.; Lämmle, B.; Wuillemin, W.A.; Hack, C.E.; Kohler, H.P. Factor XIII in severe sepsis and septic shock. *Thromb. Res.* **2007**, *119*, 311–318. [CrossRef] [PubMed]
23. Soendergaard Ch Kvist, P.H.; Seidelin, J.B.; Pelzer, H.; Nielsen, O.H. Systemic and intestinal levels of factor XIII-A: The impact of inflammation on expression in macrophage subtypes. *J. Gastroenterol.* **2016**, *51*, 796–807. [CrossRef] [PubMed]
24. Kucher, N.; Schroeder, V.; Kohler, H.P. Role of Blood Coagulation Factor XIII in Patients with Acute Pulmonary Embolism. Correlation of Factor XIII Antigen Levels with Pulmonary Occlusion Rate, Fibrinogen, D-dimer, and Clot Firmness. *Thromb. Haemost.* **2003**, *90*, 434–438. [CrossRef] [PubMed]
25. Słomka, A.; Korbal, P.; Piekuś, A.; Pawliszak, W.; Anisimowicz, L.; Żekanowska, E. Plasma Levels of the A Subunit of Factor XIII in Patients Undergoing Off-Pump Coronary Artery Bypass Surgery. *Pol. Arch. Intern. Med.* **2017**, *127*, 550–553. [CrossRef] [PubMed]
26. Kohler, H.P.; Ariëns, R.A.S.; Catto, A.J.; Carter, A.M.; Miller, G.J. Factori XIII A-subunit Concentration Predicts Outcome in Stroke Subjects and Vascular Outcome in Healthy, Middle-Aged Men. *J. Haematol.* **2002**, *118*, 825–832. [CrossRef] [PubMed]
27. Weger, M.; Renner, W.; Stanger, O.; Schmut, O.; Deutschmann, H.; Wascher, T.C.; Haas, A. Role of factor XIII Val34Leu polymorphism in retinal artery occlusion. *Stroke* **2001**, *32*, 2759–2761. [CrossRef] [PubMed]
28. Hacke, W.; Kaste, M.; Fieschi, C.; Toni, D.; Lesaffre, E.; Von Kummer, R.; Boysen, G.; Bluhmki, E.; Höxter, G.; Mahagne, M.H.; et al. Intravenous thrombolysis with recombinant tissue Plasminogen activator for acute hemispheric stroke. The European Cooperative Acute Stroke Study (ECASS). *JAMA* **1995**, *274*, 1017–1025. [CrossRef]
29. Sun, X.; Berthiller, J.; Derex, L.; Trouillas, P.; Diallo, L.; Hanss, M. Post-thrombolysis Haemostasis Changes after rt-PA treatment in Acute Cerebral Infarct. Correlations with Cardioembolic Aetiology and Outcome. *J. Neurol. Sci.* **2015**, *349*, 77–83. [CrossRef]
30. Warach, S.J.; Dula, A.N.; Milling, T.J. Tenecteplase Thrombolysis for Acute Ischemic Stroke. *Stroke* **2020**, *51*, 3440–3451. [CrossRef]
31. Vitturi, B.K.; Gagliardi, R.J. The prognostic significance of the lipid profile after an ischemic stroke. *Neurol. Res.* **2021**, *44*, 139–145. [CrossRef]

neurology international

MDPI

Review

The Effectiveness of Paired Associative Stimulation on Motor Recovery after Stroke: A Scoping Review

Andrea Baroni [1,2], Annibale Antonioni [1,2,3,*], Giulia Fregna [1,3], Nicola Lamberti [1], Fabio Manfredini [1,2], Giacomo Koch [1,4,5], Alessandro D'Ausilio [1,4] and Sofia Straudi [1,2]

1 Department of Neuroscience and Rehabilitation, University of Ferrara, 44121 Ferrara, Italy; brnndr3@unife.it (A.B.); giulia.fregna@unife.it (G.F.); nicola.lamberti@unife.it (N.L.); fabio.manfredini@unife.it (F.M.); giacomo.koch@unife.it (G.K.); alessandro.dausilio@unife.it (A.D.); strsfo@unife.it (S.S.)

2 Department of Neuroscience, Ferrara University Hospital, 44124 Ferrara, Italy

3 Doctoral Program in Translational Neurosciences and Neurotechnologies, University of Ferrara, 44121 Ferrara, Italy

4 Center for Translational Neurophysiology of Speech and Communication (CTNSC), Italian Institute of Technology (IIT), 44121 Ferrara, Italy

5 Non Invasive Brain Stimulation Unit, Istituto di Ricovero e Cura a Carattere Scientifico Santa Lucia, 00179 Rome, Italy

* Correspondence: annibale.antonioni@edu.unife.it; Tel.: + 39-3407364839

Abstract: Paired associative stimulation (PAS) is a non-invasive brain stimulation technique combining transcranial magnetic stimulation and peripheral nerve stimulation. PAS allows connections between cortical areas and peripheral nerves (C/P PAS) or between cortical regions (C/C PAS) to be strengthened or weakened by spike-timing-dependent neural plasticity mechanisms. Since PAS modulates both neurophysiological features and motor performance, there is growing interest in its application in neurorehabilitation. We aimed to synthesize evidence on the motor rehabilitation role of PAS in stroke patients. We performed a literature search following the PRISMA Extension for Scoping Reviews Framework. Eight studies were included: one investigated C/C PAS between the cerebellum and the affected primary motor area (M1), seven applied C/P PAS over the lesional, contralesional, or both M1. Seven studies evaluated the outcome on upper limb and one on lower limb motor recovery. Although several studies omit crucial methodological details, PAS highlighted effects mainly on corticospinal excitability, and, more rarely, an improvement in motor performance. However, most studies failed to prove a correlation between neurophysiological changes and motor improvement. Although current studies seem to suggest a role of PAS in post-stroke rehabilitation, their heterogeneity and limited number do not yet allow definitive conclusions to be drawn.

Keywords: paired associative stimulation (PAS); stroke; neurorehabilitation; non-invasive brain stimulation (NIBS); plasticity; neurophysiology

1. Introduction

Physiological reactions are frequent following stroke and aim to repair the damaged tissue. Plasticity refers to the ability of the brain to modify its structure and function in response to experience and environmental demand [1]. This enhanced plasticity following brain damage leads to new axon sprouting, new synapse formation, and the remapping of sensory–motor areas [2]. Several studies confirm a close relationship between neuroplasticity and functional recovery following stroke [3]. Changes in the activity and connection between neurons can be identified around the lesion up to remote areas or in the contralateral hemisphere, explaining spontaneous recovery after cerebral damage [4]. Post-stroke rehabilitation aims to improve functional recovery and promote neuroplasticity, supporting this dynamic process in rebuilding connections between neurons [5]. Non-invasive brain stimulation (NIBS) techniques are a promising adjuvant strategy for enhancing post-stroke recovery through the modulation of cortical excitability and neuronal plasticity [6]. The combination of NIBS and motor or behavioral intervention has gained substantial interest over the last years due to the promising potentiality that the combined approach offers [3]. Several studies on post-stroke patients combined NIBS and rehabilitative approaches such as intensive physiotherapy or occupational therapy [7,8], robot-assisted training [9–11], virtual reality rehabilitation [12–15], and task-oriented training [16] for promoting motor recovery. Between the available NIBS techniques, transcranial magnetic stimulation (TMS) has been used to investigate and induce plasticity in the human brain [17]. The combination of TMS and peripheral nerve electrical stimulation (PNS) is known as paired associative stimulation (PAS). PAS is an emerging NIBS approach, introduced by Stefan et al. [18], which uses the Cell Assembly Theory first formulated by Hebb in 1949 [19,20]. Hebb postulated that repeated activation of a presynaptic cell immediately before the activation of a postsynaptic cell induces synaptic strengthening, so-called long-term potentiation (LTP). Hebb did not propose an opposite activity-dependent reduction in synaptic strength or long-term depression (LTD). Indeed, later work described a heterosynaptic LTD, when a presynaptic cell repeatedly and persistently fails to excite the postsynaptic cell [21], and a homosynaptic mechanism based on low-frequency stimulation of the presynaptic element [22]. Studies on animal models have shown how PAS can influence motor cortex excitability, whereas TMS or PNS, commonly used in rehabilitation, showed no significant effect when used alone [23]. PAS's effect on the human brain was first studied on healthy subjects, and the observed increase in Motor Evoked Potential (MEP), the response induced by a TMS pulse over the Primary Motor Cortex (M1), suggested the plasticity of brain structures [24,25]. Many single-session studies explored the effects of different PAS protocols in stroke patients [26–28]. Even though promising results were found on cortical excitability and motor performance, no results were found on repeated sessions of PAS, particularly when combined with rehabilitative treatment. Several mechanisms may justify the use of PAS-empowered rehabilitative approaches: First, the increase in corticomuscular excitability induced by PAS may favor the subsequent response to neurorehabilitation treatment [28,29]. Furthermore, PAS protocols act on circuits involved in use-dependent plasticity, reinforcing connections useful for performing a specific motor task during rehabilitation [30,31]. However, although the effectiveness of PAS in stroke rehabilitation is still unclear, the emerging interest in this NIBS technique makes it necessary to summarize the current evidence on PAS-empowered motor rehabilitation. Thus, this work aims to review the available literature on PAS for motor rehabilitation following stroke. Moreover, we aimed to provide information about parameters and sites of stimulation, as well as outcomes and patients who could benefit from PAS. Due to the heterogeneity of evidence in this field, we applied a scoping review approach following the Preferred Reporting Systems for Systematic Reviews and Meta-Analyses Extension for Scoping Reviews (PRISMA-ScR) Framework [32].

2. Materials and Methods

The protocol of this scoping review has been redacted following the PRISMA-ScR Framework and has been pre-registered on an Open Science Framework (OSF) with the following doi: https://doi.org/10.17605/OSF.IO/86UAC (accessed on 20 October 2022).

2.1. Search Strategy

The PICO framework was used to define the research question. Articles published in peer-reviewed journals and pre-peer-reviewed web publications were potentially eligible for inclusion. The literature search was performed in the following electronic bibliographic databases: PubMed, Web of Science, Science Direct, and Embase. The database search was completed on 21 September 2022 and frequently updated until 31 December 2023. The search strategy included a controlled vocabulary and keywords adapted to the characteristics of the single database. A comprehensive description of the search strategy is available as a Supplemental Material (see Supplementary Materials). All the studies carried out on post-stroke adult patients where a PAS treatment was applied for the rehabilitation of motor function were considered. Only studies applying more than a single session on consecutive days were included. No restrictions on rehabilitation settings were used.

2.2. Study Inclusion/Exclusion Criteria

The study population includes adult stroke patients, without regard to the type of lesion (ischemic or hemorrhagic), time from injury, and site of brain damage, who underwent PAS as a rehabilitation treatment, in combination or not with other rehabilitation techniques. We considered eligible multi-session clinical trials (RCT, nRCT, and pre–post studies) with or without a comparator. Inclusion criteria were (i) reference in English; (ii) study subjects and setting as described above; and (iii) studies that describe the application of PAS as a rehabilitative approach for upper or lower limb in stroke patients. Exclusion criteria were (i) studies regarding PAS in patients with different pathologies other than stroke; (ii) studies evaluating the effects of PAS on non-motor outcomes in stroke patients (i.e., dysphagia).

2.3. Study Selection

Duplicate articles were excluded. Two independent reviewers (A.A. and G.F.) screened the title and abstract, and disagreement between them was solved by a third reviewer (A.B.). A.A. and G.F. reviewed the full text of the selected studies, and discordance was solved by A.B. and/or S.S. (see Figure 1).

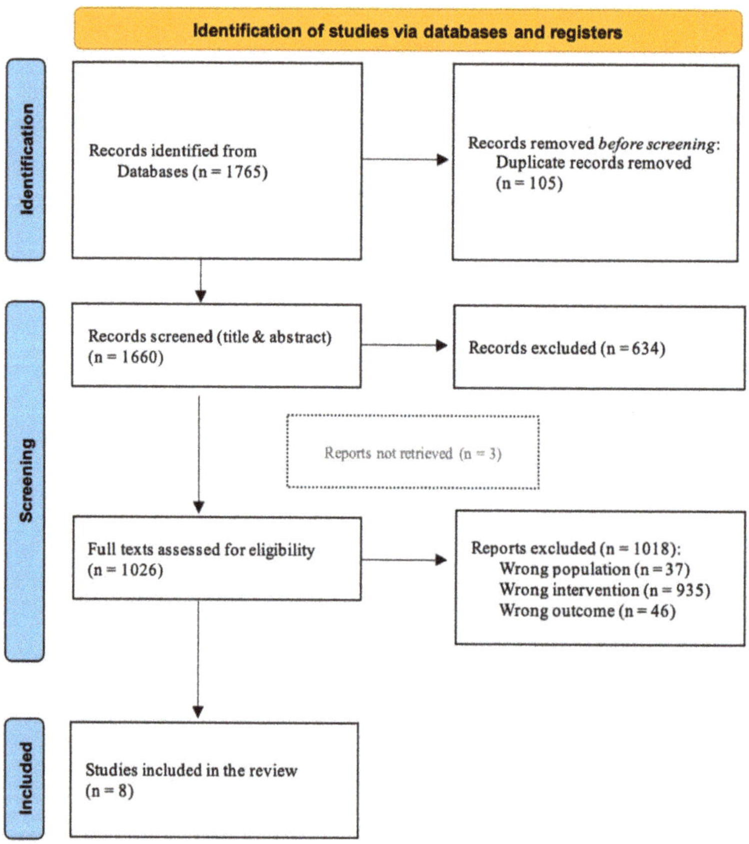

Figure 1. PRISMA review flow-chart.

2.4. Data Extraction

Two authors (A.B. and A.A.) independently extracted data using a pre-defined framework. The data framework included a field for the author(s), year of publication, country of origin, study design, sample size, type of stroke, time from stroke, PAS parameters (type, points of application, intensity and frequency of stimulation, ISI, and time of application), associated treatments, comparator details, outcome measures, and possible adverse effects related to treatment. The critical appraisal of the included papers was performed using the Cochrane Risk of Bias Tool (RoB) for RCTs [33]; nRCT and pre–post studies were evaluated using the Joanna Briggs Institute (JBI) critical appraisal tool [34]. Due to the nature of the project and the heterogeneity of the included studies, a narrative collection of results was planned.

3. Results

Database searching identified 1765 records. After removing duplicates, 1660 records were screened for the title and abstract and 634 records were excluded. Of the 1026 remaining papers, 1018 did not meet inclusion criteria and were excluded from the collection: 37 studies applied PAS to a different population, 935 studies applied a different stimulation protocol, and 46 studies evaluated a different outcome. Among the eight remaining studies [35–42], three full texts were unavailable [39–41]; however, we decided to include them in the scoping review due to the poor literature on the topic (Figure 1). Seven included studies were RCTs [34–40], and only one was a case series study [42]. All the studies were published in the last 20 years with a wide geographic distribution: four in France and one each in China, Turkey, Australia, and Ukraine. Most of the studies involved patients with ischemic stroke [35,36,38,39], two studies involved stroke patients without distinction between hemorrhagic or ischemic etiology [37,42], and two studies did not specify stroke origin [40,41]. Two studies involved chronic stroke patients (>six months post-stroke) [35,42], five studies involved subacute patients (1 to 6 months post-stroke) [36–38,40,41], and one study did not specify the time from stroke onset [39]. A total of 288 subjects were recruited. The median number of patients involved in the studies was 27.5 patients (IQR 24.75–40.25); of them, 16 (IQR 13.5–20.0) were male (one study did not specify this data [34]). Using available data, a median number of 13 (IQR 11.5–15) patients received real stimulation with different PAS protocols (one study did not specify patients' distribution among treatment groups [39]). A detailed description of the included studies is reported in Tables 1 and 2.

3.1. PAS Procedures

C/P stimulation was the PAS type most used in the included studies [36–42]. Only one study adopted a C/C protocol [35]. Most studies used the PAS protocol for upper limb treatment [35–41]. Only one of the C/P studies aimed to improve lower limb function and gait [42].

3.1.1. Cortico-Peripheral PAS

TMS was applied over the lesioned M1 in two of the included C/P studies [37,40]. One study stimulated the contralesional M1 [38]. One study stimulated both lesional and contralesional M1 in two different groups of treatment [36]. Three studies did not specify the TMS point of application [39,41,42]. Only two studies specified TMS intensity and stimulation frequency [36,38]. PNS was applied to the affected upper [37,41] or lower [42] extremity in three cases. One study used PNS in both hands in two different groups of treatment [36]. Two studies specified the site of stimulation but not the side [38,40]. One study did not specify the PNS point of application [39]. Only four studies defined both these parameters regarding the intensity and frequency of PNS [36–38,42]. Only four studies specified the ISI [36,37,40,42]: three of them applied the two stimuli with an ISI of 25 ms or 35 ms for an LTP effect [37,40,42]; and one used an ISI of 25 ms or 10 ms in two different groups of treatment to achieve LTP or LTD, respectively [36].

Table 1. Characteristics of included studies. Abbreviations: PAS = paired associative stimulation; TMS = transcranial magnetic stimulation; RCT = randomized controlled trial; RoB = risk of bias; * = conference abstract.

Unique Identifying Number	Title	Author	Year of Publication	Study Design	Country of Origin	RoB
1	Cerebello-Motor Paired Associative Stimulation and Motor Recovery in Stroke: a Randomized, Sham-Controlled, Double-Blind Pilot Trial	Rosso et al.	2022	RCT	France	6/7
2	Effect of PAS with different stimulation position on motor cortex excitability and upper limb motor function in patients with cerebral infarction	Sui et al.	2021	RCT	China	2/7
3	Five-day course of paired associative stimulation fails to improve motor function in stroke patients	Tarri et al.	2018	RCT	France	2/7
4	Effects of low-frequency repetitive transcranial magnetic stimulation and neuromuscular electrical stimulation on upper extremity motor recovery in the early period after stroke	Tosun et al.	2017	RCT	Turkey	5/7
5 *	Enhancement of cortical excitability in stroke patients after combined repetitive transcranial and peripheral magnetic stimulation	Kuznietsova et al.	2016	RCT	Ukraine	-
6 *	Study of the effects of a 5-day brain stimulation with Paired Associative Stimulation (PAS) against placebo in 28 hemiplegic patients	Tarri et al.	2015	RCT	France	-

Table 1. *Cont.*

Unique Identifying Number	Title	Author	Year of Publication	Study Design	Country of Origin	RoB
7 *	Trial of a daily program of cerebral stimulation by TMS using a PAS paradigm in the recovery phase of stroke patients	Mohamed et al.	2013	RCT	France	-
8	Does induction of plastic change in motor cortex improve leg function after stroke?	Uy et al.	2003	Case series	Australia	3/10

Table 2. Clinical characteristics of the study populations, and PAS parameters and types of treatment. Abbreviations: PAS = paired associative stimulation; TMS = transcranial magnetic stimulation; PNS = peripheral nerve stimulation; ISI = Interstimulus Interval; C/C = Cortico-Cortical; C/P = Cortico-Peripheral; M1 = Primary Motor Cortex; RMT = Resting Motor Threshold; MEP = Motor Evoked Potential; ECR = Extensor Carpi Radialis; EDC = Extensor Digitorum Communis; NMES = Neuromuscular Electrical Stimulation; CS = Conditioning Stimulus; TS = Test Stimulus; FENS = Functional Electrical Stimulation; FMA-UE = Fugl-Meyer Assessment Upper Extremity; fMRI = functional Magnetic Resonance Imaging; MI = Motricity Index; BRS-UE = Brunnstrom Recovery Stages Upper Extremity; JHFT = Jebsen Hand Function Test; GS = Grip Strength; STEF = Simple Test for Evaluating and Function; BRS-H = Brunnstrom Recovery Stages Hand; MAS = Modified Ashworth Scale; MCAS = Motor Club Assessment Scale; BI = Barthel Index; M(BI) = Modified Barthel Index; MVC = Maximum voluntary contraction; ROM = Range of Motion; * = Conference abstract.

N°	Sample Size	Type of Stroke	Time from Stroke	PAS Type	Point of Application		Parameters TMS		Parameters PNS		ISI	Time of Application	Associated Treatment	Control Group Treatment	Outcome Measures
							Intensity	Frequency	Intensity	Frequency					
1	Total n = 27 Active group n = 14 (11 males, age 63 ± 14) Sham group n = 13 (10 males, age 60 ± 11)	Ischemic	Active group 202 ± 355 months Sham group 374 ± 481 months	C/C	Contralesional cerebellum (CS)	Ipsilesional M1 (TS)	CS = 90% RMT TS = 140% RMT If MEP could not be elicited: CS = 50% RMT TS = 50% RMT	0.2 Hz	-	-	2 ms	120 paired stimuli 5 sessions (1 session/day for 5 days)	Physical therapy (45 min)	Sham PAS + physical therapy (45 min)	MEP, fMRI, JHFT, GS

Table 2. *Cont.*

N°	Sample Size	Type of Stroke	Time from Stroke	PAS Type	Point of Application		Parameters TMS		Parameters PNS		ISI	Time of Application	Associated Treatment	Control Group Treatment	Outcome Measures
							Intensity	Frequency	Intensity	Frequency					
2	Total n = 120 Ipsilateral stimulation group n = 30 (14 males, age 44.15 ± 4.76) Contralateral stimulation group n = 30 (13 males, age 43.53 ±4.88) Bilateral stimulation group n = 30 (14 males, age 45.35 ± 5.36) Control group n = 30 (15 males, age 44.83 ± 5.18)	Ischemic	Ipsilateral stimulation group 2.0 ± 0.73 months Contralateral stimulation group 1.8 ± 0.69 months Bilateral stimulation group 1.9 ± 0.78 months Control group 1.5 ± 0.71 months	C/P	Ipsilesional stimulation group (PAS25): Ipsilesional M1 Contralesional stimulation group (PAS10): contralesional M1 Bilateral stimulation group: PAS10 (contralesional M1) followed by PAS25 (ipsilesional M1)	Ipsilesional stimulation group (PAS25): median wrist nerves innervated by ipsilesional M1 Contralesional stimulation group (PAS10): median wrist nerves innervated by contralesional M1 Bilateral stimulation group: PAS10 (contralesional median wrist nerves) followed by PAS25 (ipsilesional median wrist nerves)	120% RMT	0.05 Hz	300% of the sensory threshold	0.2 ms	Ipsilesional stimulation group (PAS25): 25 ms Contralesional stimulation group (PAS10): 10 ms Bilateral stimulation group: PAS10 followed by PAS25	90 paired stimuli 28 sessions (1 session/day for 28 days)	-	Physical therapy	MEP, RMT, FMA-UE, STEF, (M)BI
3	Total n = 24 PAS group n = 13 (9 males, age 48.6 ± 12.3) Sham group n = 11 (7 males, age 51.8± 12.2)	Ischemic/ hemorrhagic	PAS group 9.8 ± 5.1 weeks Sham group 10.4 ± 5.8 weeks	C/P	Lesioned M1	ECR muscle of the paretic limb	Adjusted to obtain an ECR MEP with peak-to-peak amplitude of about 1 mV	0.1 Hz	150% of the motor threshold	5 hz	25 ms	30 min PAS 5 sessions (1 session/day for 5 days)	Physical therapy (2 h)	Sham PAS + physical therapy (2 h)	MEP, FMA-UE
4	Total n = 25 TMS group n = 9 (6 males, age 57.6 ± 12.6) TMS + NMSE group n = 7 (3 males, age 56 ± 10.1) Control group n = 9 (5 males, age 61.3 ± 10.1)	Ischemic	TMS group 49.3 ± 43.6 days TMS + NMSE group 59.6 ± 58.3 days Control group 47.2 ± 41.1 days	C/P	Contralesional M1	Wrist extensors and extensor digitorum communis	90% RMT	1 Hz	Adjusted to produce the extension of wrist and fingers (90% RMT)	50 Hz	Not specified	20 min PAS 10 sessions (5 sessions/week for 2 weeks)	Physical therapy (duration not specified)	Physical therapy (duration not specified)	fMRI, FMA-UE, MI-UE, BRS-UE, BRS-H, MAS, BI
5 *	Total n = 77 (age 63.02 ± 1.21)	Ischemic	Not specified	C/P	Not specified	Not specified	Not specified	1 Hz	Not specified	Not specified	Not specified	Not specified PAS duration 10 consecutive days	Not specified	Sham PAS	MEP, RMT, MCAS
6 *	Total n = 28 (19 males, age 49.9 ± 13.5) Analyzed n = 24 PAS group n = 13 Sham group n = 11	Not specified	10.0 ± 5.1 weeks	C/P	Wrist area (Not specified side)	Wrist extensor muscle	Not specified	0.1 Hz	Not specified	Not specified	25 ms	30 min PAS 5 sessions (1 session/day for 5 days)	-	Sham TMS	MEP, FMA-UE

Table 2. *Cont.*

N°	Sample Size	Type of Stroke	Time from Stroke	PAS Type	Point of Application		Parameters TMS		Parameters PNS		ISI	Time of Application	Associated Treatment	Control Group Treatment	Outcome Measures
							Intensity	Frequency	Intensity	Frequency					
7 *	Total n = 18 (13 males, age 47.3 ± 12.7) PAS group n = 10 Placebo n = 8	Not specified	<6 months	C/P	Not specified	ECR	Not specified	0.1 Hz	Not specified	0.1 Hz	Not specified	30 min PAS 5 sessions (1 session/day for 5 days)	Not specified	Placebo	MEP, FMA-UE
8	Total n = 9 (6 males, age 60.6 ± 10.5)	Ischemic/ hemorrhagic	3.6 ± 10.9 years	C/P	Not specified	Common peroneal nerve in the weak limb	Intensity evoking a just-visible motor response in tibialis anterior and peroneus longus	Not specified	Intensity evoking a just-visible motor response in tibialis anterior and peroneus longus	10 Hz	35 ms	30 min PAS 1 session/day for 4 weeks	-	-	MEP, MVC, ROM, GAIT PARAMETERS

3.1.2. Cortico-Cortical PAS

The study that applied C/C PAS stimulated the contralesional cerebellum and the lesional M1, specifying intensity, the frequency of stimulation, and ISI [35].

3.2. *Treatment Duration*

A median number of 10 (IQR 5–19.5) sessions of PAS was applied in the included study, with a minimum of 5 [35,37,40,41] and a maximum of 28 [36,42]. Only two studies quantified PAS duration in 30 min [41,42].

3.3. *Associated Treatments*

Four studies combined PAS with motor rehabilitation [35–38]. Specifically, three studies applied PAS before rehabilitation treatment [35,36,38], whereas one did not specify the order of the combined treatment [37]. The motor rehabilitation consisted of active-assisted range of motion exercises combined with motor imagery, strength training against gravity, and task-specific training [35]; good limb placement, bed movement, transfer training, operation treatment, daily life activity training, and other comprehensive rehabilitation treatment [36]; and activities to improve strength, flexibility, transfers, posture, balance, coordination, and activities of daily living [38]. Four studies did not specify these data [39–42].

3.4. *Comparators*

The most used comparator treatment was sham stimulation [35,37,39,40]: two studies delivered it through a sham coil applied following the same procedures used for real stimulation [35,37]; and two studies applied sham stimulation without describing the sham procedure [39,40]. One study applied a not-specified placebo as a treatment comparator [41]. Two studies used only physical therapy for the patients assigned to the control group [36,38].

3.5. *Outcome Measures*

3.5.1. Neurophysiological Measures

MEP, Resting Motor Threshold (RMT), and functional Magnetic Resonance Imaging (fMRI) were used to assess the effects of PAS on corticospinal excitability. MEP was the most used outcome measure of PAS efficacy [35–37,39–42]. Six studies reported an increase in MEP amplitude [35,36,39,42] and/or surface area (the level of corticospinal projections excitability of the target muscle) [37,39,41] in the experimental group compared to the control; however, out of all these, only four studies reported quantitative results [35,37,41,42] but no statistically significative results in both within- and between-group comparisons. One study reported a significant increase in the MEP amplitude of groups who received PAS in different protocols without reporting quantitative data (only graphs available) [36]. RMT is the amount of TMS machine output necessary to produce an MEP that exceeds an established peak-to-peak amplitude (usually 50 μV) 50% of the time in a finite number of trials [43,44]. RMT was recorded in two studies [36,39]: one study reported a significant reduction in RMT of the lesioned side in the groups who received real stimulation, only through qualitative and graphical results [36]. Kuznietsova et al. described a reduction in RMT in the experimental group without reporting numerical data [39]. fMRI was recorded in two studies and revealed increased activation of the affected hemisphere without reaching statistical significance [35,38].

3.5.2. Clinical Measures

The efficacy of PAS on upper limb function was evaluated using the Fugl-Meyer Assessment for Upper Extremity (FMA-UE) [35–37,39,40], the Motricity Index (Upper Extremity section) (MI-UE) [38], and the upper extremity section of the Brunnstrom Recovery Stages (BRS-UE) [38]. One study reported a significant improvement in FMA-UE score in the experimental group compared to the control, reporting only qualitative data and graphical results [36]. All other studies evaluating upper limb function did not record significant differences between the experimental and the control group [37,38,40,41]. Changes in hand function following PAS were evaluated using the Jebsen Hand Function Test (JHFT) [35], the grip strength (GS) [35], the Simple Test for Evaluating Hand Function (STEF) [36], and the hand section of the Brunnstrom Recovery Stages (BRS-H) [38]. No significant changes were recorded between the study groups. No effects of PAS on muscle tone [38], cognitive, and emotional function [36] were documented. Two studies assessed the efficacy of PAS on the ability to perform activities of daily living [36,38]. Of these, one study reported a significant improvement in the Barthel Index score only in the group that received real stimulation [36]. The efficacy of PAS on lower limb function (maximum voluntary contraction and range of motion) and activities (walking) was evaluated, and no changes were found between pre- and post-treatment [42]. Table 3 summarizes these data.

3.6. Adverse Effects

Possible adverse effects of PAS were recorded only in three of the included studies [35,37,38]. Two subjects showed temporary headaches after stimulation: both received C/C PAS and were allocated one in the real stimulation group and one in the control group [35]. One subject showed reflex syncope immediately after the end of the C/C sham PAS [35]. Two studies did not report adverse effects [37,38].

3.7. Quality Assessment

Considering the risk of bias evaluation of the included studies for which the full text was available [34–37,41], a heterogeneous methodological quality was noticed (Table 1). Particularly, among the RCTs involved, two studies showed an overall low risk of bias [35,38], while in the other two [36,37], the absence of explicit information on different methodological key points did not allow a precise estimation of the related methodological quality [32].

Table 3. Outcome measures and study results following the International Classification of Functioning (ICF) model. Abbreviations: PAS = paired associative stimulation; PT = physical therapy; TMS = transcranial magnetic stimulation; PNS = peripheral nerve stimulation; ISI = Interstimulus Interval; FMA-UE = Fugl-Meyer Assessment Upper Extremity; fMRI = functional Magnetic Resonance Imaging; FAr = Fractional anisotropy ratio; CST = Corticospinal tract; DTCT = Dentate-thalamo-cortical tracts; MI = Motricity Index; BRS-UE = Brunnstrom Recovery Stages Upper Extremity; JHFT = Jebsen Hand Function Test; GS = Grip strength; STEF = Simple Test for Evaluating and Function; BRS-H = Brunnstrom Recovery Stages Hand; MAS = Modified Ashworth Scale; MCAS = Motor Club Assessment Scale; BI = Barthel Index; M(BI) = Modified Barthel Index; MVC = Maximum Voluntary Contraction; ROM = Range of Motion; MEP-TA = Motor Evoked Potentials recorded from Tibialis Anterior muscle; MEP-PL = Motor Evoked Potentials recorded from Peroneus Longus; MVC-TA = Maximum Voluntary Contraction measured from Tibialis Anterior; MVC-PL = Maximum Voluntary Contraction measured from Peroneus Longus; MRC = Medical Research Council Scale; C/C = Cortico-Cortical; C/P = Cortico-Peripheral; M1 = Primary Motor Cortex; RMT = Resting Motor Threshold; MEP = Motor Evoked Potential; ECR = Extensor Carpi Radialis; EDC = Extensor Digitorum Communis; NMES = Neuromuscular Electrical Stimulation; CS = Conditioning Stimulus; TS = Test Stimulus; FENS = Functional Electrical Stimulation; D5 = Day five; D30 = Day thirty.

	Study	Intervention	Results		
UPPER EXTREMITY					
BRAIN STRUCTURE					
MEP	Rosso, 2022	Active PAS + PT vs Sham PAS + PT	*Experimental group* pre: 0.44 ± 0.62 post: 0.45 ± 0.65 fu: 0.55 ± 1.09	*Control group* pre: 0.27 ± 0.51 post: 0.33 ± 0.59 fu: 0.27 ± 0.44	*Significance* No differences within and between groups
	Sui, 2021	Ipsilateral PAS vs Contralateral PAS vs Bilateral PAS vs PT	Decrease in MEP amplitude on the contralesional side compared to before treatment. Increase in MEP amplitude on the ipsilesional side compared to before treatment. [significative differences $p < 0.05$] Decrease in MEP amplitude on the contralesional side compared to PT group. Increase in MEP amplitude on the ipsilesional side compared to PT group. [no significative differences between them] Decrease in MEP amplitude on the contralesional side and increase in MEP amplitude in the ipsilesional side in the bilateral group compared to contralesional and ipsilesional group [significative differences between them].		*Significance* Significative differences within group for the stimulation groups ($p < 0.05$) Significative differences between groups for the stimulation groups compared to PT group ($p < 0.05$) Significative differences between groups for the ipsilateral PAS25 and the contralateral PAS10 group compared to bilateral PAS group ($p < 0.05$)

Table 3. *Cont.*

	Study	Intervention	Results		
	Tarri, 2018	Active PAS + PT vs Sham PAS + PT	*Experimental group* Mean (SD) surface area of MEP was 239% (230) of baseline	*Control group* Mean (SD) surface area of MEP was 154% (81) of baseline	*Significance* No differences within and between groups
	Kuznietsova, 2016	Active PAS vs Sham PAS	Reduction of latency and increase in amplitude and area in the experimental group compared to control		
	Tarri, 2015	Active PAS vs Sham PAS	No significant differences between the two groups		
	Mohamed, 2013	Active PAS vs Placebo	*Experimental group* Increase of MEP surface of 168 ± 268%	*Control group* Increase of MEP surface of 0.1 ± 48%	*Significance* No differences between groups
RMT	Sui, 2021	Ipsilateral PAS vs Contralateral PAS vs Bilateral PAS vs PT	Increase in RMT on the contralesional side compared to before treatment. Decrease in RMT on the ipsilesional side compared to before treatment. [significative differences $p < 0.05$] Increase in RMT on the contralesional side compared to PT group. Decrease in RMT on the ipsilesional side compared to PT group. [no significative differences between them] Increase in RMT on the contralesional side and decrease in RMT in the ipsilesional side in the bilateral group compared to contralesional and ipsilesional group [significative differences between them]		*Significance* Significative differences within group for the stimulation groups ($p < 0.05$) Significative differences between groups for the stimulation groups compared to PT group ($p < 0.05$) Significative differences between groups for the ipsilateral PAS25 and the contralateral PAS10 group compared to bilateral PAS group ($p < 0.05$)
	Kuznietsova, 2016	Active PAS vs Sham PAS	Reduction in RMT in the experimental group compared to control.		

Table 3. *Cont.*

	Study	Intervention	Results			
fMRI	Rosso, 2022	Active PAS + PT vs Sham PAS + PT	*Experimental group* Ipsilesional M1 activity: pre: 4.3 ± 1.3 post: 3.9 ± 1.6 fu: 4.1 ± 0.8 Far_{CST}: pre: 0.91 ± 0.17 post: - fu: - Far_{DTCT}: pre: 0.94 ± 0.12 post: - fu: -	*Control group* Ipsilesional M1 activity: pre: 3.25 ± 1.17 post: 3.64 ± 1.45 fu: 3.62 ± 1.82 Far_{CST}: pre: 0.95 ± 0.35 post: - fu: - Far_{DTCT}: pre: 0.96 ± 0.15 post: - fu: -		*Significance* Not reported
	Tosun, 2017	Active TMS + PT vs Active PAS + PT vs PT	*Active TMS + PT group* Affected M1: Increased activation during the movements of the paretic hand in 66.7% of the group	*Active PAS + PT group* Affected M1: Increased activation during the movements of the paretic hand in 57.1% of the group	*PT group* Affected M1: 42.9% of the group revealed no change	*Significance* Not performed
BODY FUNCTION						
Upper limb function						
FMA-UE	Sui, 2021	Ipsilateral PAS vs Contralateral PAS vs Bilateral PAS vs PT	Increase in FMA-UE in stimulation groups compared to PT group			*Significance* Significative differences within group for the stimulation groups ($p < 0.05$) Significative differences between groups for the stimulation groups compared to PT group ($p < 0.05$) Significative differences between groups for the ipsilateral PAS25 and the contralateral PAS10 group compared to bilateral PAS group ($p < 0.05$)
	Tarri, 2018	Active PAS + PT vs Sham PAS + PT	No significant differences were found for time or group ($p = 0.99$). ANCOVA adjusted to the initial FMA-UE score failed to reveal a significant difference between the two groups ($p = 0.66$, 95% CI [-2.26%, 3.51%]).			

Table 3. *Cont.*

	Study	Intervention	Results			
	Tosun, 2017	Active TMS + PT vs Active PAS + PT vs PT	*Active TMS + PT group* pre: 28.8 ± 14.9 post: 51.0 ± 11.1 $p = 0.008$	*Active PAS + PT group* pre: 17.3 ± 11.6 post: 30.0 ± 14.3 $p = 0.018$	*PT group* pre: 28.5 ± 18.2 post: 33.2 ± 19.9 $p = 0.011$	*Significance* Not performed between groups comparison
	Tarri, 2015	Active PAS vs Sham PAS	No significant differences between the two groups			
	Mohamed, 2013	Active PAS vs Placebo	*Experimental group* Increase of FMA-UE score: 6.1 ± 4.5	*Control group* Increase of FMA-UE score: 4.6 ± 4.1		*Significance* Not reported
MI-UE	Tosun, 2017	Active TMS + PT vs Active PAS + PT vs PT	*Active TMS + PT group* pre: 48.4 ± 22.8 post: 78.0 ± 17.5 $p = 0.008$	*Active PAS + PT group* pre: 28.5 ± 11.1 post: 56.8 ± 18.9 $p = 0.018$	*PT group* pre: 43.9 ± 27.0 post: 51.2 ± 27.6 $p = 0.018$	*Significance* Between-group comparison not performed
BRS-UE	Tosun, 2017	Active TMS + PT vs Active PAS + PT vs PT	*Active TMS + PT group* pre: 3.4 ± 1.2 post: 4.3 ± 1.1 $p = 0.01$	*Active PAS + PT group* pre: 2.3 ± 0.8 post: 4.0 ± 1.3 $p = 0.016$	*PT group* pre: 3.2 ± 1.5 post: 3.89 ± 1.6 $p = 0.034$	*Significance* Between-group comparison not performed
Hand function						
JHFT	Rosso, 2022	Active PAS + PT vs Sham PAS + PT	*Experimental group* pre: 5.92 ± 6.95 post: 6.00 ± 7.28 fu: 5.31 ± 6.66	*Control group* pre: 9.03 ± 11.7 post: 9.71 ± 10.59 fu: 10.14 ± 12.38	*Significance* Significant GROUP*TIME interaction (F (1, 26): 3.27, p: 0.04). There was no effect of TIME (F (2, 50): 0.6, p: 0.55) and GROUP (F (1, 25): 1.1, p: 0.29). The change in JHFT score between the active and the sham group was not significant at D5 (p: 0.16) but was at D30 (p: 0.01)		
GS	Rosso, 2022	Active PAS + PT vs Sham PAS + PT	*Experimental group* pre: 0.37 ± 0.27 post: 0.48 ± 0.24 fu: 0.53 ± 0.27	*Control group* pre: 0.37 ± 0.26 post: 0.38 ± 0.26 fu: 0.41 ± 0.29	*Significance* No effect of treatment (GROUP*TIME interaction: F (1.25): 0.60; p: 0.54)		

Table 3. *Cont.*

	Study	Intervention	Results			
						Significance
STEF	Sui, 2021	Ipsilateral PAS vs Contralateral PAS vs Bilateral PAS vs PT	Increase in STEF in stimulation groups compared to PT group			Significative differences within group for the stimulation groups ($p < 0.05$) Significative differences between groups for the stimulation groups compared to PT group ($p < 0.05$) Significative differences between groups for the ipsilateral PAS25 and the contralateral PAS10 group compared to bilateral PAS group ($p < 0.05$)
			Active TMS + PT group	*Active PAS + PT group*	*PT group*	*Significance*
BRS-H	Tosun, 2017	Active TMS + PT vs Active PAS + PT vs PT	pre: 3.3 ± 1.4 post: 4.7 ± 1.2 $p = 0.006$	pre: 2.2 ± 0.4 post: 3.6 ± 0.9 $p = 0.014$	pre: 3.44 ± 1.3 post: 3.89 ± 1.5 $p = 1.02$	Not performed between groups comparison
Muscle tone						
			Active TMS + PT group	*Active PAS + PT group*	*PT group*	*Significance*
MAS	Tosun, 2017	Active TMS + PT vs Active PAS + PT vs PT	pre: 0.7 ± 0.9 post: 1.5 ± 1.0 $p = 0.102$	pre: 1 ± 0.8 post: 1.0±0.5 $p = 0.083$	pre: 0.7 ± 1.0 post: 1.0 ± 1.0 $p = 0.180$	Not performed between groups comparison
ADL						
						Significance
(M)BI	Sui, 2021	Ipsilateral PAS vs Contralateral PAS vs Bilateral PAS vs PT	Increase in MBI in stimulation groups compared to PT group			Significative differences within group for the stimulation groups ($p < 0.05$) Significative differences between groups for the stimulation groups compared to PT group ($p < 0.05$) Significative differences between groups for the ipsilateral PAS25 and the contralateral PAS10 group compared to bilateral PAS group ($p < 0.05$)
			Active TMS + PT group	*Active PAS + PT group*	*PT group*	*Significance*
BI	Tosun, 2017	Active TMS + PT vs Active PAS + PT vs PT	pre: 66.6 ± 22.7 post: 93.3 ± 6.1 $p = 0.008$	pre: 55.0 ± 22.1 post: 81.4 ± 20.1 $p = 0.017$	pre: 39.4 ± 22.3 post: 50.5 ± 32.1 $p = 0.043$	Not performed between groups comparison

Table 3. *Cont.*

	Study	Intervention	Results		
OTHER					
MCAS	Kuznietsova, 2016	Active PAS vs Sham PAS	*Experimental group* Increase of MAS score: 40.4%	*Control group* Increase of MAS score: 17.1%	*Significance* Not reported
LOWER EXTREMITY					
BRAIN STRUCTURE					
MEP	Uy, 2003	Active PAS	$MEP\text{-}TA_{relaxed}$ pre: 0.12 $MEP\text{-}TA_{relaxed}$ post: 0.19 $MEP\text{-}PL_{relaxed}$ pre: 0.8 $MEP\text{-}PL_{relaxed}$ post: 0.8	$MEP\text{-}TA_{active}$ pre: 0.74 $MEP\text{-}TA_{active}$ post: 0.87 $MEP\text{-}PL_{active}$ pre: 0.30 $MEP\text{-}PL_{active}$ post: 0.34	*Significance* No significance for grouped data
BODY FUNCTION					
Lower limb function					
MVC	Uy, 2003	Active PAS	MVC-TA pre: 0.043 MVC-TA post: 0.055 MVC-PL pre: 0.014 MVC-PL post: 0.022		*Significance* No significance for grouped data
ROM	Uy, 2003	Active PAS	No data reported		
ACTIVITIES					
Walking					
GAIT PARAMETERS	Uy, 2003	Active PAS	No changes for 10 m timed walk, step and stride length, and cadence		

4. Discussion

Although recent studies have provided a better understanding of the neurophysiological mechanism underlying PAS, and supporting its contribution to stroke recovery, few studies have specifically investigated its role in rehabilitation. This lack may be due to their relatively recent introduction in the clinical setting, which makes further studies necessary to evaluate the applicability of this technique for patient recovery [45]. Moreover, the technologies required to implement PAS into practice are extremely expensive and require specific skills not always available outside the research settings [46]. However, considering their potential from a neurorehabilitation perspective, here, to the best of our knowledge, we have gathered evidence on PAS-empowered post-stroke motor rehabilitation. In our review, we aimed to synthesize the state-of-the-art of PAS as an adjuvant to stroke rehabilitation, identifying parameters, sites of stimulation, and patients who can benefit from this combined stimulation. The implications of our results will be discussed, considering the C/P and C/C PAS studies separately.

4.1. C/P PAS

M1 represents the most frequently stimulated area due to its relatively easy accessibility with NIBS techniques, as well as the possibility of measuring the effects of its modulation (e.g., RMT and MEPs recorded from target muscles) [43,44]. Moreover, M1 is a crucial part of a wide network responsible for the regulation of motor acts where sensory stimuli, exogenous and endogenous, play a key role [47].

Several studies have shown that PNS can inhibit the subsequent homotopic muscle response evoked by a TMS pulse on M1, leading to a decrease in MEP amplitude, depending on the specific temporal interval between the sensory and the motor stimulus [48]. This phenomenon is referred to as short-latency afferent inhibition and highlights a close coupling between sensory and motor networks, dependent on the modulation of inhibitory circuits exerted by excitatory cholinergic thalamocortical afferents [48]. Therefore, considering the importance of sensorimotor integration in motor control, it is unsurprising that PAS protocols target M1 in combination with accessible peripheral regions [28,49]. Consistently, most of the included studies stimulated the impaired hand, with a particular focus on the extensor muscles, frequently impaired after stroke. By contrast, few studies stimulated the median or, generically, the whole paretic hand [50]. Interestingly, only one study applied PAS stimulation to both hands, using an excitatory protocol on the paretic one and an inhibitory protocol on the healthy one [36]. This study design is present (when the information is available) in most of the other included studies. However, limited to the paretic hand, it is based on the model developed by Di Pino et al. on post-stroke interhemispheric disequilibrium: after stroke, the normal reciprocal inhibition between the two hemispheres is altered and the damaged hemisphere is no longer able to adequately counteract the healthy one, which therefore exerts a marked inhibition on the injured hemisphere hindering the recovery of impaired functions [51]. Although recent models have considered the role of other factors in addition to the mere distinction between the injured and healthy hemisphere, this interpretation has been widely used showing remarkable efficacy in the recovery of common symptoms after stroke [52,53]. Consistently, studies that exploited this interpretative model, like the one of Sui et al. [36], have demonstrated an improvement in neurophysiological parameters, i.e., an increase in MEPs' amplitude and a decrease in the RMT of the damaged M1 (and changes in the opposite direction on healthy M1 when stimulated using an inhibitory protocol), and in motor and functional recovery.

However, it is crucial to note that statistically significant changes following PAS were observed only for the FMA score [36], showing a dissociation between neurophysiological and clinical measures. This phenomenon could reflect that functional changes observed in the subjects cannot be solely attributed to physiological modifications and clinical measures may be too coarse to detect these changes.

Regarding the quality of the reporting, some papers did not show their results except in the abstract or graphical form, making it complex to evaluate what was achieved [36,39].

In contrast, other studies used outcome measurement scales that are scarcely used internationally (e.g., Motor Club Assessment Scale, MCAS), making the generalization of the obtained results difficult [39]. Looking at the stroke timeframe, half of the studies evaluated patients in the subacute phase (between 1 and 6 months), three studies evaluated the chronic phase (>6 months), and one considered patients in the acute/subacute phase (<6 months). This choice may depend on the need to reconcile, on the one hand, the clinical stabilization of the patient (normally difficult to achieve in the acute phase) and, on the other, to exploit the interval of increased cortical plasticity that gradually decreases over time [54,55]. In this sense, the subacute phase seems to be the most suitable to reconcile these needs [56]. The study of Sui et al. showed significant changes in sub-acute patients following PAS [36], offering insights into applying this protocol to improve motor function even after the acute injury.

We cannot draw definitive conclusions about the number of treatment sessions due to the heterogeneity of the studies. Sui et al. found a functional improvement after 28 sessions of stimulation [36]; therefore, fewer sessions may not be sufficient to achieve this goal.

Only one study used PAS to improve lower limb function; the combined stimulation of M1 and the common peroneal nerve of the affected leg showed neurophysiological changes in the related brain areas and a functional improvement in gait [42]. The reason why the lower limb is much less investigated surely depends on its mesial area of cortical representation, less accessible with NIBS techniques. In addition, upper limb impairment most frequently afflicts stroke survivors' daily autonomy, making the evaluation of lower limb recovery less frequent in the literature [57–59]. Functional improvements are observed, even in this case, in chronic patients, making the application of PAS of great interest in patients with gait impairment following stroke.

4.2. C/C PAS

PAS protocols aimed at improving connections between two or more brain areas (C/C PAS) have been more recently exploited to strengthen or weaken connections based on the timing of the stimulations [60]. Indeed, cortical areas are interconnected by extensive fiber bundles, both intra- and inter-hemispheric, and these reciprocal connections are crucial for the modulation of numerous activities and, in particular, motor actions [61,62]. Recent studies have shown that C/C PAS on areas involved in motor control induces significant changes, not only in neurophysiological parameters but also in motor actions [63,64], making its application of growing interest in the rehabilitation field. Consistently, several RCTs are underway to evaluate its potential in combination with various rehabilitative approaches like upper limb robot-assisted therapy [65].

Recently, several advanced PAS techniques involving combined trans-modality stimulation (e.g., between motor cortical areas and visual or acoustic ones) have been employed, remaining tied to research contexts despite promising results [66]. Therefore, although these techniques are likely to become part of stroke rehabilitation in the future, at the moment, we have limited our discussion to the study that we included in our review. Rosso et al. used a more explored "within" motor system protocol that exploited the long-range connections of the cerebellum and M1 [35], like the dentate nucleus–thalamus–cortical pathway [67]. Indeed, the contralesional cerebellum plays a significant role in the reorganization of the motor network and during the recovery process following stroke [68]. Notably, stroke patients often need to relearn basic motor strategies, a process actively governed by the cerebellum [69] that can be empowered by the simultaneous application of NIBS techniques [70].

Consistently, Rosso et al. found a significant improvement in hand function at one-month follow-up [35], leading to the hypothesis that cerebellar modulation influences motor output through morphological modifications and LTP mechanisms in the motor areas. These changes are possible in the presence of integral afferent and efferent circuits of M1 as necessary substrates for functional improvements [71]. Improvements in motor outcomes were achieved with only five days of stimulation in a group of patients who had suffered from stroke years earlier; therefore, C/C PAS seems to offer an attractive rehabilitative opportunity in chronic patients, even with a small number of sessions, probably due to its focus on the CNS, free from the influences of various peripheral factors that could reduce the effectiveness of the treatment [72].

In line with our findings, a promising role for PAS has been demonstrated in other neurological disorders, such as spinal cord injury (SCI). Neuromodulatory effects of PAS have been proposed to improve functional outcomes because of cortico-spinal and cortico-peripheral stimulation protocols [73,74]. C/P PAS, particularly, has demonstrated excellent results in terms of corticospinal transmission and functional outcome, reasonably exploiting the spike-timing-dependent plasticity principles similar to those described in this review [75,76]. However, while SCI affects areas of the nervous system that are remote from the higher brain centers, stroke damages a network of closely related and mutually influencing areas, making treatment outcomes more complex and difficult to predict [77,78]. Summarizing, although the low number of included studies makes it difficult to generalize the results, it is possible to highlight some crucial aspects. Firstly, clinical studies about PAS are highly heterogeneous in terms of stimulation protocol and parameters, stroke timeframe, session duration, number, comparators, and the motor and functional assessment scale. Moreover, PAS-empowered rehabilitation is widely used for upper limb recovery, compared to only one study using PAS for lower limb rehabilitation. Finally, the traditional C/P PAS paradigm, as described in the literature, is the most used for rehabilitative purposes, compared to the more recent C/C technique. A substantial number of studies omitted crucial information about the generalizability of the intervention, like the site and parameters of stimulation, or the specific site of the stroke lesion. Many studies evaluated the effects of PAS in terms of neurophysiological changes like MEPs and RMT, while only a few studies showed changes in clinical or functional scales used to evaluate the clinical correlation of neurophysiological aspects. Less than half of the studies reported no adverse effects or provided information to analyze their characteristics, which were in any case rare and generally minor and transient. Considering that the presence of adverse effects has not been evaluated in most of the included studies, the improvement in the reporting quality appears to be essential for a thorough analysis of the obtained results, and future work needs to adopt a methodology capable of addressing this issue.

Considering the above, it is necessary to underline that our scoping review aimed to synthesize evidence on PAS protocols in stroke rehabilitation, without considering the possible neurophysiological limitations of this technique, which, at present, require further investigation. Consequently, we were unable to quantitatively evaluate the precise interactions between PAS and rehabilitative intervention. Furthermore, it is worth noting that spinal cord circuitry gating, along with the functional status of polysynaptic descending pathways regulating the interaction between M1 and peripheral effectors, could potentially influence the outcomes of PAS protocols. However, no information is currently available on these fundamental issues. Understanding these crucial aspects could significantly impact the outcomes of PAS protocols, informing their use in neurorehabilitation.

5. Conclusions

The small number and the heterogeneity of the studies included in our review make it challenging to identify the role of PAS in motor rehabilitation following stroke. Despite the fact that several outcome measures have been used to quantify the efficacy of PAS in stroke rehabilitation, the study of neurophysiological modifications such as MEP and RMT is the most frequent. Our data may provide valuable information about the current use of PAS in neurorehabilitation, becoming a reference point for future studies on tailored PAS protocols identifying patients where combined stimulation can be added to motor rehabilitation to reach the best rehabilitative outcome.

Supplementary Materials: The following supporting information can be downloaded at: https://www.mdpi.com/article/10.3390/neurolint16030043/s1, Supplementary Materials: the details about the search strategy.

Author Contributions: Conceptualization, A.B., A.A., G.F. and S.S.; methodology, A.B., A.A., G.F. and S.S.; investigation, A.B., A.A., G.F. and S.S.; data curation, A.B., A.A., G.F. and S.S.; writing—original draft preparation, A.B. and A.A.; writing—review and editing, A.B., A.A., N.L., F.M., G.K., A.D. and S.S; supervision, N.L., F.M., G.K., A.D. and S.S; funding acquisition, F.M., A.D. and S.S. All authors have read and agreed to the published version of the manuscript.

Funding: This work was supported by a grant from the 2010–2012 Research Programme of Emilia Romagna Region [grant number 1786/2012] and by #NEXTGENERATIONEU (NGEU) and funded by the Ministry of University and Research (MUR), National Recovery and Resilience Plan (NRRP), project MNESYS (PE0000006)—A Multiscale integrated approach to the study of the nervous system in health and disease (DN. 1553 11.10.2022); Ricerca Finalizzata 2018—Giovani Ricercatori (GR-2018-12366027); EU project PRIMI (HORIZON-CL4-2022-DIGITAL-EMER- GING-02-101120727).

Institutional Review Board Statement: Not applicable.

Informed Consent Statement: Not applicable.

Data Availability Statement: No new data were created or analyzed in this study. Data sharing is not applicable to this article. All the literature used for this review is listed in the bibliography.

Acknowledgments: We thank Matilde Piva and Alessandra Zanetti for their help during the screening process.

Conflicts of Interest: The authors declare no conflicts of interest.

References

1. Freed, W.J.; de Medinaceli, L.; Wyatt, R.J. Promoting Functional Plasticity in the Damaged Nervous System. *Science* **1985**, *227*, 1544–1552. [CrossRef] [PubMed]
2. Joy, M.T.; Carmichael, S.T. Encouraging an Excitable Brain State: Mechanisms of Brain Repair in Stroke. *Nat. Rev. Neurosci.* **2021**, *22*, 38–53. [CrossRef] [PubMed]
3. Ahmed, I.; Mustafaoglu, R.; Rossi, S.; Cavdar, F.A.; Agyenkwa, S.K.; Pang, M.Y.C.; Straudi, S. Non-Invasive Brain Stimulation Techniques for the Improvement of Upper Limb Motor Function and Performance in Activities of Daily Living After Stroke: A Systematic Review and Network Meta-Analysis. *Arch. Phys. Med. Rehabil.* **2023**, *104*, 1683–1697. [CrossRef] [PubMed]
4. Su, F.; Xu, W. Enhancing Brain Plasticity to Promote Stroke Recovery. *Front. Neurol.* **2020**, *11*, 554089. [CrossRef] [PubMed]
5. Gulyaeva, N.V. Molecular Mechanisms of Neuroplasticity: An Expanding Universe. *Biochemistry* **2017**, *82*, 237–242. [CrossRef] [PubMed]
6. Wessel, M.J.; Zimerman, M.; Hummel, F.C. Non-Invasive Brain Stimulation: An Interventional Tool for Enhancing Behavioral Training after Stroke. *Front. Hum. Neurosci.* **2015**, *9*, 265. [CrossRef]
7. Bornheim, S.; Croisier, J.-L.; Maquet, P.; Kaux, J.-F. Transcranial Direct Current Stimulation Associated with Physical-Therapy in Acute Stroke Patients—A Randomized, Triple Blind, Sham-Controlled Study. *Brain Stimul.* **2020**, *13*, 329–336. [CrossRef] [PubMed]
8. Kim, W.-S.; Kwon, B.; Seo, H.; Park, J.; Paik, N.-J. Low-Frequency Repetitive Transcranial Magnetic Stimulation Over Contralesional Motor Cortex for Motor Recovery in Subacute Ischemic Stroke: A Randomized Sham-Controlled Trial. *Neurorehabilit. Neural Repair* **2020**, *34*, 856–867. [CrossRef]
9. Bressi, F.; Cinnera, A.M.; Morone, G.; Campagnola, B.; Cricenti, L.; Santacaterina, F.; Miccinilli, S.; Zollo, L.; Paolucci, S.; Di Lazzaro, V.; et al. Combining Robot-Assisted Gait Training and Non-Invasive Brain Stimulation in Chronic Stroke Patients: A Systematic Review. *Front. Neurol.* **2022**, *13*, 795788. [CrossRef]
10. Mazzoleni, S.; Tran, V.-D.; Dario, P.; Posteraro, F. Effects of Transcranial Direct Current Stimulation (tDCS) Combined with Wrist Robot-Assisted Rehabilitation on Motor Recovery in Subacute Stroke Patients: A Randomized Controlled Trial. *IEEE Trans. Neural Syst. Rehabil. Eng.* **2019**, *27*, 1458–1466. [CrossRef]
11. Straudi, S.; Fregni, F.; Martinuzzi, C.; Pavarelli, C.; Salvioli, S.; Basaglia, N. tDCS and Robotics on Upper Limb Stroke Rehabilitation: Effect Modification by Stroke Duration and Type of Stroke. *BioMed Res. Int.* **2016**, *2016*, 5068127. [CrossRef] [PubMed]
12. Cassani, R.; Novak, G.S.; Falk, T.H.; Oliveira, A.A. Virtual Reality and Non-Invasive Brain Stimulation for Rehabilitation Applications: A Systematic Review. *J. Neuroeng. Rehabil.* **2020**, *17*, 147. [CrossRef] [PubMed]
13. Chen, Y.-H.; Chen, C.-L.; Huang, Y.-Z.; Chen, H.-C.; Chen, C.-Y.; Wu, C.-Y.; Lin, K. Augmented Efficacy of Intermittent Theta Burst Stimulation on the Virtual Reality-Based Cycling Training for Upper Limb Function in Patients with Stroke: A Double-Blinded, Randomized Controlled Trial. *J. Neuroeng. Rehabil.* **2021**, *18*, 91. [CrossRef] [PubMed]

14. Meng, J.; Yan, Z.; Gu, F.; Tao, X.; Xue, T.; Liu, D.; Wang, Z. Transcranial Direct Current Stimulation with Virtual Reality versus Virtual Reality Alone for Upper Extremity Rehabilitation in Stroke: A Meta-Analysis. *Heliyon* **2023**, *9*, e12695. [CrossRef] [PubMed]
15. Yao, X.; Cui, L.; Wang, J.; Feng, W.; Bao, Y.; Xie, Q. Effects of Transcranial Direct Current Stimulation with Virtual Reality on Upper Limb Function in Patients with Ischemic Stroke: A Randomized Controlled Trial. *J. Neuroeng. Rehabil.* **2020**, *17*, 73. [CrossRef]
16. Baroni, A.; Magro, G.; Martinuzzi, C.; Brondi, L.; Masiero, S.; Milani, G.; Zani, G.; Bergonzoni, A.; Basaglia, N.; Straudi, S. Combined Effects of Cerebellar tDCS and Task-Oriented Circuit Training in People with Multiple Sclerosis: A Pilot Randomized Control Trial. *Restor. Neurol. Neurosci.* **2022**, *40*, 85–95. [CrossRef] [PubMed]
17. Hernandez-Pavon, J.C.; San Agustín, A.; Wang, M.C.; Veniero, D.; Pons, J.L. Can We Manipulate Brain Connectivity? A Systematic Review of Cortico-Cortical Paired Associative Stimulation Effects. *Clin. Neurophysiol.* **2023**, *154*, 169–193. [CrossRef]
18. Stefan, K.; Kunesch, E.; Cohen, L.G.; Benecke, R.; Classen, J. Induction of Plasticity in the Human Motor Cortex by Paired Associative Stimulation. *Brain* **2000**, *123 Pt 3*, 572–584. [CrossRef]
19. Hebb, D.O. *The Organization of Behavior: A Neuropsychological Theory*; Wiley: Oxford, UK, 1949; p. 335.
20. Classen, J.; Wolters, A.; Stefan, K.; Wycislo, M.; Sandbrink, F.; Schmidt, A.; Kunesch, E. Paired Associative Stimulation. *Suppl. Clin. Neurophysiol.* **2004**, *57*, 563–569.
21. Stent, G.S. A Physiological Mechanism for Hebb's Postulate of Learning. *Proc. Natl. Acad. Sci. USA* **1973**, *70*, 997–1001. [CrossRef]
22. Kirkwood, A.; Dudek, S.M.; Gold, J.T.; Aizenman, C.D.; Bear, M.F. Common Forms of Synaptic Plasticity in the Hippocampus and Neocortex in Vitro. *Science* **1993**, *260*, 1518–1521. [CrossRef] [PubMed]
23. Zhang, X.; Sui, Y.; Guo, T.; Wang, S.; Hu, Y.; Lu, Y. Effect of Paired Associative Stimulation on Motor Cortex Excitability in Rats. *Curr. Med. Sci.* **2018**, *38*, 903–909. [CrossRef] [PubMed]
24. Costanzo, M.; Leodori, G.; Cutrona, C.; Marchet, F.; De Bartolo, M.I.; Mancuso, M.; Belvisi, D.; Conte, A.; Berardelli, A.; Fabbrini, G. Motor Cortical Correlates of Paired Associative Stimulation Induced Plasticity: A TMS-EEG Study. *Brain Sci.* **2023**, *13*, 921. [CrossRef] [PubMed]
25. Müller, J.F.M.; Orekhov, Y.; Liu, Y.; Ziemann, U. Homeostatic Plasticity in Human Motor Cortex Demonstrated by Two Consecutive Sessions of Paired Associative Stimulation. *Eur. J. Neurosci.* **2007**, *25*, 3461–3468. [CrossRef] [PubMed]
26. Castel-Lacanal, E.; Gerdelat-Mas, A.; Marque, P.; Loubinoux, I.; Simonetta-Moreau, M. Induction of Cortical Plastic Changes in Wrist Muscles by Paired Associative Stimulation in Healthy Subjects and Post-Stroke Patients. *Exp. Brain Res.* **2007**, *180*, 113–122. [CrossRef] [PubMed]
27. Castel-Lacanal, E.; Marque, P.; Tardy, J.; de Boissezon, X.; Guiraud, V.; Chollet, F.; Loubinoux, I.; Moreau, M.S. Induction of Cortical Plastic Changes in Wrist Muscles by Paired Associative Stimulation in the Recovery Phase of Stroke Patients. *Neurorehabilit. Neural Repair* **2009**, *23*, 366–372. [CrossRef]
28. Palmer, J.A.; Wolf, S.L.; Borich, M.R. Paired Associative Stimulation Modulates Corticomotor Excitability in Chronic Stroke: A Preliminary Investigation. *Restor. Neurol. Neurosci.* **2018**, *36*, 183–194. [CrossRef]
29. Tsuji, T.; Suzuki, K.; Masakado, Y.; Ota, T.; Kimura, A.; Liu, M.; Chino, N. Long-Lasting Effects of Paired Associative Stimulation in Hemiparetic Stroke Patients. *Int. Congr. Ser.* **2005**, *1278*, 280–283. [CrossRef]
30. Koganemaru, S.; Fukuyama, H.; Mima, T. Two Is More Than One: How to Combine Brain Stimulation Rehabilitative Training for Functional Recovery? *Front. Syst. Neurosci.* **2015**, *9*, 154. [CrossRef]
31. Bestmann, S.; Krakauer, J.W. The Uses and Interpretations of the Motor-Evoked Potential for Understanding Behaviour. *Exp. Brain Res.* **2015**, *233*, 679–689. [CrossRef]
32. Tricco, A.C.; Lillie, E.; Zarin, W.; O'Brien, K.K.; Colquhoun, H.; Levac, D.; Moher, D.; Peters, M.D.J.; Horsley, T.; Weeks, L.; et al. PRISMA Extension for Scoping Reviews (PRISMA-ScR): Checklist and Explanation. *Ann. Intern. Med.* **2018**, *169*, 467–473. [CrossRef] [PubMed]
33. Higgins, J.P.T.; Altman, D.G.; Gøtzsche, P.C.; Jüni, P.; Moher, D.; Oxman, A.D.; Savovic, J.; Schulz, K.F.; Weeks, L.; Sterne, J.A.C.; et al. The Cochrane Collaboration's Tool for Assessing Risk of Bias in Randomised Trials. *BMJ* **2011**, *343*, d5928. [CrossRef] [PubMed]
34. Joanna Briggs Institute. *Checklist for Systematic Reviews and Research Syntheses*; Joanna Briggs Institute: Adelaide, Australia, 2017.
35. Rosso, C.; Moulton, E.J.; Kemlin, C.; Leder, S.; Corvol, J.-C.; Mehdi, S.; Obadia, M.A.; Obadia, M.; Yger, M.; Meseguer, E.; et al. Cerebello-Motor Paired Associative Stimulation and Motor Recovery in Stroke: A Randomized, Sham-Controlled, Double-Blind Pilot Trial. *Neurotherapeutics* **2022**, *19*, 491–500. [CrossRef] [PubMed]
36. Sui, Y.-F.; Tong, L.-Q.; Zhang, X.-Y.; Song, Z.-H.; Guo, T.-C. Effects of Paired Associated Stimulation with Different Stimulation Position on Motor Cortex Excitability and Upper Limb Motor Function in Patients with Cerebral Infarction. *J. Clin. Neurosci.* **2021**, *90*, 363–369. [CrossRef] [PubMed]
37. Tarri, M.; Brihmat, N.; Gasq, D.; Lepage, B.; Loubinoux, I.; De Boissezon, X.; Marque, P.; Castel-Lacanal, E. Five-Day Course of Paired Associative Stimulation Fails to Improve Motor Function in Stroke Patients. *Ann. Phys. Rehabil. Med.* **2018**, *61*, 78–84. [CrossRef] [PubMed]
38. Tosun, A.; Türe, S.; Askin, A.; Yardimci, E.U.; Demirdal, S.U.; Kurt Incesu, T.; Tosun, O.; Kocyigit, H.; Akhan, G.; Gelal, F.M. Effects of Low-Frequency Repetitive Transcranial Magnetic Stimulation and Neuromuscular Electrical Stimulation on Upper Extremity Motor Recovery in the Early Period after Stroke: A Preliminary Study. *Top. Stroke Rehabil.* **2017**, *24*, 361–367. [CrossRef] [PubMed]

39. Kuznietsova, S.; Skachkova, N.; Semonova, O. ID 243—Enhancement of Cortical Excitability in Stroke Patients after Combined Repetitive Transcranial and Peripheral Magnetic Stimulation. *Clin. Neurophysiol.* **2016**, *127*, e121. [CrossRef]
40. Tarri, M.; Simonetta-Moreau, M.; Loubinoux, I.; De Boissezon, X.; Gasq, D.; Marque, P.; Castel-Lacanal, E. Study of the Effects of a 5-Day Brain Stimulation with Paired Associative Stimulation (PAS) against Placebo in 28 Hemiplegic Patients. *Ann. Phys. Rehabil. Med.* **2015**, *58*, e2. [CrossRef]
41. Mohamed, T.; Marion, S.-M.M.; Isabelle, L.; Xavier, D.B.; David, G.; Phillipe, M.; Evelyne, C.-L. P 123. CIPASS: Trial of a Daily Program of Cerebral Stimulation by TMS Using a PAS Paradigm in the Recovery Phase of Stroke Patients. *Clin. Neurophysiol.* **2013**, *124*, e123. [CrossRef]
42. Uy, J.; Ridding, M.C.; Hillier, S.; Thompson, P.D.; Miles, T.S. Does Induction of Plastic Change in Motor Cortex Improve Leg Function after Stroke? *Neurology* **2003**, *61*, 982–984. [CrossRef]
43. Rossi, S.; Antal, A.; Bestmann, S.; Bikson, M.; Brewer, C.; Brockmöller, J.; Carpenter, L.L.; Cincotta, M.; Chen, R.; Daskalakis, J.D.; et al. Safety and Recommendations for TMS Use in Healthy Subjects and Patient Populations, with Updates on Training, Ethical and Regulatory Issues: Expert Guidelines. *Clin. Neurophysiol.* **2021**, *132*, 269–306. [CrossRef] [PubMed]
44. Rossini, P.M.; Barker, A.T.; Berardelli, A.; Caramia, M.D.; Caruso, G.; Cracco, R.Q.; Dimitrijević, M.R.; Hallett, M.; Katayama, Y.; Lücking, C.H. Non-Invasive Electrical and Magnetic Stimulation of the Brain, Spinal Cord and Roots: Basic Principles and Procedures for Routine Clinical Application. Report of an IFCN Committee. *Electroencephalogr. Clin. Neurophysiol.* **1994**, *91*, 79–92. [CrossRef] [PubMed]
45. Silverstein, J.; Cortes, M.; Tsagaris, K.Z.; Climent, A.; Gerber, L.M.; Oromendia, C.; Fonzetti, P.; Ratan, R.R.; Kitago, T.; Iacoboni, M.; et al. Paired Associative Stimulation as a Tool to Assess Plasticity Enhancers in Chronic Stroke. *Front. Neurosci.* **2019**, *13*, 792. [CrossRef]
46. Edwards, J.D.; Dominguez-Vargas, A.U.; Rosso, C.; Branscheidt, M.; Sheehy, L.; Quandt, F.; Zamora, S.A.; Fleming, M.K.; Azzollini, V.; Mooney, R.A.; et al. A Translational Roadmap for Transcranial Magnetic and Direct Current Stimulation in Stroke Rehabilitation: Consensus-Based Core Recommendations from the Third Stroke Recovery and Rehabilitation Roundtable. *Int. J. Stroke* **2023**, *19*, 145–157. [CrossRef]
47. Strick, P.L.; Dum, R.P.; Rathelot, J.-A. The Cortical Motor Areas and the Emergence of Motor Skills: A Neuroanatomical Perspective. *Annu. Rev. Neurosci.* **2021**, *44*, 425–447. [CrossRef] [PubMed]
48. Turco, C.V.; El-Sayes, J.; Savoie, M.J.; Fassett, H.J.; Locke, M.B.; Nelson, A.J. Short- and Long-Latency Afferent Inhibition; Uses, Mechanisms and Influencing Factors. *Brain Stimul.* **2018**, *11*, 59–74. [CrossRef] [PubMed]
49. Edwards, L.L.; King, E.M.; Buetefisch, C.M.; Borich, M.R. Putting the "Sensory" Into Sensorimotor Control: The Role of Sensorimotor Integration in Goal-Directed Hand Movements After Stroke. *Front. Integr. Neurosci.* **2019**, *13*, 16. [CrossRef] [PubMed]
50. Raghavan, P. The Nature of Hand Motor Impairment after Stroke and Its Treatment. *Curr. Treat. Options Cardiovasc. Med.* **2007**, *9*, 221–228. [CrossRef]
51. Di Pino, G.; Pellegrino, G.; Assenza, G.; Capone, F.; Ferreri, F.; Formica, D.; Ranieri, F.; Tombini, M.; Ziemann, U.; Rothwell, J.C.; et al. Modulation of Brain Plasticity in Stroke: A Novel Model for Neurorehabilitation. *Nat. Rev. Neurol.* **2014**, *10*, 597–608. [CrossRef]
52. Koch, G.; Bonnì, S.; Giacobbe, V.; Bucchi, G.; Basile, B.; Lupo, F.; Versace, V.; Bozzali, M.; Caltagirone, C. θ-Burst Stimulation of the Left Hemisphere Accelerates Recovery of Hemispatial Neglect. *Neurology* **2012**, *78*, 24–30. [CrossRef]
53. Motolese, F.; Capone, F.; Di Lazzaro, V. New Tools for Shaping Plasticity to Enhance Recovery after Stroke. *Handb. Clin. Neurol.* **2022**, *184*, 299–315. [CrossRef] [PubMed]
54. Dimyan, M.A.; Cohen, L.G. Neuroplasticity in the Context of Motor Rehabilitation after Stroke. *Nat. Rev. Neurol.* **2011**, *7*, 76–85. [CrossRef] [PubMed]
55. Morotti, A.; Poli, L.; Costa, P. Acute Stroke. *Semin. Neurol.* **2019**, *39*, 61–72. [CrossRef] [PubMed]
56. Johansson, B.B. Current Trends in Stroke Rehabilitation. A Review with Focus on Brain Plasticity. *Acta Neurol. Scand.* **2011**, *123*, 147–159. [CrossRef] [PubMed]
57. Anwer, S.; Waris, A.; Gilani, S.O.; Iqbal, J.; Shaikh, N.; Pujari, A.N.; Niazi, I.K. Rehabilitation of Upper Limb Motor Impairment in Stroke: A Narrative Review on the Prevalence, Risk Factors, and Economic Statistics of Stroke and State of the Art Therapies. *Healthcare* **2022**, *10*, 190. [CrossRef] [PubMed]
58. Jang, S.H. The Recovery of Walking in Stroke Patients: A Review. *Int. J. Rehabil. Res.* **2010**, *33*, 285–289. [CrossRef] [PubMed]
59. Schieber, M.H. Modern Coordinates for the Motor Homunculus. *J. Physiol.* **2020**, *598*, 5305–5306. [CrossRef] [PubMed]
60. Koch, G.; Ponzo, V.; Di Lorenzo, F.; Caltagirone, C.; Veniero, D. Hebbian and Anti-Hebbian Spike-Timing-Dependent Plasticity of Human Cortico-Cortical Connections. *J. Neurosci.* **2013**, *33*, 9725–9733. [CrossRef]
61. Ashe, J.; Lungu, O.V.; Basford, A.T.; Lu, X. Cortical Control of Motor Sequences. *Curr. Opin. Neurobiol.* **2006**, *16*, 213–221. [CrossRef]
62. Mesulam, M. The Evolving Landscape of Human Cortical Connectivity: Facts and Inferences. *Neuroimage* **2012**, *62*, 2182–2189. [CrossRef]
63. Casarotto, A.; Dolfini, E.; Cardellicchio, P.; Fadiga, L.; D'Ausilio, A.; Koch, G. Mechanisms of Hebbian-like Plasticity in the Ventral Premotor—Primary Motor Network. *J. Physiol.* **2023**, *601*, 211–226. [CrossRef] [PubMed]

64. Casarotto, A.; Dolfini, E.; Fadiga, L.; Koch, G.; D'Ausilio, A. Cortico-Cortical Paired Associative Stimulation Conditioning Superficial Ventral Premotor Cortex-Primary Motor Cortex Connectivity Influences Motor Cortical Activity during Precision Grip. *J. Physiol.* **2023**, *601*, 3945–3960. [CrossRef]
65. Cinnera, A.M.; Bonnì, S.; D'Acunto, A.; Maiella, M.; Ferraresi, M.; Casula, E.P.; Pezzopane, V.; Tramontano, M.; Iosa, M.; Paolucci, S.; et al. Cortico-Cortical Stimulation and Robot-Assisted Therapy (CCS and RAT) for Upper Limb Recovery after Stroke: Study Protocol for a Randomised Controlled Trial. *Trials* **2023**, *24*, 823. [CrossRef]
66. Guidali, G.; Roncoroni, C.; Bolognini, N. Paired Associative Stimulations: Novel Tools for Interacting with Sensory and Motor Cortical Plasticity. *Behav. Brain Res.* **2021**, *414*, 113484. [CrossRef] [PubMed]
67. Spampinato, D.A.; Block, H.J.; Celnik, P.A. Cerebellar-M1 Connectivity Changes Associated with Motor Learning Are Somatotopic Specific. *J. Neurosci.* **2017**, *37*, 2377–2386. [CrossRef]
68. Rehme, A.K.; Eickhoff, S.B.; Rottschy, C.; Fink, G.R.; Grefkes, C. Activation Likelihood Estimation Meta-Analysis of Motor-Related Neural Activity after Stroke. *Neuroimage* **2012**, *59*, 2771–2782. [CrossRef]
69. Celnik, P. Understanding and Modulating Motor Learning with Cerebellar Stimulation. *Cerebellum* **2015**, *14*, 171–174. [CrossRef] [PubMed]
70. Koch, G.; Bonnì, S.; Casula, E.P.; Iosa, M.; Paolucci, S.; Pellicciari, M.C.; Cinnera, A.M.; Ponzo, V.; Maiella, M.; Picazio, S.; et al. Effect of Cerebellar Stimulation on Gait and Balance Recovery in Patients with Hemiparetic Stroke: A Randomized Clinical Trial. *JAMA Neurol.* **2018**, *76*, 170–178. [CrossRef]
71. Krakauer, J.W.; Shadmehr, R. Consolidation of Motor Memory. *Trends Neurosci.* **2006**, *29*, 58–64. [CrossRef]
72. Rossini, P.M.; Burke, D.; Chen, R.; Cohen, L.G.; Daskalakis, Z.; Di Iorio, R.; Di Lazzaro, V.; Ferreri, F.; Fitzgerald, P.B.; George, M.S.; et al. Non-Invasive Electrical and Magnetic Stimulation of the Brain, Spinal Cord, Roots and Peripheral Nerves: Basic Principles and Procedures for Routine Clinical and Research Application. An Updated Report from an I.F.C.N. Committee. *Clin. Neurophysiol.* **2015**, *126*, 1071–1107. [CrossRef]
73. Mazwi, N.L.; Adeletti, K.; Hirschberg, R.E. Traumatic Spinal Cord Injury: Recovery, Rehabilitation, and Prognosis. *Curr. Trauma Rep.* **2015**, *1*, 182–192. [CrossRef]
74. Ling, Y.T.; Alam, M.; Zheng, Y.-P. Spinal Cord Injury: Lessons about Neuroplasticity from Paired Associative Stimulation. *Neuroscientist* **2020**, *26*, 266–277. [CrossRef] [PubMed]
75. Bunday, K.L.; Perez, M.A. Motor Recovery after Spinal Cord Injury Enhanced by Strengthening Corticospinal Synaptic Transmission. *Curr. Biol.* **2012**, *22*, 2355–2361. [CrossRef] [PubMed]
76. Debanne, D.; Inglebert, Y. Spike Timing-Dependent Plasticity and Memory. *Curr. Opin. Neurobiol.* **2023**, *80*, 102707. [CrossRef] [PubMed]
77. Ji, J.L.; Spronk, M.; Kulkarni, K.; Repovš, G.; Anticevic, A.; Cole, M.W. Mapping the Human Brain's Cortical-Subcortical Functional Network Organization. *Neuroimage* **2019**, *185*, 35–57. [CrossRef] [PubMed]
78. Guggisberg, A.G.; Koch, P.J.; Hummel, F.C.; Buetefisch, C.M. Brain Networks and Their Relevance for Stroke Rehabilitation. *Clin. Neurophysiol.* **2019**, *130*, 1098–1124. [CrossRef]

neurology international

MDPI

Article

Communicating Arteries and Leptomeningeal Collaterals: A Synergistic but Independent Effect on Patient Outcomes after Stroke

Sara Sablić [1], Krešimir Dolić [1,2,3], Danijela Budimir Mršić [1,2], Mate Čičmir-Vestić [4], Antonela Matana [3], Sanja Lovrić Kojundžić [1,2,3] and Maja Marinović Guić [1,2,3,*]

1 Clinical Department of Diagnostic and Interventional Radiology, University Hospital of Split, 21000 Split, Croatia; sarasablic@gmail.com (S.S.); kdolic79@gmail.com (K.D.); danijelabudimir@gmail.com (D.B.M.); lovric.sanja@gmail.com (S.L.K.)
2 School of Medicine, University of Split, 21000 Split, Croatia
3 University Department of Health Studies, University of Split, 21000 Split, Croatia; antonela.matana@gmail.com
4 Department of Neurology, University Hospital of Split, 21000 Split, Croatia; mate.c-v@hotmail.com
* Correspondence: maja.marinovic.guic@gmail.com

Abstract: The collateral system is a compensatory mechanism activated in the acute phase of an ischemic stroke. It increases brain perfusion to the hypoperfused area. Arteries of the Willis' circle supply antegrade blood flow, while pial (leptomeningeal) arteries direct blood via retrograde flow. The aim of our retrospective study was to investigate the relationship between both collateral systems, computed tomography perfusion (CTP) values, and functional outcomes in acute stroke patients. Overall, 158 patients with anterior circulation stroke who underwent mechanical thrombectomy were included in the study. We analyzed the presence of communicating arteries and leptomeningeal arteries on computed tomography angiography. Patients were divided into three groups according to their collateral status. The main outcomes were the rate of functional independence 3 months after stroke (modified Rankin scale score, mRS) and mortality rate. Our study suggests that the collateral status, as indicated by the three groups (unfavorable, intermediate, and favorable), is linked to CT perfusion parameters, potential recuperation ratio, and stroke outcomes. Patients with favorable collateral status exhibited smaller core infarct and penumbra volumes, higher mismatch ratios, better potential for recuperation, and improved functional outcomes compared to patients with unfavorable or intermediate collateral status.

Keywords: cerebral infarction; stroke; mechanical thrombectomy; endovascular treatment; collateral circulation

1. Introduction

Acute ischemic stroke is a medical condition that poses a threat to life and disability [1]. Numerous advances have been made in research on its pathogenesis and treatment. Activation of the collateral system has been extensively studied in the past few years as a contributing factor to a better prognosis following stroke [2–7], regardless of whether intravenous thrombolytic agent administration was the primary treatment or if it was followed by endovascular treatment (EVT) [8–12]. While mechanical thrombectomy (MT) is unquestionably an effective therapy approach, it is not always associated with favorable functional outcomes. The time window for MT was extended from 6 h to even 24 h from the onset of the stroke due to the great results of this treatment method [13–15]. However, some patients do not benefit from this treatment, even when the EVT is administered rapidly after stroke onset. Several factors contribute to this, such as older age, diabetes, a higher NIHSS score, an occlusion site (internal carotid artery occlusion), a lower ASPECT score, and distal clot migration [16–18]. The term "fast progressors" refers to patients who quickly deteriorate due to fast infarct progression [19,20]. These patients usually have terminal carotid occlusions and impaired collateral flow via the anterior or posterior cerebral artery, which indicates the importance of collateral rerouting of blood flow [19,20]. Numerous studies that examined pial arteries using CT angiography (CTA) discovered that a healthy collateral system "saves" the hypoperfused area of the brain and can even extend the time window for potential mechanical thrombectomy treatment [21]. Arteries of the Willis circle (mainly communicating arteries) as branches of the internal carotid artery are considered primary collaterals. They abruptly redirect blood flow to the site of occlusion. Small arterial branches and leptomeningeal arteries (branches of the external carotid artery) are considered secondary collaterals, and they retrogradely supply branches of the occluded artery [22,23]. Few studies have evaluated anatomic variants of the Willis' circle (CoW), while other studies analyzed secondary collateral scores on CT angiography and their impact on outcomes after ischemic stroke [9,10,24–27]. These studies concluded that patients with better collateral status had a smaller computed tomography perfusion (CTP) core and penumbra, a larger mismatch ratio, and better functional outcomes. In addition, baseline collateral status is significantly associated with clinical and safety outcomes in acute ischemic stroke patients treated with reperfusion therapy [28]. The most recent guidelines from the American Heart Association/American Stroke Association indicate that it would be appropriate to take collateral flow status into account when making clinical decisions about which individuals are eligible for mechanical thrombectomy [29].

The aim of our study was to make a comprehensive analysis of both collateral systems on CTA and investigate the relationship between collateral status, CT perfusion parameters, and functional outcomes in anterior circulation acute ischemic stroke patients treated with endovascular therapy.

2. Materials and Methods

2.1. Study Design and Participants

This study complied with the ethical guidelines of the 1975 Declaration of Helsinki and was approved by the institutional ethics committee at the University Hospital in Split, Croatia (protocol code 2181-147-01/06/LJ.Z.-23.02). Informed consent was waived due to the retrospective nature of the study. We examined patients with acute ischemic stroke in the anterior circulation who were eligible for endovascular therapy at our comprehensive stroke center from 1 January 2021 to 1 July 2023. The inclusion criteria included the following: patients aged 18 years and older, onset of neurological symptoms less than six hours before presentation to the Neurology Emergency Department, and radiological confirmation of the occlusion. Exclusion criteria included the following: patients with acute ischemic stroke in the posterior circulation (vertebral or basilar artery) and patients with stroke caused by carotid artery dissection. Our inclusion criteria were met by 158 patients. We gathered clinical data (National Institutes of Health Stroke Scale (NIHSS) on admission, onset-to-door time, door-to-needle time, and onset-to-groin puncture time) as well as patient demographics (age, sex, stroke risk factors, comorbidities, and anticoagulant treatment) from the hospital information system (HIS).

2.2. Radiological Data

Prior to endovascular treatment, every patient underwent a standardized stroke protocol that includes emergency nonenhanced brain CT, cerebral CT angiography, and CT perfusion on a 128-slice MSCT Siemens Somatom Definition AS, Erlangen, Germany. All scans were analyzed using Syngovia software (version VB60A, Hofix08). Cerebral arteries were evaluated using the advanced vessel analysis software (syngo.via CT vascular, version VB60A, Hofix08), while perfusion analysis was performed in syngo.via Perfusion CT Neuro software (version VB60A, Hofix08). Two experienced neuroradiologists evaluated cerebral CT angiography to determine the presence of different anatomical variants of CoW. The presence of anterior and posterior communicating arteries was observed (present/absent) without measuring their diameters. Primary collaterals were evaluated as described: Patients were categorized as absent (no communicating arteries present) or present (anterior and/or posterior communicating arteries). Anterior communicating artery (AcoA) was considered positive only if an ipsilateral A1 segment was present on the side of the occluded vessel.

If the patient had ACoA but aplasia of the A1 segment on the ipsilateral side, ACoA was considered absent. Present posterior communicating artery (PCoA) was considered positive only if it was present on the side of the occluded vessel.

The collateral grade score of secondary collaterals was evaluated on CT angiography with the collateral score proposed by Tan et al. [30], which uses a 4-point grading system (0–3) and grades vessel filling in the territory of the occluded artery to assess collateral circulation. A Tan score of 0 indicates absent filling, with no visible vessels within the occluded middle cerebral artery (MCA) territory. A Tan score of 1 indicates arterial contrast filling of $\leq$50% of the occluded MCA territory. A score of 2 was given when contrast filling was present in >50% but <100% of the occluded MCA territory, and a score of 3 was given when contrast filling was present in 100% of the occluded MCA territory. Low Tan scores range from 0 to 1, while high Tan scores range from 2 to 3.

2.3. Patient Grouping

We categorized patients into three groups: Group 1 (unfavorable collateral group) consisted of patients who had absent communicating arteries and a low Tan score (0 or 1); group 2 (intermediate collateral group) consisted of patients who had at least one communicating artery present and a low Tan score or expressed a high Tan score (2 or 3) but had absent communicating arteries; group 3 (favorable collateral group) consisted of patients who had at least one communicating artery and who had a high Tan score (2 or 3).

2.4. Patient Management

Mechanical thrombectomy was performed under local or general anesthesia with approved endovascular devices (aspiration catheters, stent retrievers, or a combination of both).

2.5. Outcomes and Measures

The main outcome was the rate of functional independence measured by the modified Rankin scale score (0–6) 90 days after a stroke incident. The favorable mRS score was considered 0–2.

Additional outcomes that were automatically calculated by software were volumes of core infarct and penumbra on CT perfusion, mismatch ratio (penumbra divided by the infarct core volume), and the potential recuperation ratio, which is calculated using a formula [31]. The recanalization grade on mechanical thrombectomy was scored using the TICI scale [32], and TICI scores 2B, 2C, and 3 were considered a successful recanalization. We further evaluated the hypodense area on follow-up unenhanced brain CT scans that were obtained approximately 24 h after the endovascular treatment. A score of 0 was assigned to patients who had no infarction detected; a score of 1 included patients with small/lacunar ischemic lesion; a score of 2 was assigned to patients with larger ischemic lesions that occupied less than 50% of the vascular territory of the occluded vessel; and a score of 3 included patients with ischemic lesions that occupied more than 50% of the vascular territory of the affected vessel.

2.6. Statistical Analysis

For normality checking, the Kolmogorov–Smirnov test was used. Continuous variables with normal distribution were presented as mean (standard deviation, SD); non-normal variables were reported as median (interquartile range, IQR). Categorical variables are expressed with frequencies (percentages). Differences in categorical variables were analyzed using Pearson chi-square test. Kruskal–Wallis test was used for not normally distributed continuous variables, while ANOVA was used for normally distributed continuous variables. Statistically significance was considered for two-sided p-values less than 0.05. Statistical analysis was conducted using Statistical Package Software for Social Science, version 28 (SPSS Inc., Chicago, IL, USA).

3. Results

This single-center retrospective study included 158 patients with acute ischemic stroke in the anterior circulation. All groups had comparable age and sex distribution. The unfavorable collateral group had the highest incidence of arterial hypertension ($p = 0.023$) and atrial fibrillation ($p = 0.029$). The favorable collateral group had the lowest NIHSS median on admittance ($p = 0.011$). Systolic and diastolic blood pressure values were similar among groups. All three groups had comparable onset-to-door, door-to-needle, door-to-puncture, and onset-to-groin puncture times (Table 1). Twenty-seven patients had wake-up strokes, or the onset of the neurological deficit was unknown. Groups did not differ in thrombus localization ($p = 0.176$), and the most common site of occlusion was the M1 segment of the middle cerebral artery (Table 1).

Table 1. Baseline characteristics.

	Unfavorable Collateral Group (n = 24)	Intermediate Collateral Group (n = 69)	Favorable Collateral Group (n = 65)	p-Value
Age, years (median, IQR)	82.5 (11.5)	78 (17)	75 (18)	0.051 *
Male sex (n, %)	13 (54.2%)	46 (66.7%)	39 (60%)	0.503 †
Arterial hypertension (n,%)	22 (91.7%)	49 (72.1%)	39 (61.9%)	0.023 †
Diabetes mellitus (n,%)	5 (20.8%)	15 (22.1%)	15 (23.8%)	0.948 †
Stroke cause Atrial fibrillation (n,%) Atherosclerosis (n,%)	 15 (65.2%) 8 (34.8%)	 43 (63.2%) 25 (36.8%)	 26 (41.9%) 36 (58.1%)	0.029 †
Anticoagulant/antithrombotic therapy (n,%)	13 (54.2%)	28 (41.2%)	19 (30.2%)	0.104 †
NIHSS on admittance, median (IQR)	15 (3)	15 (3.8)	12 (6.5)	0.011 *
Systolic blood pressure (mmHg, median IQR)	150 (53.5)	140 (40)	150 (45)	0.952 *
Diastolic blood pressure (mmHg, median IQR)	80 (28.8)	80 (20)	80 (20)	0.988 *
Onset-to-door (minutes, median IQR)	87.5 (101.5)	84 (74)	80 (47)	0.747 *
Door-to-needle (minutes, median IQR)	101.14 (29.8)	113.9 (38.4)	107.5 (29.3)	0.277 ‡
Onset-to-puncture (minutes, median IQR)	189 (84)	189 (88)	191 (59)	0.833 *
Intravenous thrombolysis (n,%)	9 (37.5%)	30 (43.5%)	24 (36.9%)	0.717 †
Thrombus localization MCA occlusion ICA + MCA occlusion	 16 (66.7%) 8 (33.3%)	 54 (78.2%) 15 (21.7%)	 55 (84.6%) 10 (15.4%)	0.176 †

* Kruskal–Wallis test. † chi-square test. ‡ ANOVA. NIHSS, The National Institutes of Health Stroke Scale; MCA, middle cerebral artery; ICA, internal carotid artery.

Patients in the favorable collateral group had the lowest volume of core infarct ($p = 0.001$) and penumbra on CT perfusion maps ($p = 0.033$) (Table 2). The mismatch ratio increased with a more developed collateral system and was the highest in the favorable collateral group ($p \leq 0.001$). During mechanical thrombectomy procedures, similar devices were used in all groups ($p = 0.915$); only the aspiration catheters were used in most patients. Unsuccessful placement of endovascular devices was similar in all groups, predominantly because of anatomic or groin puncture difficulties. The potential recuperation ratio was the lowest in patients with poorly developed collaterals, with the highest percentage in patients with developed both primary and secondary collaterals ($p \leq 0.001$).

Table 2. CT perfusion analysis and stroke outcomes.

	Unfavorable Collateral Group (n = 24)	Intermediate Collateral Group (n = 69)	Favorable Collateral Group (n = 65)	p-Value
Core infarct, cm^3 (median, IQR)	25.75 (27.9)	23.11 (29.2)	11.64 (12.1)	0.001 *
Penumbra, cm^3 (median, IQR)	49.69 (45.1)	54.74 (39.3)	39.69 (27.2)	0.033 *
Mismatch ratio (median, IQR)	1.72 (1.1)	2.3 (1.7)	3.55 (3.2)	<0.001 *
PRR% (median, IQR)	66.8 (10.6)	70.52 (9.3)	77.57 (10.3)	<0.001 †
Mechanical thrombectomy technique: Aspiration only Aspiration catheter + stent retriever Unsuccessful placement	 17 (70.8%) 2 (33.3%) 5 (20.8%)	 46 (66.7%) 11 (15.9%) 12 (17.4%)	 44 (67.7%) 10 (15.4%) 11 (16.9%)	0.915 ‡
Successful recanalization (n,%)	7 (29.2%)	36 (53.7%)	32 (49.2%)	0.219 ‡
TICI 3 (n,%)	11 (45.8%)	40 (59.7%)	39 (59%)	0.476 ‡
Number of passes (median, IQR)	2 (2)	1 (2)	1 (1.8)	0.430 *
First-pass recanalization (n,%)	7 (36.8%)	36 (59%)	32 (57.1%)	0.219 ‡
Control CT area of ischemia None Lacunar <50% >50%	 1 (47%) 6 (25%) 10 (41.7%) 7 (29.2%)	 9 (13.2%) 23 (33.8%) 20 (29.4%) 16 (23.5%)	 10 (15.4%) 32 (49.2%) 16 (24.6%) 7 (10.7%)	 0.366 ‡ 0.056 ‡ 0.291 † 0.072 ‡
SAH or ICH (n,%)	7 (29.2%)	13 (19.1%)	16 (24.6%)	0.551 ‡
Favorable mRS score (0–2) (n,%)	4 (17.4%)	24 (44.4%)	31 (60.8%)	0.002 ‡
Death (n,%)	14 (58.3%)	23 (37.7%)	12 (20.3%)	0.003 ‡

* Kruskal–Wallis test. † ANOVA. ‡ chi-square test. PRR, potential recuperation ratio; TICI, thrombolysis in cerebral infarction scale; SAH, subarachnoid hemorrhage; ICH; intracerebral hemorrhage; mRS, modified Rankin scale score.

Successful recanalization (TICI 2B-C) and first-pass recanalization rates were achieved similarly in all groups. We found no difference in the incidence of postprocedural complications among groups ($p = 0.551$) (Table 2). The rate of good functional recovery increased among groups; patients with poor collateral activation had the lowest percentage of mRS 0–2 ($p = 0.002$) and the highest mortality rate ($p = 0.003$). Patients with the most collaterals present had the best functional outcomes (Table 2).

We further evaluated if the presence of primary collaterals affected recruiting secondary collaterals, and we found no statistical significance ($p = 0.313$). A good functional outcome was associated with good primary collaterals ($p = 0.04$) as well as a good TAN score on CT angiography ($p = 0.001$).

4. Discussion

Our study findings highlight the influence of collateral circulation on patients' prognosis following an acute ischemic stroke. Our key finding is that acute stroke patients who have a more developed collateral system (a more closed circle of Willis and well-activated leptomeningeal arteries) have better functional recovery and experience a lower mortality rate.

We assessed the effects of primary and secondary collaterals on the radiological and clinical outcomes of patients by proposing a classification that covered both types of collaterals. We analyzed whether the communicating arteries (both ACoA and PCoA) were visible on CTA and did not measure the diameters, contrary to a study by Zhou et al. [26].

While some studies based on the analysis of the circle of Willis variants have not proven a relationship between anatomy and stroke outcomes [33,34], our results are consistent with other earlier research [24,26,35]. We think the primary cause of this discrepancy between studies is that various classification systems were employed by the authors in the analysis of CoW arteries. These conflicting results show that additional research on the CoW anatomy is needed.

Patients in our three groups had comparable age and sex distribution; however, when it comes to stroke causes, the patients in the unfavorable collateral group had a higher incidence of arterial hypertension and atrial fibrillation. Chronic hypertension is a well-known risk factor for worse outcomes in stroke patients. According to a review by Noor et al., hypertension, peripheral artery disease, and younger age correlate with poor functional outcomes in patients undergoing EVT [36]. Furthermore, Alobaida et al. concluded that atrial fibrillation was also linked to a worse functional recovery but not with intracerebral hemorrhage or a higher death rate [37]. Worse outcomes of stroke connected to atrial fibrillation were found in a study by Rebello et al., where they explained that atheroembolic strokes appear to be associated with improved collateral flow in comparison to patients with atrial fibrillation-related strokes [38]. This was further supported by other studies [39,40].

Studies have shown that patients with higher collateral activation have lower NIHSS values on admission [41,42]. This is in accordance with our study, explaining that higher NIHSS reflects larger ischemic tissue volume [43,44], as confirmed in our results. The admission blood pressure is another crucial aspect. Research has revealed that higher blood pressure levels prior to endovascular treatment decreased favorable 3-month functional independence [43,44]. Furthermore, lower post-treatment blood pressure was associated with better outcomes [45–47]. Our patients in all three groups had comparable systolic blood pressure values on admission, so we were unable to correlate that.

In our study, we also evaluated the onset-to-door time in order to assess any variations in the time between symptom onset and groin puncture. Our findings indicated that our patients had comparable onset times, and there were no delays in receiving endovascular therapy for either group. This information is particularly significant, as our patients were referred from distant primary stroke centers, highlighting the effectiveness of our well-established protocols and trained healthcare professionals in managing acute stroke patients. Numerous published studies have explored the impact of treatment delays on the prognosis of stroke patients. For instance, Snyder et al. [48] demonstrated that patients who presented within 6 h of stroke onset and underwent mechanical thrombectomy achieved better outcomes. Additionally, it is important to consider whether patients are directly admitted to comprehensive stroke centers, as this can lead to a shorter time to recanalization and consequently better outcomes [49]. Furthermore, additional research has shown that patients admitted during off-hours may experience poorer functional outcomes due to treatment delays [50]. These findings emphasize the importance of timely and efficient management of stroke patients, which can significantly influence their overall prognosis and functional outcomes.

There was no significant difference in the type of mechanical thrombectomy devices used among the three collateral groups. The majority of patients underwent the procedure using aspiration catheters. This indicates that the choice of mechanical thrombectomy technique was not influenced by the collateral status. With regards to collateral status and functional outcomes, the favorable collateral group had a higher percentage of patients with favorable mRS scores (indicating better functional outcomes) and a lower mortality rate compared to the other groups. This suggests that better collateral status may be associated with improved stroke outcomes. There were no significant differences between the collateral groups in terms of unsuccessful placement of endovascular devices, successful recanalization, TICI scores, number of passes during thrombectomy, control CT area of ischemia, and the presence of subarachnoid or intracerebral hemorrhage. These factors did not appear to be strongly influenced by the collateral status.

Our study revealed distinct profiles of CT perfusion parameters and stroke outcomes among the three groups (unfavorable, intermediate, and favorable) based on collateral status. The favorable collateral group had the lowest volume of core infarct and penumbra on CT perfusion maps compared to the other groups. Additionally, the mismatch ratio, which indicates the extent of mismatch between core infarct and penumbra, was the highest in the favorable collateral group, suggesting better perfusion patterns. The potential recuperation ratio, which represents the likelihood of recovery, was lowest in patients with poorly developed collaterals and highest in patients with well-developed primary and secondary collaterals. This suggests that better collateral status may contribute to a higher potential for recuperation.

Several studies correlated the presence of collaterals with automatically calculated values on CT perfusion. Regarding penumbra, some studies found that pial collaterals directly affect the proportion of recoverable penumbra [21], which is in accordance with our results. In addition, we also demonstrated that, as several CT perfusion studies have found, collateral status is related to core volume [51–53]. Our study showed that patients with the best collateral activation tended to have a higher rate of small/lacunar infarcts as a final core volume in comparison with low-collateral patients.

A higher recanalization rate [9] is correlated with collateral status, according to certain research evaluating leptomeningeal collaterals; however, other studies have not reached this conclusion [2,54]. According to our research, there was no correlation between successful recanalization and the degree of the collateral system. We also collected information on first-pass recanalization since it is known to be a predictor of better outcomes [55–59]. Our groups of patients had similar first-pass recanalization rates, which means that all our patients had similar treatment conditions.

Each of these studies underscores the significance of collateral circulation as a crucial factor in predicting the efficacy of endovascular therapy. Our study aligns with previous research, but we provide a significant added value by demonstrating the combined yet independent effect of both primary and secondary collaterals on patient survival and recovery.

Our study has several limitations. Firstly, we focused on patients within the "early" time window, which may limit the generalizability of our findings to patients in later stages of stroke. Additionally, our total sample size was limited, which could impact the statistical power and generalizability of our results. Nonetheless, we mitigated these limitations by conducting thorough examinations of comorbidities, meticulously analyzing radiological data, and collecting comprehensive information on thrombectomy procedures. Another limitation of our study is that the relationship between collateral status and stroke etiology was not assessed as analyzed in recent meta-analysis [58,59]. Despite these limitations, our study provides valuable insights into the impact of collateral circulation on endovascular therapy outcomes, offering a unique perspective on the independent contributions of primary and secondary collaterals to patient survival and recovery.

5. Conclusions

In conclusion, all our patients had equal conditions for achieving a successful treatment effect with mechanical thrombectomy, including arrival time and technical aspects of treatment. However, they differed in the cause of the ischemic stroke itself and the status of collaterals.

Our study revealed a significant relationship between collateral system status and CT perfusion values on emergency CT scans. The analysis demonstrated that both primary and secondary collateral systems work synergistically, with the circle of Willis playing a more prominent role in occlusions involving the distal internal carotid artery compared to M1 occlusions. These findings suggest a potential relationship between collateral circulation and CT perfusion parameters.

Our study's findings, which demonstrated better functional outcomes and higher survival rates in patients with good collateral circulation, provide valuable assistance to radiologists in making decisions about endovascular treatment. This is particularly relevant for patients who are located far from comprehensive stroke centers, where longer transfer times may be experienced, as well as for those who receive radiological examinations in centers without access to sophisticated analyses like CT perfusion. Additionally, our results may be useful in cases of wake-up strokes or strokes with unknown onset times. Further studies with larger patient cohorts have the potential to enhance and expand the current recommendations for mechanical thrombectomy. By carefully analyzing patient age, comorbidities, and collateral circulation, interventional neuroradiologists can make informed decisions about the use of invasive procedures such as mechanical thrombectomy.

Author Contributions: Conceptualization, S.S. and M.M.G.; methodology, S.L.K.; software, S.S.; validation, K.D., D.B.M., and S.L.K.; formal analysis, A.M.; investigation, S.S.; resources, M.Č.-V.; data curation, M.Č.-V.; writing—original draft preparation, S.S., M.M.G., and D.B.M.; writing—review and editing, K.D. and S.L.K.; visualization, A.M.; supervision, M.M.G. All authors have read and agreed to the published version of the manuscript.

Funding: This research received no external funding.

Institutional Review Board Statement: The study was conducted in accordance with the Declaration of Helsinki and approved by the Institutional Review Board (or Ethics Committee) of University hospital Split (protocol code 2181-147-01/06/LJ.Z.-23.02).

Informed Consent Statement: Patient consent was waived due to retrospective study.

Data Availability Statement: The datasets analyzed during the study are available from the corresponding author on reasonable request.

Acknowledgments: The authors thank to the Radiology and Neurology Department team included in the treatment of acute ischemic stroke patients at the University Hospital of Split.

Conflicts of Interest: The authors declare no conflicts of interest.

References

1. GBD 2019 Stroke Collaborators. Global, regional, and national burden of stroke and its risk factors, 1990–2019: A systematic analysis for the Global Burden of Disease Study 2019. *Lancet Neurol.* **2021**, *20*, 795–820. [CrossRef] [PubMed]
2. Nordmeyer, H.; Webering, N.; Chapot, R.; Hadisurya, J.; Heddier, M.; Stracke, P.; Berger, K.; Isenmann, S.; Weber, R. The association between collateral status, recanalization and long term outcome in stroke patients treated with stent retrievers—Are there indications not to perform thrombectomy based on CT angiography? *J. Neuroradiol.* **2017**, *44*, 217–222. [CrossRef]
3. Sheth, S.A.; Liebeskind, D.S. Collaterals in endovascular therapy for stroke. *Curr. Opin. Neurol.* **2015**, *28*, 10–15. [CrossRef]
4. Ginsberg, M.D. The cerebral collateral circulation: Relevance to pathophysiology and treatment of stroke. *Neuropharmacology* **2018**, *134*, 280–292. [CrossRef]
5. Martinon, E.; Lefevre, P.H.; Thouant, P.; Osseby, G.V.; Ricolfi, F.; Chavent, A. Collateral circulation in acute stroke: Assessing methods and impact: A literature review. *J. Neuroradiol.* **2014**, *41*, 97–107. [CrossRef] [PubMed]
6. Bang, O.Y.; Saver, J.L.; Kim, S.J.; Kim, G.M.; Chung, C.S.; Ovbiagele, B.; Lee, K.H.; Liebeskind, D.S. Collateral flow predicts response to endovascular therapy for acute ischemic stroke. *Stroke* **2011**, *42*, 693–699. [CrossRef] [PubMed]
7. Piedade, G.S.; Schirmer, C.M.; Goren, O.; Zhang, H.; Aghajanian, A.; Faber, J.E.; Griessenauer, C.J. Cerebral collateral circulation: A review in the context of ischemic stroke and mechanical thrombectomy. *World Neurosurg.* **2019**, *122*, 33–42. [CrossRef]
8. Leng, X.; Lan, L.; Liu, L.; Leung, T.W.; Wong, K.S. Good collateral circulation predicts favorable outcomes in intravenous thrombolysis: A systematic review and meta-analysis. *Eur. J. Neurol.* **2016**, *23*, 1738–1749. [CrossRef]
9. Liebeskind, D.S.; Tomsick, T.A.; Foster, L.D.; Yeatts, S.D.; Carrozzella, J.; Demchuk, A.M.; Jovin, T.G.; Khatri, P.; von Kummer, R.; Sugg, R.M.; et al. Collaterals at angiography and outcomes in the Interventional Management of Stroke (IMS) III trial. *Stroke* **2014**, *45*, 759–764. [CrossRef]
10. Maas, M.B.; Lev, M.H.; Ay, H.; Singhal, A.B.; Greer, D.M.; Smith, W.S.; Harris, G.J.; Halpern, E.; Kemmling, A.; Koroshetz, W.J.; et al. Collateral vessels on CT angiography predict outcome in acute ischemic stroke. *Stroke* **2009**, *40*, 3001–3005. [CrossRef]
11. Sallustio, F.; Motta, C.; Pizzuto, S.; Diomedi, M.; Giordano, A.; D'Agostino, V.C.; Samà, D.; Mangiafico, S.; Saia, V.; Legramante, J.M.; et al. CT angiography-based collateral flow and time to reperfusion are strong predictors of outcome in endovascular treatment of patients with stroke. *J. NeuroInterventional Surg.* **2017**, *9*, 940–943. [CrossRef] [PubMed]
12. Gariel, F.; Lapergue, B.; Bourcier, R.; Berge, J.; Barreau, X.; Mazighi, M.; Kyheng, M.; Labreuche, J.; Fahed, R.; Blanc, R.; et al. Mechanical thrombectomy outcomes with or without intravenous thrombolysis. *Stroke* **2018**, *49*, 2383–2390. [CrossRef]
13. Bhan, C.; Koehler, T.J.; Elisevich, L.; Singer, J.; Mazaris, P.; James, E.; Combs, J.; Dejesus, M.; Tubergen, T.; Packard, L.; et al. Mechanical thrombectomy for acute stroke: Early versus late time window outcomes. *J. Neuroimaging* **2020**, *30*, 315–320. [CrossRef] [PubMed]

14. de Castro-Afonso, L.H.; Fornazari, V.R.; Machado, J.P.; Nakiri, G.S.; Abud, T.G.; Monsignore, L.M.; Pontes-Neto, O.M.; Abud, D.G. Thrombectomy for anterior circulation stroke in a witnessed late time window versus early time window. *Neurohospitalist* **2023**, *13*, 243–249. [CrossRef] [PubMed]
15. Ni, H.; Liu, X.; Hang, Y.; Jia, Z.; Cao, Y.; Shi, H.; Liu, S.; Zhao, L. Predictors of futile recanalization in patients with acute ischemic stroke undergoing mechanical thrombectomy in late time windows. *Front. Neurol.* **2022**, *13*, 958236. [CrossRef]
16. Mohammaden, M.H.; Stapleton, C.J.; Brunozzi, D.; Hussein, A.E.; Khedr, E.M.; Atwal, G.; Alaraj, A. Predictors of poor outcome despite successful mechanical thrombectomy of anterior circulation large vessel occlusions within 6 h of symptom onset. *Front. Neurol.* **2020**, *11*, 907. [CrossRef] [PubMed]
17. van Horn, N.; Kniep, H.; Leischner, H.; McDonough, R.; Deb-Chatterji, M.; Broocks, G.; Thomalla, G.; Brekenfeld, C.; Fiehler, J.; Hanning, U.; et al. Predictors of poor clinical outcome despite complete reperfusion in acute ischemic stroke patients. *J. NeuroInterventional Surg.* **2021**, *13*, 14–18. [CrossRef] [PubMed]
18. Linfante, I.; Starosciak, A.K.; Walker, G.R.; Dabus, G.; Castonguay, A.C.; Gupta, R.; Sun, C.H.; Martin, C.; Holloway, W.E.; Mueller-Kronast, N.; et al. Predictors of poor outcome despite recanalization: A multiple regression analysis of the NASA registry. *J. NeuroInterventional Surg.* **2016**, *8*, 224–229. [CrossRef] [PubMed]
19. Ospel, J.M.; Hill, M.D.; Kappelhof, M.; Demchuk, A.M.; Menon, B.K.; Mayank, A.; Dowlatshahi, D.; Frei, D.; Rempel, J.L.; Baxter, B.; et al. Which acute ischemic stroke patients are fast progressors?: Results from the ESCAPE trial control arm. *Stroke* **2021**, *52*, 1847–1850. [CrossRef]
20. Mohammaden, M.H.; Haussen, D.C.; Pisani, L.; Al-Bayati, A.R.; Bhatt, N.R.; Jillella, D.V.; Bianchi, N.A.; Belagaje, S.R.; Frankel, M.R. Nogueira RG. Characterizing fast and slow progressors in anterior circulation large vessel occlusion strokes. *Interv. Neuroradiol.* **2023**, *29*, 379–385. [CrossRef]
21. Rusanen, H.; Saarinen, J.T.; Sillanpää, N. Collateral circulation predicts the size of the infarct core and the proportion of salvageable penumbra in hyperacute ischemic stroke patients treated with intravenous thrombolysis. *Cerebrovasc. Dis.* **2015**, *40*, 182–190. [CrossRef]
22. Liebeskind, D.S. Collateral circulation. *Stroke* **2003**, *34*, 2279–2284. [CrossRef]
23. Maguida, G.; Shuaib, A. Collateral circulation in ischemic stroke: An updated review. *J. Stroke* **2023**, *25*, 179–198. [CrossRef] [PubMed]
24. Lin, E.; Kamel, H.; Gupta, A.; RoyChoudhury, A.; Girgis, P.; Glodzik, L. Incomplete circle of Willis variants and stroke outcome. *Eur. J. Radiol.* **2022**, *153*, 110383. [CrossRef] [PubMed]
25. Sadeh-Gonik, U.; Budylev, A.; Krivitzky, D.; Molad, J.; Halevi, H.; Jonas-Kimchi, T.; Yashar, H.; Ben Assayag, E.; Seyman, E. Circle of Willis integrity in acute middle cerebral artery occlusion: Does the posterior communicating artery matter? *J. NeuroInterventional Surg.* **2023**, jnis-2023-020326. [CrossRef]
26. Zhou, H.; Sun, J.; Ji, X.; Lin, J.; Tang, S.; Zeng, J.; Fan, Y.H. Correlation between the integrity of the circle of Willis and the severity of initial noncardiac cerebral infarction and clinical prognosis. *Medicine* **2016**, *95*, e2892. [CrossRef]
27. Gerber, J.C.; Petrova, M.; Krukowski, P.; Kuhn, M.; Abramyuk, A.; Bodechtel, U.; Dzialowski, I.; Engellandt, K.; Kitzler, H.; Pallesen, L.P.; et al. Collateral state and the effect of endovascular reperfusion therapy on clinical outcome in ischemic stroke patients. *Brain Behav.* **2016**, *6*, e00513. [CrossRef]
28. Sinha, A.; Stanwell, P.; Killingsworth, M.C.; Bhaskar, S.M.M. Prognostic accuracy and impact of cerebral collateral status on clinical and safety outcomes in acute ischemic stroke patients receiving reperfusion therapy: A systematic meta-analysis. *Acta Radiol.* **2023**, *64*, 698–718. [CrossRef] [PubMed]
29. Powers, W.J.; Rabinstein, A.A.; Ackerson, T.; Adeoye, O.M.; Bambakidis, N.C.; Becker, K.; Biller, J.; Brown, M.; Demaerschalk, B.M.; Hoh, B.; et al. Guidelines for the early management of patients with acute ischemic stroke: 2019 update to the 2018 guidelines for the early management of acute ischemic stroke: A guideline for healthcare professionals from the American Heart Association/American Stroke Association. *Stroke* **2019**, *50*, e344–e418.
30. Tan, J.C.; Dillon, W.P.; Liu, S.; Adler, F.; Smith, W.S.; Wintermark, M. Systematic comparison of perfusion-CT and CT-angiography in acute stroke patients. *Ann. Neurol.* **2007**, *61*, 533–543. [CrossRef]
31. Meuli, R.A. Imaging Viable Brain Tissue with CT Scan during Acute Stroke. *Cerebrovasc. Dis.* **2018**, *17* (Suppl. 3), 28. [CrossRef] [PubMed]
32. Higashida, R.T.; Furlan, A.J.; Roberts, H.; Tomsick, T.; Connors, B.; Barr, J.; Dillon, W.; Warach, S.; Broderick, J.; Tilley, B.; et al. Trial design and reporting standards for intra-arterial cerebral thrombolysis for acute ischemic stroke. *Stroke* **2003**, *34*, e109–e137. [CrossRef] [PubMed]
33. Oumer, M.; Alemayehu, M.; Muche, A. Association between circle of Willis and ischemic stroke: A systematic review and meta-analysis. *BMC Neurosci.* **2021**, *22*, 3. [CrossRef] [PubMed]
34. Seifert-Held, T.; Eberhard, K.; Christensen, S.; Hofer, E.; Enzinger, C.; Albers, G.W.; Lansberg, M. Circle of Willis variants are not associated with thrombectomy outcomes. *Stroke Vasc. Neurol.* **2021**, *6*, 310–313. [CrossRef] [PubMed]
35. Širvinskas, A.; Lengvenis, G.; Ledas, G.; Mosenko, V.; Lukoševičius, S. Circle of Willis configuration and thrombus localization impact on ischemic stroke patient outcomes: A systematic review. *Medicina* **2023**, *59*, 2115. [CrossRef] [PubMed]
36. Noor, A.; Bennett, G.; Arndt, S.; Vidal, G. Comorbidities related to clinical outcomes in patients with acute ischemic stroke undergoing mechanical thrombectomy: Review of literature and experience at a single comprehensive stroke center. *Ochsner J.* **2019**, *19*, 13–16. [CrossRef] [PubMed]

37. Alobaida, M.; Harrison, S.L.; Lane, D.A.; Underhill, P.; Hill, A.; Lip, G.Y.H. Outcomes in patients with ischaemic stroke undergoing endovascular thrombectomy: Impact of atrial fibrillation. *J. Stroke Cerebrovasc. Dis.* **2023**, *32*, 106917. [CrossRef] [PubMed]
38. Rebello, L.C.; Bouslama, M.; Haussen, D.C.; Grossberg, J.A.; Dehkharghani, S.; Anderson, A.; Belagaje, S.R.; Bianchi, N.A.; Grigoryan, M.; Frankel, M.R.; et al. Stroke etiology and collaterals: Atheroembolic strokes have greater collateral recruitment than cardioembolic strokes. *Eur. J. Neurol.* **2017**, *24*, 762–767. [CrossRef]
39. Kobeissi, H.; Ghozy, S.; Seymour, T.; Gupta, R.; Bilgin, C.; Kadirvel, R.; Rabinstein, A.A.; Kallmes, D.F. Outcomes of patients with atrial fibrillation following thrombectomy for stroke: A systematic review and meta-analysis. *JAMA Netw. Open* **2023**, *6*, e2249993. [CrossRef]
40. Yang, J.; Wu, Y.; Gao, X.; Shang, Q.; Xu, Y.; Han, Q.; Li, J.; Chen, C.; Bivard, A.; Parsons, M.W.; et al. Poor collateral flow with severe hypoperfusion explains worse outcome in acute stroke patients with atrial fibrillation. *Int. J. Stroke* **2023**, *18*, 689–696. [CrossRef]
41. Sperti, M.; Arba, F.; Acerbi, A.; Busto, G.; Fainardi, E.; Sarti, C. Determinants of cerebral collateral circulation in acute ischemic stroke due to large vessel occlusion. *Front. Neurol.* **2023**, *14*, 1181001. [CrossRef] [PubMed]
42. Mangiafico, S.; Saia, V.; Nencini, P.; Romani, I.; Palumbo, V.; Pracucci, G.; Pracucci, G.; Consoli, A.; Rosi, A.; Renieri, L.; et al. Effect of the interaction between recanalization and collateral circulation on functional outcome in acute ischaemic stroke. *Interv. Neuroradiol.* **2014**, *20*, 704–714. [CrossRef] [PubMed]
43. Samuels, N.; van de Graaf, R.A.; Mulder, M.J.H.L.; Brown, S.; Roozenbeek, B.; van Doormaal, P.J.; Goyal, M.; Campbell, B.C.V.; Muir, K.W.; Agrinier, N.; et al. Admission systolic blood pressure and effect of endovascular treatment in patients with ischaemic stroke: An individual patient data meta-analysis. *Lancet Neurol.* **2023**, *22*, 312–319. [CrossRef] [PubMed]
44. Liu, C.H.; Wei, Y.C.; Lin, J.R.; Chang, C.H.; Chang, T.Y.; Huang, K.L.; Chang, Y.J.; Ryu, S.J.; Lin, L.C.; Lee, T.H. Initial blood pressure is associated with stroke severity and is predictive of admission cost and one-year outcome in different stroke subtypes: A SRICHS registry study. *BMC Neurol.* **2016**, *16*, 27. [CrossRef] [PubMed]
45. Anadani, M.; Orabi, M.Y.; Alawieh, A.; Goyal, N.; Alexandrov, A.V.; Petersen, N.; Kodali, S.; Maier, I.L.; Psychogios, M.N.; Swisher, C.B.; et al. Blood pressure and outcome after mechanical thrombectomy with successful revascularization. *Stroke* **2019**, *50*, 2448–2454. [CrossRef] [PubMed]
46. Matusevicius, M.; Cooray, C.; Holmin, S.; Bottai, M.; Ahmed, N. Association between systolic blood pressure course and outcomes after stroke thrombectomy. *BMJ Neurol. Open* **2021**, *3*, e000183. [CrossRef] [PubMed]
47. Katsanos, A.H.; Malhotra, K.; Ahmed, N.; Seitidis, G.; Mistry, E.A.; Mavridis, D.; Kim, J.T.; Veroniki, A.A.; Maier, I.; Matusevicius, M.; et al. Blood pressure after endovascular thrombectomy and outcomes in patients with acute ischemic stroke: An individual patient data meta-analysis. *Neurology* **2022**, *98*, e291–e301. [CrossRef] [PubMed]
48. Snyder, T.; Agarwal, S.; Huang, J.; Ishida, K.; Flusty, B.; Frontera, J.; Lord, A.; Torres, J.; Zhang, C.; Rostanski, S.; et al. Stroke treatment delay limits outcome after mechanical thrombectomy: Stratification by arrival time and ASPECTS. *J. Neuroimaging* **2020**, *30*, 625–630. [CrossRef] [PubMed]
49. MacKenzie, I.E.R.; Arusoo, T.; Sigounas, D. Impact of direct admission versus interfacility transfer on endovascular treatment outcomes for acute ischemic stroke: Systematic review and meta-analysis. *World Neurosurg.* **2021**, *152*, e387–e397. [CrossRef]
50. Ghozy, S.; Hasanzadeh, A.; Kobeissi, H.; Abdelghaffar, M.; Shafie, M.; Beizavi, Z.; Hassankhani, A.; Azzam, A.Y.; Dmytriw, A.A.; Kadirvel, R.; et al. The impact of off-hour mechanical thrombectomy therapy on outcomes for acute ischemic stroke: A systematic review and meta-analysis. *J. Neurol. Sci.* **2023**, *453*, 120802. [CrossRef]
51. de Havenon, A.P.; Mlynash, M.; Kim-Tenser, M.A.; Lansberg, M.G.; Leslie-Mazwi, T.; Christensen, S.; McTaggart, R.A.; Alexander, M.; Albers, G.; Broderick, J.; et al. Results from DEFUSE 3: Good collaterals are associated with reduced ischemic core growth but not neurologic outcome. *Stroke* **2019**, *50*, 632–638. [CrossRef] [PubMed]
52. Nannoni, S.; Cereda, C.W.; Sirimarco, G.; Lambrou, D.; Strambo, D.; Eskandari, A.; Dunet, V.; Wintermark, M.; Michel, P. Collaterals are a major determinant of the core but not the penumbra volume in acute ischemic stroke. *Neuroradiology* **2019**, *61*, 971–978. [CrossRef] [PubMed]
53. Lin, L.; Yang, J.; Chen, C.; Tian, H.; Bivard, A.; Spratt, N.J.; Levi, C.R.; Parsons, M.W. Association of collateral status and ischemic core growth in patients with acute ischemic stroke. *Neurology* **2021**, *96*, e161–e170. [CrossRef] [PubMed]
54. Uniken Venema, S.M.; Dankbaar, J.W.; Wolff, L.; van Es, A.C.G.M.; Sprengers, M.; van der Lugt, A.; Dippel, D.W.J.; van der Worp, H.B. Collateral status and recanalization after endovascular treatment for acute ischemic stroke. *J. NeuroInterventional Surg.* **2023**, *15*, 531–538. [CrossRef] [PubMed]
55. Bai, X.; Zhang, X.; Yang, W.; Zhang, Y.; Wang, T.; Xu, R.; Wang, Y.; Li, L.; Feng, Y.; Yang, K.; et al. Influence of first-pass effect on recanalization outcomes in the era of mechanical thrombectomy: A systemic review and meta-analysis. *Neuroradiology* **2021**, *63*, 795–807. [CrossRef] [PubMed]
56. Jang, K.M.; Choi, H.H.; Nam, T.K.; Byun, J.S. Clinical outcomes of first-pass effect after mechanical thrombectomy for acute ischemic stroke: A systematic review and meta-analysis. *Clin. Neurol. Neurosurg.* **2021**, *211*, 107030. [CrossRef] [PubMed]
57. Huo, X.; Sun, D.; Nguyen, T.N.; Raynald; Jia, B.; Tong, X.; Ma, G.; Wang, A.; Ma, N.; Gao, F.; et al. First-pass effect of mechanical thrombectomy for anterior circulation large vessel occlusion: Incidence, predictors, and clinical impact. Insight from the ANGEL-ACT registry. *J. Neurosurg.* **2023**, *139*, 670–677. [CrossRef] [PubMed]

58. Ravindran, A.V.; Killingsworth, M.C.; Bhaskar, S. Cerebral collaterals in acute ischaemia: Implications for acute ischaemic stroke patients receiving reperfusion therapy. *Eur. J. Neurosci.* **2021**, *53*, 1238–1261. [CrossRef]
59. Sinha, A.; Stanwell, P.; Beran, R.G.; Calic, Z.; Killingsworth, M.C.; Bhaskar, S.M.M. Stroke Aetiology and Collateral Status in Acute Ischemic Stroke Patients Receiving Reperfusion Therapy-A Meta-Analysis. *Neurol. Int.* **2021**, *13*, 608–621. [CrossRef]

neurology international

Article

Redefining Infarction Size for Small-Vessel Occlusion in Acute Ischemic Stroke: A Retrospective Case–Control Study

Yen-Chu Huang [1,2,*], Hsu-Huei Weng [3], Leng-Chieh Lin [4], Jiann-Der Lee [1,2], Jen-Tsung Yang [2,5], Yuan-Hsiung Tsai [2,3] and Chao-Hui Chen [1]

1 Department of Neurology, Chang Gung Memorial Hospital at Chiayi, Chiayi 613, Taiwan; jdlee540908@cgmh.org.tw (J.-D.L.); trianglescorpio@gmail.com (C.-H.C.)
2 College of Medicine, Chang-Gung University, Taoyuan 333, Taiwan; jents716@ms32.hinet.net (J.-T.Y.); russell.tsai@gmail.com (Y.-H.T.)
3 Department of Diagnostic Radiology, Chang Gung Memorial Hospital at Chiayi, Chiayi 613, Taiwan; hweng@post.harvard.edu
4 Department of Emergency Medicine, Chang Gung Memorial Hospital at Chiayi, Chiayi 613, Taiwan; a3456711@ms65.hinet.net
5 Department of Neurosurgery, Chang Gung Memorial Hospital at Chiayi, Chiayi 613, Taiwan
* Correspondence: yenchu.huang@msa.hinet.net

Abstract: Background/Objectives: Small-vessel occlusion, previously referred to as lacunar infarcts, accounts for approximately one-third of all ischemic strokes, using an axial diameter of less than 20 mm on diffusion-weighted imaging. However, this threshold may not adequately differentiate small-vessel occlusion from other pathologies, such as branch atheromatous disease (BAD) and embolism. This study aimed to assess the clinical significance and pathological implications of acute small subcortical infarctions (SSIs) based on infarct diameter. Methods: We conducted a retrospective case–control study using data from stroke patients recorded between 2016 and 2021 of the Stroke Registry in Chang Gung Healthcare System. Patients with acute SSIs in penetrating artery territories were included. Key variables such as patient demographics, stroke severity, and medical history were collected. Infarcts were categorized based on size, and the presence of early neurological deterioration (END) and favorable functional outcomes were assessed. Results: Among the 855 patients with acute SSIs, the median age was 70 years and the median National Institutes of Health Stroke Scale (NIHSS) score at arrival was four. END occurred in 97 patients (11.3%). Those who experienced END were significantly less likely to achieve a favorable functional outcome compared to those who did not (18.6% vs. 59.9%, $p < 0.001$). The incidence of END increased progressively with infarct sizes of 15 mm or larger, with the optimal threshold for predicting END identified as 15.5 mm and for BAD, it was 12.1 mm. A multiple logistic regression analysis revealed that motor tract involvement [adjusted odds ratio (aOR) 2.3; 95% confidence interval (CI) 1.1–4.7], an initial heart rate greater than 90 beats per minute (aOR 2.3; 95% CI 1.2–4.3), and a larger infarct size (15 mm to less than 20 mm vs. 10 mm to less than 15 mm; aOR 3.0; 95% CI 1.4–6.3) were significantly associated with END. Conclusions: Our findings suggest that setting the upper limit for small-vessel occlusion at 15 mm would be more effective in distinguishing it from BAD. However, these findings should be interpreted in the context of the retrospective design and study population. Further multi-center research utilizing high-resolution vessel wall imaging is necessary to refine this threshold and enhance diagnostic accuracy.

Keywords: small subcortical infarction; small-vessel disease; MRI; early neurological deterioration; small-vessel occlusion

1. Introduction

Small-vessel occlusion, formerly referred to as lacunar infarcts, comprises approximately one-third of all ischemic strokes [1]. However, lacunar infarcts are not solitary cerebrovascular events; they often constitute a part of the spectrum of manifestations seen in cerebral small-vessel disease. This condition can be initiated by various pathologies, primarily hypertensive arteriopathy and cerebral amyloid angiopathy, both associated with endothelial dysfunction, blood–brain barrier degradation, inflammation, and other contributing factors [2,3]. Moreover, acute small subcortical infarctions (SSIs) can also arise from other causes, including branch atheromatous disease (BAD), intracranial artery stenosis, and embolism [3,4].

In the Trial of Org 10,172 in Acute Stroke Treatment (TOAST) classification, acute SSIs were categorized as small-vessel occlusions when the axial diameter was less than 15 mm [5]. This upper limit was revised to 20 mm in diffusion-weighted imaging (DWI) for small-artery occlusion in the Stop Stroke Study (SSS)–TOAST classification [6] or for a recent SSI in cerebral small-vessel disease [7]. However, specific infarct patterns such as a tubular shape, larger infarct size, or proximal location may suggest a BAD pathology [8–10]. While certain SSIs with BAD pathology may fit the current definition of small-artery occlusion, from a pathological standpoint, BAD should be classified as intracranial atherosclerosis and may warrant more aggressive treatment to reduce END and improve clinical outcomes [11,12]. This inconsistency could potentially complicate stroke classification and impact clinical research and treatment strategies.

In the Chinese ischemic stroke subclassification, BAD is distinctly categorized as large-artery atherosclerosis, while small-artery occlusion, now renamed as penetrating artery disease in this classification, is defined as an acute isolated infarct in a clinically relevant territory of one penetrating artery, regardless of the size of the infarct and without evidence of atherosclerotic plaque detected by high-resolution MRI [13].

However, visualizing these microatheromas or junctional plaques requires high-resolution vessel wall imaging, which is a time-consuming and impractical procedure for routine clinical application [14]. Consequently, BAD is often characterized by a large infarct size (>15 mm), proximity to the parent artery, or a tubular shape [4]. Nevertheless, the current definition, which relies on an arbitrary cut-off of 20 mm for small-vessel occlusion, is not entirely satisfactory, as it encompasses a significant number of cases involving BAD or embolic infarction

The objective of this study is to assess the clinical relevance and possible pathologies of acute SSIs based on their infarct diameter.

2. Materials and Methods

This retrospective case–control study utilized data from stroke patients recorded between 2016 and 2021 at the Chiayi branch of the Stroke Registry in Chang Gung Healthcare System (SRICHS), Chiayi, Taiwan [15]. Clinical information for these acute ischemic stroke patients was gathered prospectively. Patients were excluded if their stroke onset time was unknown, they arrived more than 48 h post-onset, or did not undergo MRI within seven days of onset (Figure 1). The analysis focused on those with SSIs in penetrating artery territories with a maximum lesion size of 40 mm.

Key variables, including patient demographics, stroke severity, and medical history, were collected upon admission. The medical history encompassed conditions such as diabetes mellitus, hypertension, coronary artery disease, atrial fibrillation, congestive heart failure, hyperlipidemia, and prior strokes. Comprehensive MRI evaluations were performed on all patients, including extracranial and intracranial angiography via time-of-flight (TOF) techniques, and carotid and transcranial duplex scans were used to detect large-vessel pathologies. The decision to perform extended EKG monitoring or echocardiography was made by the attending physician. In addition, certain patients underwent investigations for occult atrial fibrillation using serial 12-lead electrocardiography, 24 h Holter monitoring, or echocardiography. For patients under 45, further assessments were carried out to determine potential causes of stroke at a young age. Stroke subtypes were classified based on TOAST criteria, which include large-artery atherosclerosis, cardioembolism, small-vessel occlusion, undetermined etiology, and other specific causes [5]. The small-vessel occlusion subtype was defined by an SSI within penetrating artery territories, with a maximum diameter of 20 mm on DWI. Early neurological deterioration (END) was defined as an increase of 4 or more points on the National Institutes of Health Stroke Scale (NIHSS) within 72 h of stroke onset [16]. A modified Rankin Scale score of ≤1 at 3 months was considered a favorable outcome.

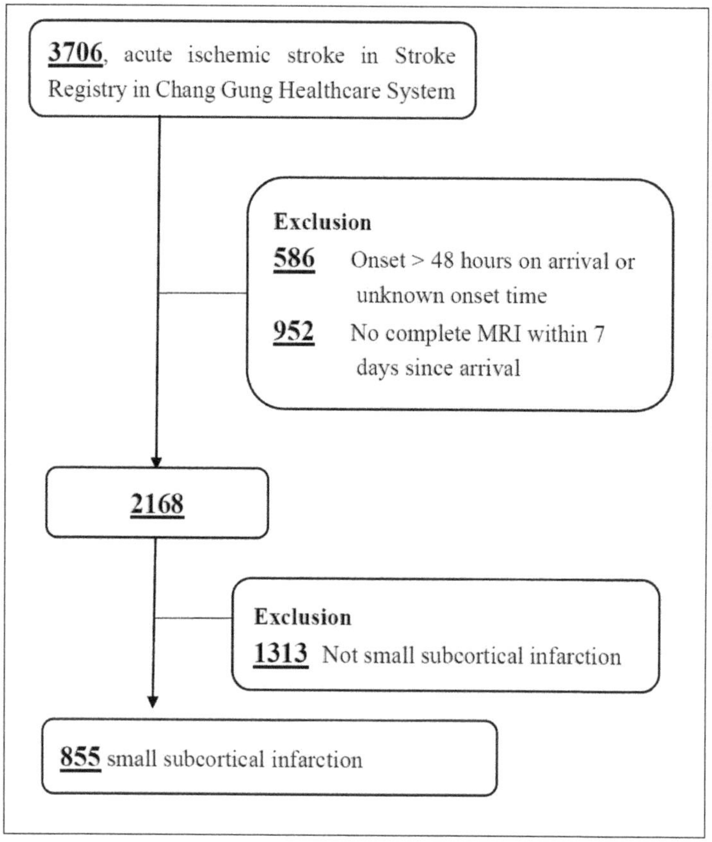

Figure 1. Study flowchart of patient selection.

BAD was diagnosed based on two criteria, namely (1) infarctions visible on four or more axial MRI slices (slice thickness 5–7 mm) within the lenticulostriate artery territory or extending from the basal surface of the pons within the paramedian pontine artery territory and (2) the absence of significant large-artery stenosis (>50%) or occlusion and no evidence of cardiogenic embolism [4,17]. Corticospinal tract involvement has been linked to the prediction of END. Therefore, we defined motor tract involvement as the presence of an acute infarct in specific regions, namely the posterior part of the corona radiata adjacent to the lateral ventricle, the posterior limb of the internal capsule, or the basal surface of the pons [18,19]. The imaging markers of cerebral small-vessel disease, such as cerebral microbleeds, enlarged perivascular spaces (EPVSs), and white matter hyperintensities (WMH), were also evaluated [20].

MRI data were obtained using either a 1.5-Tesla Philips Gyroscan Intera scanner (Philips Medical Systems, Best, The Netherlands) or a 3-Tesla Siemens Verio MRI system (Siemens Medical System, Erlangen, Germany), including standard sequences such as axial DWI, the apparent diffusion coefficient, T1- and T2-weighted images, fluid-attenuated inversion recovery images, and 3D TOF angiography covering the extracranial carotid artery and the circle of Willis.

Statistical Analyses

Statistical analyses were conducted using SPSS software (version 27, Chicago, IL, USA). For continuous variables, the Kruskal–Wallis test was applied, while Fisher's exact test or Pearson's chi-square test was used for categorical data. When significant differences were identified across the five infarct sizes, post hoc comparisons between groups were performed using Mann–Whitney or chi-square tests with Bonferroni correction. Receiver-operating characteristic (ROC) curve analysis was utilized to determine the optimal infarct diameter cut-off for predicting END, BAD, and large-artery atherosclerosis. To adjust for baseline confounders, multivariate logistic regression analysis was performed for END and favorable outcomes. A two-tailed p-value of less than 0.05 was considered statistically significant.

3. Results

A total of 3706 stroke patients were identified from the Stroke Registry in Chang Gung Healthcare System (SRICHS) at the Jiayi branch between 2016 and 2021 (Figure 1). Out of these, 586 patients whose stroke onset occurred more than 48 h before arrival, as well as 952 patients who did not undergo a complete MRI study, were excluded from the analysis. This left 2168 patients, of whom 855 had acute SSIs.

For the 855 patients with acute SSIs, the median age was 70 years (61–78) and the median NIHSS score at the time of arrival was four (2–6). Among these patients, 329 (38.5%) had brainstem strokes, 97 (11.3%) experienced END, and 36 (4.2%) received intravenous alteplase treatment. At the three-month follow-up, 465 patients (54.4%) achieved a favorable functional outcome. Patients who experienced END were significantly less likely to achieve a favorable functional outcome compared to those who did not experience END (18.6% vs. 59.9%, $p < 0.001$). According to the TOAST classification, most patients were identified as having small-vessel occlusion ($n = 582$, 68.1%), followed by large-artery atherosclerosis ($n = 104$, 12.2%), cardioembolism ($n = 72$, 8.4%), undetermined causes ($n = 94$, 11.0%), and other causes ($n = 3$, 0.4%).

Table 1 outlines the baseline clinical characteristics based on infarct size. Patients with infarcts smaller than 10 mm had the mildest initial NIHSS scores at arrival and on the third day, the lowest incidence of BAD, the lowest incidence of END, and the highest rate of favorable outcomes at three months. Stroke risk factors, including hypertension, diabetes mellitus, atrial fibrillation, congestive heart failure, or previous stroke, did not show significant differences across various infarct sizes. Similarly, the incidence of cardioembolism was consistent across different infarct sizes. However, patients with infarct sizes of 25 mm or larger had the highest rates of large-artery atherosclerosis, END, and intravenous alteplase treatment. For small-vessel disease imaging markers, patients with infarcts smaller than 10 mm had the lowest WMH scores in both the basal ganglia and periventricular areas, as well as the lowest EPVS scores in the basal ganglia. Specifically, patients with infarcts smaller than 10 mm had significantly lower WMH scores in the periventricular area compared to those with infarcts measuring 20 mm to less than 25 mm (1.8 vs. 2.3, adjusted $p = 0.044$). Similarly, WMH scores in the basal ganglia (1.6 vs. 2.2, adjusted $p < 0.001$) and EPVS scores in the basal ganglia (1.6 vs. 2.2, adjusted $p = 0.001$) were also lower in patients with infarcts smaller than 10 mm compared to those with infarcts measuring 20 mm to less than 25 mm.

Table 1. Associations between infarction sizes and clinical characteristics.

Characteristic	Diameter <10 mm	Diameter 10 mm–<15 mm	Diameter 15 mm–<20 mm	Diameter 20 mm–<25 mm	Diameter ≥25 mm	p-Value
Number	248	302	182	74	49	
Age	69.5 (60.3–77.8)	69 (61.8–78)	70 (61.8–78)	74.5 (65–79)	66 (58–78)	0.182
Female gender	72 (29.0%)	99 (32.8%)	74 (40.7%)	35 (47.3%)	14 (28.6%)	0.012
Stroke risk factors						
Diabetes mellitus	150 (60.5%)	158 (52.3%)	101 (55.5%)	37 (50.0%)	24 (49.0%)	0.251
Hypertension	206 (83.1%)	245 (81.1%)	147 (80.8%)	62 (83.8%)	41 (83.7%)	0.940
Hypercholesterolemia	82 (33.1%)	125 (41.4%)	92 (50.5%)	33 (44.6%)	26 (53.1%)	0.003
Coronary artery disease	10 (4.0%)	20 (6.6%)	11 (6.0%)	5 (6.8%)	2 (4.1%)	0.697
Atrial fibrillation	26 (10.5%)	21 (7.0%)	16 (8.8%)	4 (5.4%)	5 (10.2%)	0.506
Congestive heart failure	4 (1.6%)	7 (2.3%)	3 (1.6%)	1 (1.4%)	1 (2.0%)	0.965
Old stroke	61 (24.6%)	72 (23.8%)	29 (15.9%)	12 (16.2%)	8 (16.3%)	0.101
TOAST stroke subtypes						<0.001
Large-artery atherosclerosis	30 (12.1%)	24 (7.9%)	27 (14.8%)	12 (16.2%)	11 (22.4%)	
Cardioembolism	27 (10.9%)	23 (7.6%)	13 (7.1%)	5 (6.8%)	4 (8.2%)	
Small-vessel occlusion	186 (75.0%)	255 (84.4%)	137 (75.3%)	0 (0%)	0 (0%)	
Other determined	2 (0.8%)	0 (0%)	0 (0%)	1 (1.4%)	0 (0%)	
Undetermined	3 (1.2%)	0 (0%)	5 (2.7%)	56 (75.7%)	34 (69.4%)	
Stroke information						
Initial NIHSS	3 (1–4)	4 (3–5)	5 (3–6)	5 (3–7)	4 (3–9.5)	<0.001
NIHSS score on 3rd day	2 (1–4)	4 (2.5–6)	5 (3–7)	6 (4–9)	7.5 (4–10)	<0.001
END (ΔNIHSS ≥ 4)	16 (6.5%)	22 (7.3%)	30 (16.5%)	16 (21.6%)	13 (26.5%)	<0.001
Axial diameter (mm)	7 (6–8.5)	12 (11–13.1)	17 (16–18.1)	22 (21–23)	29 (26–32)	<0.001
Intravenous alteplase	6 (2.4%)	11 (3.6%)	8 (4.4%)	3 (4.1%)	8 (16.3%)	<0.001
mRS ≥ 1 at 3 months	179 (72.2%)	165 (54.6%)	83 (45.6%)	20 (27.0%)	18 (36.7%)	<0.001
Mortality	2 (0.8%)	2 (0.7%)	0 (0%)	1 (1.4%)	2 (4.1%)	0.551
Neuroimaging information						
Onset–MRI duration (hour)	26.9 (14.8–56.2)	29.2 (18.2–56.2)	33.2 (18.6–57.8)	41.8 (14.9–76.3)	38.3 (25.8–64.1)	0.072
Stroke locations						<0.001
Brainstem	110 (44.4%)	105 (34.8%)	83 (45.6%)	28 (37.8%)	2 (4.1%)	
Thalamus	24 (9.7%)	19 (6.3%)	1 (0.6%)	0 (0%)	0 (0%)	
MCA territory	114 (46.0%)	178 (58.9%)	96 (53.0)	45 (60.8)	47 (95.9)	
Branch atheromatous disease	46 (18.5%)	137 (45.4%)	141 (77.5%)	67 (90.5%)	44 (89.8%)	<0.001
Parent artery stenosis ≥ 50%	38 (15.3%)	32 (10.6%)	32 (17.6%)	14 (18.9%)	11 (22.4%)	0.006
WMH scores of periventricular area	1.8 ± 1.1	2.0 ± 1.1	1.8 ± 1.0	2.3 ± 1.2	2.3 ± 0.9	0.004
WMH scores of basal ganglia	1.6 ± 0.9	1.8 ± 0.9	1.6 ± 0.9	2.2 ± 0.9	2.3 ± 0.8	<0.001
EPVS scores of centrum semiovale	1.8 ± 0.8	1.9 ± 1.0	1.8 ± 0.8	2.2 ± 1.0	1.9 ± 0.8	0.157
EPVS scores of basal ganglia	1.6 ± 0.8	1.8 ± 0.8	1.6 ± 0.8	2.1 ± 0.9	1.6 ± 0.7	0.001
CMBs at lobar areas *	48 (23.4%)	53 (22.9%)	21 (14.6%)	8 (18.2%)	4 (10.8%)	0.120
CMBs at deep areas *	46 (22.4%)	58 (25.1%)	28 (19.4%)	12 (27.3%)	6 (12.2%)	0.549
CMBs at infratentorial areas *	38 (18.5%)	41 (17.7%)	12 (8.3%)	5 (11.4%)	3 (8.1%)	0.034

* CMB data were unavailable in 194 patients. Data were presented as *n* (%), median (interquartile range), or mean ± standard deviation. Abbreviations: CMBs, cerebral microbleeds; END, early neurological deterioration; EPVS, enlarged perivascular space; MCA, middle cerebral artery; MRI, magnetic resonance imaging; mRS, modified Rankin Scale; NIHSS, National Institutes of Health Stroke Scale; WMH, white matter hyperintensity.

Figure 2 presents the percentage of patients with END, BAD, large-artery atherosclerosis, and cardioembolism across different infarct sizes. The incidence of END increased progressively with infarct sizes of 15 mm or larger. Patients with infarct sizes between 15 mm and less than 20 mm had a higher rate of END compared to those with infarct sizes between 10 mm and less than 15 mm (16.5% vs. 7.3%, adjusted $p = 0.02$). The optimal threshold for predicting END was found to be 15.5 mm. The incidence of BAD also rose with increasing infarct size, with significantly higher rates in the following comparisons: less than 10 mm vs. 10 mm to less than 15 mm (15.8% vs. 45.4%, adjusted $p < 0.001$) and 10 mm to less than 15 mm vs. 15 mm to less than 20 mm (45.4% vs. 77.5%, adjusted $p < 0.001$). The optimal threshold for predicting BAD was determined to be 12.1 mm. Patients with infarct sizes of 25 mm or larger had the highest incidence of large-artery atherosclerosis, significantly higher than those with infarct sizes between 10 mm and less than 15 mm (22.4% vs. 7.9%, adjusted $p < 0.04$). The optimal threshold for predicting large-artery atherosclerosis was 14.8 mm. The incidence of cardioembolism did not vary significantly across different infarct sizes.

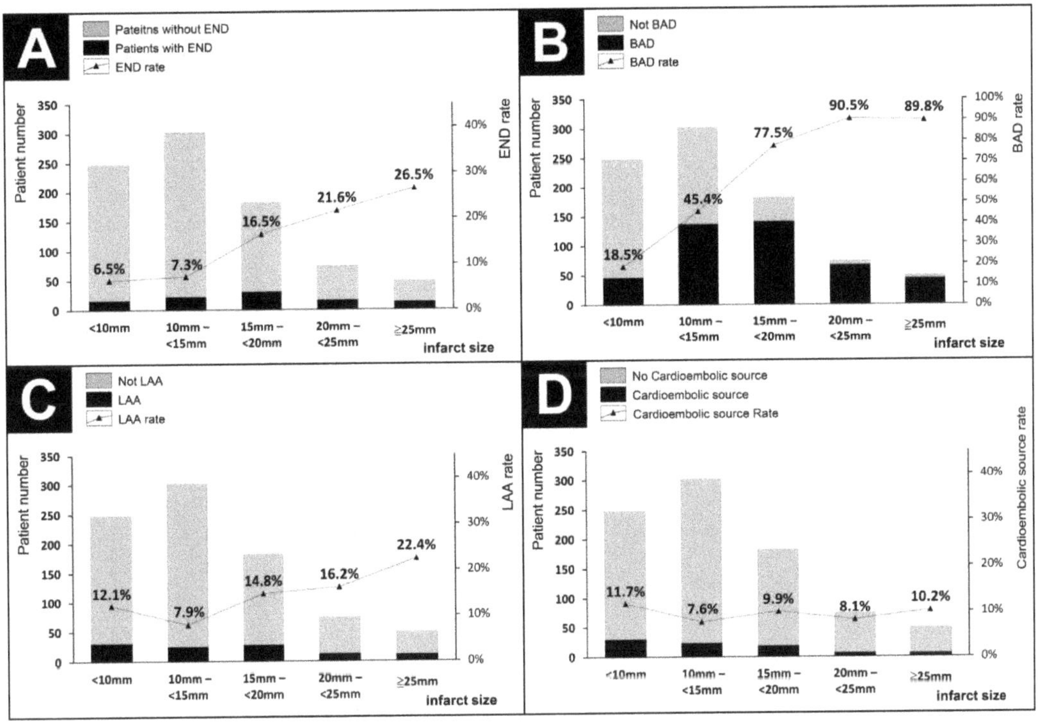

Figure 2. Number and percentage of patients with early neurological deterioration, branch atheromatous disease, large-artery atherosclerosis, and cardioembolism across different infarct sizes. (**A**) Early neurological deterioration (END). (**B**) Branch atheromatous disease (BAD). (**C**) Large-artery atherosclerosis (LAA). (**D**) Cardioembolism.

Table 2 presents the results of the multiple logistic regression analysis, identifying variables significantly associated with END. These include an age over 70 years [adjusted odds ratio (aOR) 2.2; 95% confidence interval (CI) 1.2–4.0], motor tract involvement (aOR 2.3; 95% CI 1.1–4.7), a heart rate above 90 beats per minute (aOR 2.3; 95% CI 1.2–4.3), and a larger infarct size (15 mm to less than 20 mm vs. 10 mm to less than 15 mm; aOR 3.0; 95% CI 1.4–6.3).

Table 2. Logistic regression analysis of predictors for early neurological deterioration in patients with small subcortical infarction.

	Simple Logistic Regression			Multiple Logistic Regression		
	Odds Ratio	95% CI	p	Adjusted Odds Ratio	95% CI	p
Age > 70 years	2.2	1.4–3.4	<0.001	2.2	1.2–4.0	0.008
Female gender	1.4	0.9–2.1	0.133			
Diabetes mellitus	1.3	0.8–1.9	0.311			
Hypertension	1.5	0.8–2.7	0.212			
Hypercholesterolemia	0.8	0.5–1.2	0.221			
Coronary artery disease	1.4	0.6–3.2	0.468			
Previous stroke	1.0	0.6–1.6	0.864			
Atrial fibrillation	1.3	0.6–2.6	0.478			
Congestive heart failure	1.1	0.3–5.0	0.883			
Motor tract involvement	2.2	1.4–3.6	0.001	2.3	1.1–4.7	0.021
Intravenous alteplase	2.0	0.8–4.6	0.124			

Table 2. *Cont.*

	Simple Logistic Regression			Multiple Logistic Regression		
	Odds Ratio	95% CI	*p*	Adjusted Odds Ratio	95% CI	*p*
Sugar > 180 mg/dL	1.8	1.1–2.9	0.028			
Heart rate > 90/min	2.0	1.2–3.2	0.007	2.3	1.2–4.3	0.009
SBP > 140 mmHg	2.3	1.1–4.9	0.031			
DBP > 90 mmHg	0.9	0.6–1.5	0.776			
Parent artery stenosis	1.5	0.9–2.5	0.146			
Diameter *			<0.001			0.023
<10 mm	1.1	0.6–2.2	0.702	0.7	0.3–1.6	0.335
15 mm–<20 mm	2.9	1.5–5.4	0.001	3.0	1.4–6.3	0.004
20 mm–<25 mm	4.0	1.9–8.5	<0.001	2.8	1.0–7.7	0.045
≥25 mm	5.2	2.3–11.8	<0.001	3.9	1.3–11.6	0.015

* With reference to infarct diameter 10 mm–<15 mm. Abbreviations: SBP, systolic blood pressure; DBP, diastolic blood pressure.

Table 3 shows that the multiple logistic regression analysis identified factors significantly associated with a good outcome. These included an age over 70 years (aOR 0.4; 95% CI 0.3–0.5), previous stroke (aOR 0.4; 95% CI 0.3–0.6), atrial fibrillation (aOR 0.6; 95% CI 0.3–1.0), congestive heart failure (aOR 0.2; 95% CI 0.03–0.8), motor tract involvement (aOR 0.6; 95% CI 0.4–0.9), a heart rate above 90 beats per minute (aOR 0.6; 95% CI 0.4–0.9), parent artery stenosis (aOR 0.5; 95% CI 0.3–0.8), and a larger infarct size (10 mm to less than 15 mm vs. less than 10 mm; aOR 0.4; 95% CI 0.3–0.6).

Table 3. Logistic regression analysis of predictors for good functional outcomes with mRS ≤ 1 at 3 months in patients with small subcortical infarction.

	Simple Logistic Regression			Multiple Logistic Regression		
	Odds Ratio	95% CI	*p*	Adjusted Odds Ratio	95% CI	*p*
Age > 70	0.4	0.3–0.6	<0.001	0.4	0.3–0.5	<0.001
Female gender	0.7	0.5–1.0	0.022			
Diabetes mellitus	1.0	0.8–1.3	0.958			
Hypertension	0.7	0.5–1.0	0.068			
Hypercholesterolemia	1.2	0.9–1.6	0.152			
Coronary artery disease	0.6	0.4–1.1	0.131			
Previous stroke	0.5	0.3–0.7	<0.001	0.4	0.3–0.6	<0.001
Atrial fibrillation	0.5	0.3–0.8	0.003	0.6	0.3–1.0	0.047
Congestive heart failure	0.1	0.03–0.5	0.005	0.2	0.03–0.8	0.025
Motor tract involvement	0.5	0.4–0.6	<0.001	0.6	0.4–0.9	0.004
Intravenous alteplase	0.8	0.4–1.6	0.590			
Sugar > 180 mg/dL	0.8	0.6–1.2	0.331			
Heart rate > 90/min	0.7	0.5–1.0	0.078	0.6	0.4–0.9	0.012
SBP > 140 mmHg	0.9	0.6–1.2	0.404			
DBP > 90 mmHg	1.2	0.9–1.5	0.311			
Parent artery stenosis	0.5	0.3–0.7	<0.001	0.5	0.3–0.8	0.003
Diameter *			<0.001			<0.001
10 mm–<15 mm	0.5	0.3–0.7	<0.001	0.4	0.3–0.6	<0.001
15 mm–<20 mm	0.3	0.2–0.5	<0.001	0.3	0.2–0.5	<0.001
20 mm–<25 mm	0.1	0.1–0.3	<0.001	0.2	0.1–0.3	<0.001
≥25 mm	0.2	0.1–0.4	<0.001	0.2	0.1–0.3	<0.001

* With reference to infarct diameter < 10 mm. Abbreviations: SBP, systolic blood pressure; DBP, diastolic blood pressure.

4. Discussion

Our study suggests that the sizes of acute SSIs could indicate distinct underlying pathologies. However, the current definition of 20 mm on DWI as the upper limit for small-vessel occlusion is not entirely satisfactory, as it includes a significant number of cases of BAD. A more practical definition of small-vessel disease would be to set the upper limit to 15 mm, but further high-resolution vessel wall imaging is warranted to refine the definition and diagnosis.

Our study demonstrated that an infarct diameter greater than 15 mm was independently linked to a higher incidence of END and a reduced likelihood of favorable functional outcomes. These findings align with previous research indicating that BAD is strongly associated with END, especially in cases of progressive motor deficits, which often result in poor functional recovery [21]. The suggested threshold for the onset of END is an infarct diameter exceeding 15 mm, while the diagnostic threshold for BAD is set to 12 mm. Aligned with a previous review suggesting 15 mm or larger as the definition of BAD, the majority of SSIs measuring 15 mm or more were associated with BAD in our study [4]. A recent study utilizing high-resolution vessel wall imaging for BAD reported a mean infarct diameter of 16.7 $\pm$ 4.9 mm in patients with visible atherosclerotic plaques in the parent arteries [22]. These findings underscore the necessity of re-evaluating the diameter criterion for small-vessel occlusion, given that the existing 20 mm threshold falls short. Up to now, various treatment combinations have been explored and suggested to mitigate END in BAD [12]. However, there is currently no therapy with established efficacy for BAD. Therefore, achieving a precise diagnosis of BAD distinct from lacunar infarction is crucial for advancing research and determining the most effective treatment strategies.

BAD is primarily caused by atherosclerosis, particularly when large-caliber penetrating arteries become occluded or narrowed at their origins due to microatheromas or junctional plaques [3]. Several studies have associated intracranial atherosclerotic disease with BAD, identifying it by mild stenosis and a distinct infarct pattern that extends to the basal surface of the parent artery [23]. Even with traditional DWI diagnostic criteria, the detection rate of atherosclerotic plaques along parent artery perforators can reach up to 74% when using high-resolution vessel wall imaging [22]. Importantly, the incidence of parent artery stenosis increases significantly when the arterial diameter exceeds 15 mm. This underscores the value of high-resolution vessel wall imaging in cases where the diameter is greater than 15 mm.

A previous study revealed that perfusion defects in SSIs were associated with BAD, as well as with both EPVS and WMH in the basal ganglia, suggesting a shared underlying pathology [20]. In our study, a larger infarct size was correlated with higher EPVS and WMH scores in the basal ganglia. This may be due to the increased arteriopathy burden in the basal ganglia caused by BAD or intracranial atherosclerosis. This finding aligns with previous research that demonstrated an association between intracranial atherosclerosis and larger WMH volumes [24].

An embolic stroke of undetermined source (ESUS) is typically characterized by a non-lacunar brain infarct without significant proximal arterial stenosis or identifiable cardioembolic sources [25]. According to the TOAST classification, lesions with a diameter of $\geq$20 mm are currently considered cryptogenic strokes or ESUSs when neither arterial stenosis nor cardioembolic sources are present [5,26]. However, no significant differences in cardioembolism rates were observed across patient groups with different infarct sizes. Prior studies have suggested that only separated subcortical lesions are associated with cardioembolic strokes rather than larger infarct sizes [27]. Thus, using the ESUS classification for lacunar infarctions with a diameter of $\geq$20 mm may be inappropriate.

The findings of our study suggest that a larger infarction size in acute SSIs is an independent factor that reduces the likelihood of achieving favorable functional outcomes. Additionally, several stroke risk factors—such as previous stroke, atrial fibrillation, and congestive heart failure—also diminished the chances of favorable outcomes, consistent with previous studies [28,29]. These results underscore the importance of the comprehensive evaluation and management of underlying risk factors, even in cases traditionally diagnosed as lacunar infarction. Our study also aligns with earlier research indicating that initial in-hospital heart rate is associated with major adverse cardiovascular events [30].

Although numerous blood biomarkers for stroke have been identified, none have demonstrated adequate sensitivity and specificity for routine clinical use [31,32]. Currently,

MRI is the standard tool in clinical practice for identifying acute ischemic stroke rather than blood biomarkers. However, further research is needed to identify reliable serum biomarkers associated with small-vessel disease and to enhance the diagnosis of underlying pathology by combining these biomarkers with MRI.

There are several limitations to the present study. First, we did not utilize high-resolution vessel walls to evaluate the parent arteries of SSIs, which may lead to an inaccurate diagnosis of BAD. While we propose lowering the threshold for small-vessel occlusion from 20 mm to 15 mm, this adjustment is based on current imaging techniques, and future studies should incorporate high-resolution imaging to refine this threshold. Second, our retrospective single-center study, limited to a Taiwanese population, affects the generalizability of our findings. Genetic and environmental factors, as well as local stroke management practices, may not apply universally. A broader multi-center study is needed to validate these results. Moreover, potential confounding factors, such as comorbidities, medication use, and stroke management, may influence the relationship between infarct size and clinical outcomes. Additionally, infarct size can evolve over time, making a single measurement insufficient to fully capture the dynamic changes. Lastly, the absence of significant differences in cardioembolism rates suggests that the ESUS may not be suitable for large lacunar infarctions. However, we did not thoroughly investigate larger infarcts for potential undiagnosed atrial fibrillation using long-term EKG monitoring, which warrants further studies. Despite these limitations, our findings contribute valuable insights to the existing knowledge regarding the association between infarct morphologies and clinical characteristics in acute SSIs.

5. Conclusions

Our study suggests that the current definition of a 20 mm axial diameter on DWI as the upper limit for small-vessel occlusion is inadequate, given the observed rates of BAD, END, and functional outcomes. A more practical threshold for small-vessel disease would be an upper limit of 15 mm. However, these findings should be interpreted in the context of the retrospective design and study population. Further multi-center research using high-resolution vessel wall imaging is needed to refine this definition and improve diagnostic accuracy.

Author Contributions: Conceptualization, Y.-C.H.; methodology, Y.-C.H. and H.-H.W.; validation, J.-D.L. and L.-C.L.; formal analysis, Y.-C.H., C.-H.C. and Y.-H.T.; investigation, J.-D.L., J.-T.Y. and L.-C.L.; data curation, Y.-C.H.; writing—original draft preparation, Y.-C.H.; writing—review and editing, H.-H.W.; supervision, J.-D.L. and Y.-H.T.; project administration, Y.-C.H. and Y.-H.T. All authors have read and agreed to the published version of the manuscript.

Funding: The study was supported by Chang Gung Memorial Hospital research grants (CORPG6L0151 and CORPG6N0241).

Institutional Review Board Statement: The study was approved by the Institutional Review Board of Chang Gung Memorial Hospital (201802107B0).

Informed Consent Statement: Written informed consent was waived from the institutional review board of Chang Gung Memorial Hospital due to the retrospective design.

Data Availability Statement: All data inquiries can be directed to the corresponding author.

Acknowledgments: We thank Yi-Chen Kuo for assisting us with this study.

Conflicts of Interest: The authors have no conflicts of interest to declare.

References

1. Pantoni, L. Cerebral small vessel disease: From pathogenesis and clinical characteristics to therapeutic challenges. *Lancet Neurol.* **2010**, *9*, 689–701. [CrossRef] [PubMed]
2. Gao, Y.; Li, D.; Lin, J.; Thomas, A.M.; Miao, J.; Chen, D.; Li, S.; Chu, C. Cerebral small vessel disease: Pathological mechanisms and potential therapeutic targets. *Front. Aging Neurosci.* **2022**, *14*, 961661. [CrossRef] [PubMed]

3. Charidimou, A.; Pantoni, L.; Love, S. The concept of sporadic cerebral small vessel disease: A road map on key definitions and current concepts. *Int. J. Stroke* **2016**, *11*, 6–18. [CrossRef]
4. Petrone, L.; Nannoni, S.; Del Bene, A.; Palumbo, V.; Inzitari, D. Branch Atheromatous Disease: A Clinically Meaningful, Yet Unproven Concept. *Cerebrovasc. Dis.* **2016**, *41*, 87–95. [CrossRef]
5. Adams, H.P., Jr.; Bendixen, B.H.; Kappelle, L.J.; Biller, J.; Love, B.B.; Gordon, D.L.; Marsh, E.E., III. Classification of subtype of acute ischemic stroke. Definitions for use in a multicenter clinical trial. TOAST. Trial of Org 10172 in Acute Stroke Treatment. *Stroke* **1993**, *24*, 35–41. [CrossRef] [PubMed]
6. Arsava, E.M.; Ballabio, E.; Benner, T.; Cole, J.W.; Delgado-Martinez, M.P.; Dichgans, M.; Fazekas, F.; Furie, K.L.; Illoh, K.; Jood, K.; et al. The Causative Classification of Stroke system: An international reliability and optimization study. *Neurology* **2010**, *75*, 1277–1284. [CrossRef]
7. Wardlaw, J.M.; Smith, E.E.; Biessels, G.J.; Cordonnier, C.; Fazekas, F.; Frayne, R.; Lindley, R.I.; O'Brien, J.T.; Barkhof, F.; Benavente, O.R.; et al. Neuroimaging standards for research into small vessel disease and its contribution to ageing and neurodegeneration. *Lancet Neurol.* **2013**, *12*, 822–838. [CrossRef] [PubMed]
8. Ryu, D.W.; Shon, Y.M.; Kim, B.S.; Cho, A.H. Conglomerated beads shape of lacunar infarcts on diffusion-weighted MRI: What does it suggest? *Neurology* **2012**, *78*, 1416–1419. [CrossRef]
9. Nah, H.W.; Kang, D.W.; Kwon, S.U.; Kim, J.S. Diversity of Single Small Subcortical Infarctions According to Infarct Location and Parent Artery Disease. *Stroke* **2010**, *41*, 2822–2827. [CrossRef]
10. Zhu, T.; Ren, L.; Zhang, L.; Shao, Y.; Wan, L.; Li, Y.; Liang, D.; Zheng, H.; Liu, X.; Zhang, N. Comparison of plaque characteristics of small and large subcortical infarctions in the middle cerebral artery territory using high-resolution magnetic resonance vessel wall imaging. *Quant. Imaging Med. Surg.* **2021**, *11*, 57–66. [CrossRef]
11. Deguchi, I.; Takahashi, S. Pathophysiology and Optimal Treatment of Intracranial Branch Atheromatous Disease. *J. Atheroscler. Thromb.* **2023**, *30*, 701–709. [CrossRef] [PubMed]
12. Al Kasab, S.; Nguyen, T.N.; Derdeyn, C.P.; Yaghi, S.; Amin-Hanjani, S.; Kicielinski, K.; Zaidat, O.O.; de Havenon, A. Emergent Large Vessel Occlusion due to Intracranial Stenosis: Identification, Management, Challenges, and Future Directions. *Stroke* **2024**, *55*, 355–365. [CrossRef]
13. Gao, S.; Wang, Y.J.; Xu, A.D.; Li, Y.S.; Wang, D.Z. Chinese ischemic stroke subclassification. *Front. Neurol.* **2011**, *2*, 6. [CrossRef]
14. Liao, S.; Deng, Z.; Wang, Y.; Jiang, T.; Kang, Z.; Tan, S.; Shan, Y.; Zou, Y.; Lu, Z. Different Mechanisms of Two Subtypes of Perforating Artery Infarct in the Middle Cerebral Artery Territory, A High-Resolution Magnetic Resonance Imaging Study. *Front. Neurol.* **2018**, *9*, 657. [CrossRef]
15. Lee, T.H.; Chang, C.H.; Chang, Y.J.; Chang, K.C.; Chung, J.; Chang Gung Medical System Stroke Registry Group. Establishment of electronic chart-based stroke registry system in a medical system in Taiwan. *J. Formos. Med. Assoc.* **2011**, *110*, 543–547. [CrossRef] [PubMed]
16. Siegler, J.E.; Martin-Schild, S. Early Neurological Deterioration (END) after stroke: The END depends on the definition. *Int. J. Stroke* **2011**, *6*, 211–212. [CrossRef]
17. Li, S.; Ni, J.; Fan, X.; Yao, M.; Feng, F.; Li, D.; Qu, J.; Zhu, Y.; Zhou, L.; Peng, B. Study protocol of Branch Atheromatous Disease-related stroke (BAD-study): A multicenter prospective cohort study. *BMC Neurol.* **2022**, *22*, 458. [CrossRef] [PubMed]
18. Yamamoto, Y.; Ohara, T.; Hamanaka, M.; Hosomi, A.; Tamura, A.; Akiguchi, I. Characteristics of intracranial branch atheromatous disease and its association with progressive motor deficits. *J. Neurol. Sci.* **2011**, *304*, 78–82. [CrossRef] [PubMed]
19. Ohara, T.; Yamamoto, Y.; Tamura, A.; Ishii, R.; Murai, T. The infarct location predicts progressive motor deficits in patients with acute lacunar infarction in the lenticulostriate artery territory. *J. Neurol. Sci.* **2010**, *293*, 87–91. [CrossRef]
20. Pan, Y.T.; Tsai, Y.H.; Lee, J.D.; Weng, H.H.; Yang, J.T.; Huang, Y.C. Evaluation of clinical relevance and underlying pathology for hemodynamic compromise in acute small subcortical infarction using MRI-based neuroimaging markers. *Biomed. J.* **2023**, *46*, 100529. [CrossRef]
21. Del Bene, A.; Palumbo, V.; Lamassa, M.; Saia, V.; Piccardi, B.; Inzitari, D. Progressive lacunar stroke: Review of mechanisms, prognostic features, and putative treatments. *Int. J. Stroke* **2012**, *7*, 321–329. [CrossRef] [PubMed]
22. Huang, Y.C.; Tsai, Y.H.; Lin, L.C.; Weng, H.H.; JDLee; Yang, J.T. Preliminary results on temporal evolution and clinical implications of atherosclerotic plaque in branch atheromatous disease after statin treatment. *Ther. Adv. Neurol. Disord.* **2024**. [CrossRef]
23. Bang, O.Y. Intracranial Atherosclerosis: Current Understanding and Perspectives. *J. Stroke* **2014**, *16*, 27. [CrossRef] [PubMed]
24. Nam, K.W.; Kwon, H.M.; Jeong, H.Y.; Park, J.H.; Kim, S.H.; Jeong, S.M.; Yoo, T.G.; Kim, S. Cerebral white matter hyperintensity is associated with intracranial atherosclerosis in a healthy population. *Atherosclerosis* **2017**, *265*, 179–183. [CrossRef] [PubMed]
25. Hart, R.G.; Diener, H.C.; Coutts, S.B.; Easton, J.D.; Granger, C.B.; O'Donnell, M.J.; Sacco, R.L.; Connolly, S.J. Embolic strokes of undetermined source: The case for a new clinical construct. *Lancet Neurol.* **2014**, *13*, 429–438. [CrossRef]
26. Chen, P.H.; Gao, S.; Wang, Y.J.; Xu, A.D.; Li, Y.S.; Wang, D. Classifying Ischemic Stroke, from TOAST to CISS. *CNS Neurosci. Ther.* **2012**, *18*, 452–456. [CrossRef]
27. Huang, Y.C.; Lee, J.D.; Lin, L.C.; Weng, H.H.; Yang, J.T.; Tsai, Y.H.; Chen, C.H. Exploring the relationship between lesion morphology and pathogenesis in acute small subcortical infarction. *Cerebrovasc. Dis.* **2023**, 1–8. [CrossRef]
28. Siedler, G.; Sommer, K.; Macha, K.; Marsch, A.; Breuer, L.; Stoll, S.; Engelhorn, T.; Dorfler, A.; Arnold, M.; Schwab, S.; et al. Heart Failure in Ischemic Stroke, Relevance for Acute Care and Outcome. *Stroke* **2019**, *50*, 3051–3056. [CrossRef]

29. Yang, H.J.; Wang, G.J.; Shuai, W.; Shen, C.J.; Kong, B.; Huang, H. The Value of the CHADS2 and CHA2DS2-VASc Score for Predicting the Prognosis in Lacunar Stroke with or without Atrial Fibrillation Patients. *J. Stroke Cerebrovasc. Dis.* **2019**, *28*, 104143. [CrossRef]
30. Lin, C.H.; Zhang, J.F.; Kuo, Y.W.; Kuo, C.F.; Huang, Y.C.; Lee, M.; Lee, J.D. Assessment of the impact of resting heart rate on the risk of major adverse cardiovascular events after ischemic stroke: A retrospective observational study. *BMC Neurol.* **2024**, *24*, 267. [CrossRef]
31. Makris, K.; Haliassos, A.; Chondrogianni, M.; Tsivgoulis, G. Blood biomarkers in ischemic stroke: Potential role and challenges in clinical practice and research. *Crit. Rev. Clin. Lab. Sci.* **2018**, *55*, 294–328. [CrossRef] [PubMed]
32. Kamtchum-Tatuene, J.; Jickling, G.C. Blood biomarkers for stroke diagnosis and management. *Neuromol. Med.* **2019**, *21*, 344–368. [CrossRef] [PubMed]

neurology international

MDPI

Brief Report

Functional Outcomes After Decompressive Surgery in Patients with Malignant Space-Occupying Cerebellar Infarction

Enayatullah Baki [1,*], Lea Baumgart [2], Victoria Kehl [3], Felix Hess [1], Andreas Wolfgang Wolff [1], Arthur Wagner [4], Moritz Roman Hernandez Petzsche [5], Tobias Boeckh-Behrens [5], Bernhard Hemmer [1,6], Bernhard Meyer [4], Jens Gempt [2] and Silke Wunderlich [1]

1 Department of Neurology, Klinikum Rechts der Isar, School of Medicine and Health, Technical University of Munich, 81675 Munich, Germany
2 Department of Neurosurgery, University Medical Center Hamburg-Eppendorf, 20251 Hamburg, Germany
3 Institute of AI and Informatics in Medicine, Klinikum Rechts der Isar, School of Medicine and Health, Technical University of Munich, 81675 Munich, Germany
4 Department of Neurosurgery, Klinikum Rechts der Isar, School of Medicine and Health, Technical University of Munich, 81675 Munich, Germany
5 Department of Diagnostic and Interventional Neuroradiology, Klinikum Rechts der Isar, School of Medicine and Health, Technical University of Munich, 81675 Munich, Germany
6 Munich Cluster for Systems Neurology (SyNergy), 81377 Munich, Germany
* Correspondence: enayatullah.baki@mri.tum.de

Abstract: Background and Purpose: Decompressive surgery is a potentially life-saving treatment in patients with malignant space-occupying cerebellar infarction. However, there is only limited literature on functional outcomes and complications after surgery. Our aim was to establish markers which predict poor outcome. **Methods**: We retrospectively analyzed data of all patients who underwent surgery due to malignant swelling of a space-occupying cerebellar infarction in our hospital between 2005 and 2023. Statistical analyses were conducted on multiple parameters to identify predictors of poor functional outcome (mRS 4–6) 90 days after surgery. Complications during hospitalization were reviewed for each patient. **Results**: In total, 58 patients received decompressive surgery. The 90-day mortality rate was 27.6% (n = 16). A good functional outcome (mRS 0–3) 90 days after surgery was achieved in 24 patients (41.4%). Multivariable analysis revealed multiple factors associated with a poor outcome on day 90 (mRS 4–6): a higher premorbid mRS score (OR 2.715 [95% CI, 1.166–6.323]; $p = 0.021$), higher NIHSS score on admission (OR 1.088 [95% CI, 1.014, 1.168]; $p = 0.019$) and the presence of an additional brainstem infarction (OR 7.035, [95% CI, 1.255, 39.424], $p = 0.027$). Hyperactive delirium was associated with good clinical outcome (OR 0.020 [95%CI, 0.001–0.623]; $p = 0.026$). Aspiration pneumonia (n = 22, 37.9%), urinary tract infection (n = 15, 25.9%), and hyperactive delirium (n = 8, 13.8%) were the most common complications during hospitalization. **Conclusions**: Decompressive surgery is a safe, life-saving treatment for malignant space-occupying cerebellar infarction. Higher premorbid mRS, higher NIHSS score on admission and the presence of brainstem infarction are associated with a poor functional outcome.

Keywords: ischemic stroke; cerebellar infarction; decompressive surgery

1. Introduction

Malignant swelling can occur as a severe complication in space-occupying cerebellar infarction. The compression of the brainstem and the obstruction of the flow of cerebral spinal fluid will become fatal if left untreated [1]. Despite decompressive surgery of the posterior fossa being the only acute life-saving treatment option, there is only a little evidence regarding the clinical benefits. The aim of this study was to analyze the functional outcomes and complications after decompressive surgery in patients with malignant space-occupying cerebellar infarction and to identify predictors of a poor functional outcome.

2. Methods

The study was conducted in accordance with the guidelines of the Declaration of Helsinki. The data that support the findings of this study are available from the corresponding author upon reasonable request.

2.1. Data Sources and Patient Selection

Patients who were admitted to our hospital and underwent decompressive surgery for malignant swelling of space-occupying cerebellar infarctions between 2005 and 2023 were eligible for inclusion in this retrospective study. Malignant swelling was defined as a mass effect in the posterior cranial fossa, accompanied by a decrease in consciousness due to the compression of the brainstem and/or the development of obstructive hydrocephalus. Other potential underlying causes of decreased consciousness were ruled out [2,3].

We retrospectively identified patients who fulfilled the predefined criteria from our stroke quality assurance registry (Bavarian Working Group for Quality Assessment, BAQ).

2.2. Baseline Demographic, Clinical and Radiological Characteristics

Patients' demographic and clinical characteristics included age, sex, premorbid functional status, vascular risk factors, modified Rankin Scale (mRS) and National Institutes of Health Stroke Scale (NIHSS) on admission and stroke subtypes according to the Trial of Org 10172 in Acute Stroke Treatment (TOAST) criteria.

Radiologic parameters, such as infarct volume, brainstem compression, hydrocephalus, side of infarction, vascular territory, basilar artery occlusion and the presence of additional brainstem infarction, were analyzed by experienced neuroradiologists. Infarct volume was measured using the well-known "ABC/2" method [4], in which A is the largest diameter of the selected slice by eye with the largest infarct area, B is the largest diameter perpendicular to the line above and C is the total number of slices of infarct seen multiplied by the slice thickness. The infarct volume was determined on the last cranial CT or MRI conducted before surgery.

Decompressive surgery was further analyzed with regard to the duration of the surgical procedure and the time span between symptom onset, respectively, malignant swelling and surgery. Other therapeutic interventions considered were the insertion of an external ventricular drain (EVD), intravenous thrombolysis with tissue plasminogen activator (tPA, alteplase) and endovascular treatment (EVT) of a basilar artery occlusion.

2.3. Endpoints

The primary endpoints were mortality and functional outcome, measured using the modified Rankin Scale (mRS) 90 days after surgery. Good functional outcome was defined as a score of 0–3 on the modified Rankin Scale and poor functional outcome as a score $\geq$4. To identify predictors of a poor functional outcome, the total cohort was divided into two groups: patients with good outcome and those with a poor outcome at day 90. Complications during hospitalization were evaluated as secondary endpoints for each patient individually.

2.4. Statistical Analysis

All statistical analyses were conducted using IBM SPSS Statistics for Windows (version 28.0.1.1). Descriptive statistics were presented as frequencies with percentages for categorical variables, as means with standard deviations for approximately normally distributed continuous variables and as medians with ranges for ordinal or not normally distributed variables, as appropriate. To determine potential predictors of poor functional outcome, we used univariate and multivariable logistic models and presented the results using the odds ratio (OR), 95% confidence interval (CI) and the p-value of the Wald statistic. Cumulative survival was analyzed using the Kaplan–Meier method. All statistical tests were performed on two sides at the global 5% significance level. Bonferroni–Holm correction was used to adjust the p-values for multiple testing.

3. Results

Our database included 58 patients who received decompressive surgery as an acute treatment for malignant space-occupying cerebellar infarction.

3.1. Survival

Overall, 16 out of 58 patients died within 90 days after surgery, resulting in a 90-day mortality rate of 27.6%. Kaplan–Meier analysis, as illustrated in Figure 1, was used to model the cumulative probability for survival.

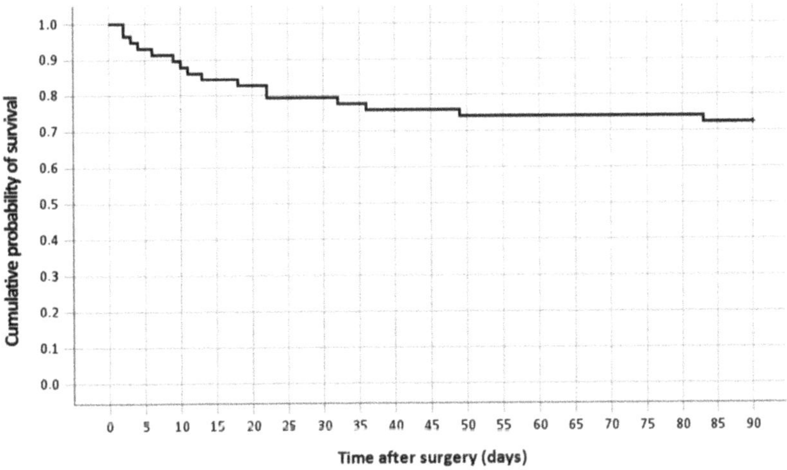

Figure 1. Kaplan–Meier curve showing the cumulative probability for survival in all patients.

3.2. Functional Outcome

Overall, 34 out of 58 patients (58.6%) had a poor outcome at 90 days following decompressive surgery. Accordingly, 24 patients (41.4%) had a good outcome. The detailed distribution of mRS scores (premorbid, on admission and 90 days after surgery) is depicted in Figure 2.

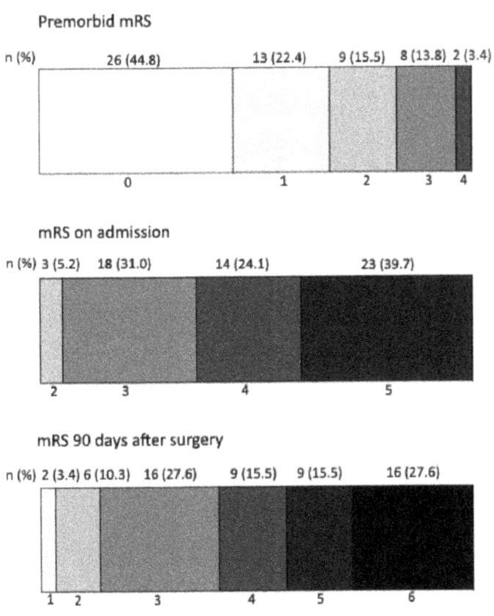

Figure 2. Distribution of mRS at premorbid, on admission and 90 days after surgery.

3.3. Complications

Aspiration pneumonia was the most frequent complication during hospitalization (n = 22, 37.9%), followed by urinary tract infection (n = 15, 25.9%), hyperactive delirium (n = 8, 13.8%), cardiac arrhythmia (n = 7, 12.1%) and ventriculitis (n = 4, 6.9%).

Three patients (5.2%) suffered acute kidney failure and two patients (3.4%) developed meningitis after surgical treatment. Postoperative wound healing disorder with surgical site infection and leak of cerebrospinal fluid was reported in two patients (3.4%). All complications are depicted in Figure 3.

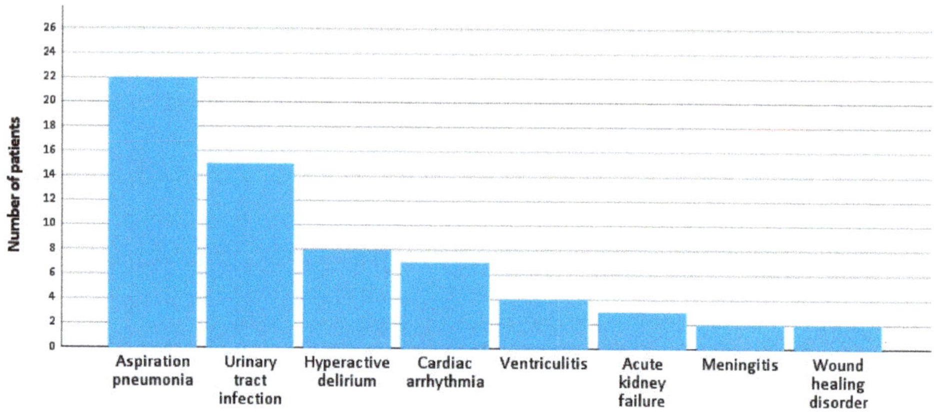

Figure 3. Complications during hospitalization.

3.4. Comparison of Patients with Good and Poor Functional Outcomes

Comparison of demographic, clinical and radiological characteristics are summarized in Table 1. The same table also gives details on the univariate logistic regression analysis.

Table 1. Comparison of demographic, clinical, radiological and therapeutic characteristics between patients with good functional outcome (mRS 0–3) and patients with poor functional outcome (mRS 4–6) at day 90, including results from the univariate logistic regression models with the dependent variable "poor functional outcome" (yes/no); *p*-values of the Wald statistic from the logistic models are given without adjusting for multiplicity.

	Total Cohort (n = 58)	Poor Functional Outcome, mRS 4–6 (n = 34)	Good Functional Outcome, mRS 0–3 (n = 24)	Odds Ratio	95% CI for OR	*p*-Value
Demographic characteristics						
Age (years), *mean (±SD)*	61.2 (±15.4)	60.3 (±17.3)	62.5 (±12.6)	0.991	(0.957, 1.026)	0.594
Female, *n (%)*	21 (36.2%)	15 (44.1%)	6 (25.0%)	2.368	(0.753, 7.446)	0.140
Premorbid status						
Premorbid mRS, *median (range)*	1 (0–4)	1 (0–4)	0 (0–2)	2.011	(1.165, 3.468)	**0.012**
Vascular risk factors, *n (%)*						
Arterial hypertension	38 (65.5%)	20 (58.8%)	18 (75.0%)	0.476	(0.151, 1.502)	0.206
Diabetes mellitus	18 (31.0%)	11 (32.4%)	7 (29.2%)	1.161	(0.373, 3.618)	0.796
Coronary heart disease	15 (25.9%)	10 (29.4%)	5 (20.8%)	1.583	(0.462, 5.421)	0.464
Hypercholesterolemia	18 (31.0%)	9 (26.5%)	9 (37.5%)	0.600	(0.195, 1.846)	0.373

Table 1. *Cont.*

	Total Cohort (n = 58)	Poor Functional Outcome, mRS 4–6 (n = 34)	Good Functional Outcome, mRS 0–3 (n = 24)	Odds Ratio	95% CI for OR	*p*-Value
Acute stroke severity index						
NIHSS on admission, *median (range)*	8 (0–38)	14 (0–38)	5 (0–34)	1.106	(1.035, 1.181)	**0.003**
Cause of infarction according to TOAST criteria, *n (%)*						
Large-artery atherosclerosis	10 (17.2%)	5 (14.7%)	5 (20.8%)	0.655	(0.167, 2.573)	0.545
Cardioembolism	22 (37.9%)	13 (38.2%)	9 (37.5%)	1.032	(0.351, 3.031)	0.955
Small-vessel occlusion	0					
Stroke of other determined etiology	11 (19.0%)	8 (23.5%)	3 (12.5%)	2.154	(0.507, 9.147)	0.298
Stroke of undetermined etiology	15 (25.9%)	8 (23.5%)	7 (29.2%)	0.747	(0.229, 2.442)	0.630
Imaging						
Infarct volume (cm^3), *mean (±SD)*	54.0 (±14.3)	55.5 (±13.0)	51.9 (±15.9)	1.019	(0.981, 1.059)	0.331
Brainstem infarction, *n (%)*	21 (36.2%)	18 (52.9%)	3 (12.5%)	7.875	(1.972, 31.444)	**0.003**
Basilar artery occlusion	10 (17.2%)	8 (23.5%)	2 (8.3%)	3.385	(0.650, 17.628)	0.148
Vascular territory, *n (%)*						
PICA	39 (67.2%)	20 (58.8%)	19 (79.2%)	0.376	(0.113, 1.247)	0.110
SCA	4 (6.9%)	4 (11.8%)	0	*	*	0.999
AICA	0					
Multiple different territories	15 (25.9%)	10 (29.4%)	5 (20.8%)	1.583	(0.462, 5.421)	0.464
Side of the cerebellum						
Bilateral	19 (32.8%)	12 (35.3%)	7 (29.2%)	1.325	(0.429, 4.087)	0.625
Left	19 (32.8%)	10 (29.4%)	9 (37.5%)	0.694	(0.229, 2.102)	0.519
Right	20 (34.5%)	12 (35.3%)	8 (33.3%)	1.091	(0.362, 3.285)	0.877
Therapy						
Operation duration (minutes), *mean (±SD)*	124 (±44.1)	121 (±44.3)	128 (±44.4)	0.996	(0.984, 1.008)	0.528
Time between symptom onset and surgery (days), *median (range)*	2 (1–10)	2 (1–9)	2 (1–10)	0.922	(0.691, 1.231)	0.581
Time between malignant swelling and surgery (hours), *median (range)*	6 (1–34)	7 (1–34)	5 (2–17)	1.068	(0.958, 1.190)	0.234
EVD, *n (%)*	46 (79.3%)	30 (88.2%)	16 (66.7%)	3.750	(0.977, 14.391)	0.054
Intravenous tPA, *n (%)*	9 (15.5%)	4 (11.8%)	5 (20.8%)	0.507	(0.121, 2.128)	0.353
EVT, *n (%)*	9 (15.5%)	7 (20.6%)	2 (8.3%)	2.852	(0.537, 15.138)	0.219
Most common complications						
Aspiration pneumonia	22 (37.9%)	15 (44.1%)	7 (29.2%)	1.917	(0.632, 5.820)	0.251
Urinary tract infection	15 (25.9%)	10 (29.4%)	5 (20.8%)	1.583	(0.462, 5.421)	0.464
Hyperactive delirium	8 (13.8%)	1 (2.9%)	7 (29.2%)	0.074	(0.008, 0.648)	**0.019**

* Not enough data for a meaningful estimate.

The multivariable logistic regression model, shown in Table 2, confirmed multiple variables associated with a poor functional outcome, as follows.

Table 2. Multivariable logistic regression predicting poor functional outcome (mRS 4–6) at day 90.

	Odds Ratio	95% CI for OR	*p*-Value
Premorbid mRS	2.715	(1.166, 6.323)	0.021
NIHSS on admission	1.088	(1.014, 1.168)	0.019
Brainstem infarction	7.035	(1.255, 39.424)	0.027
Hyperactive delirium	0.020	(0.001, 0.623)	0.026

Among clinical characteristics, a higher premorbid mRS score (OR 2.715 [95% CI, 1.166–6.323]; $p = 0.021$) and a higher NIHSS score on admission (OR 1.088 [95% CI, 1.014, 1.168]; $p = 0.019$) were associated with a poor functional outcome.

Among imaging parameters, the presence of an additional brainstem infarction was also associated with a poor functional outcome (OR 7.035, [95% CI, 1.255, 39.424], $p = 0.027$).

After analyzing the most common complications as potential predictors of poor outcome, the occurrence of postoperative hyperactive delirium was associated with a good functional outcome (OR 0.020 [95%CI, 0.001–0.623]; $p = 0.026$).

4. Discussion

Malignant swelling can emerge as a fatal complication of space-occupying cerebellar infarction. A recently published retrospective multicenter study showed that decompressive surgery is beneficial in patients with larger cerebellar infarct volumes [5].

Overall, there is only limited literature evaluating the functional outcomes and complications of this procedure. In this retrospective single-center study, we determined a 90-day mortality rate of 27.6% for 58 patients after decompressive surgery for malignant space-occupying cerebellar infarction. This mortality rate is relatively similar to those calculated in the largest single-center studies on this subject [6–8]. Pfefferkorn et al. reported a 6-month mortality rate of 28% in a cohort of 57 patients [6]. While Jüttler et al. measured a mortality rate at discharge of 25% in 56 evaluated patients [7], Tsitsopoulos et al. stated a mortality rate of 31% after long-term follow-up over five years in a cohort of 32 patients [8].

We used poor functional outcome, defined as a modified Rankin Scale (mRS) score of 4–6 at 90 days after surgery, as the primary endpoint. This definition is valid according to results from a study that used long-term data comparing quality of life across mRS subgroups [9].

Our determined rate of 41.4% of patients who achieved good functional outcome coincides with results stated in the mentioned single-center studies [6–8]. Won et al. reported a favorable functional outcome of 61% at 1-year follow-up in surgically treated patients with cerebellar infarct volumes of 35 cm^3 or greater [5].

Among clinical parameters, higher premorbid mRS and higher NIHSS scores on admission were significantly associated with a poor functional outcome.

Brainstem infarction being the most significant predictor for poor outcome in our analysis corresponds to the results of Pfefferkorn et al. [6]. Nonetheless, it is worth noting that one-seventh of all patients with brainstem infarction still achieved a good outcome. The presence of a brainstem infarction should therefore not solely determine the therapy. We suggest taking multiple individual factors into consideration in the therapy decision-making process.

Aspiration pneumonia, urinary tract infection and delirium were the most common problems during hospitalization in our study population, and are known to be the most frequent complications after acute stroke in general [10]. Unexpectedly, the occurrence of postoperative hyperactive delirium was associated with a good functional outcome. This contradicts the known negative impact of delirium on the prognosis of patients with acute stroke. Previous studies have demonstrated that patients with post-stroke delirium have a worse functional outcome and higher mortality [11,12]. We ascribe this result of our study to the effect of the so-called "survivorship bias", an error that occurs when researchers focus on groups or cases that have passed a certain selection process. Patients in the "poor functional outcome group" were often in such a bad clinical state, i.e., dead within a few days or unconscious, that they were unable to express symptoms of delirium. Thus, delirium remained an underrepresented complication in this group. However, due to the small sample size of only eight cases of reported hyperactive delirium, we emphasize that this result must be treated with caution and warrants verification in a larger prospective study.

The main limitations of this study can be attributed to the retrospective and single-center design.

Our analyses relied on data from medical records that were not originally collected for research purposes. The relatively small cohort size of 58 patients limited the statistical power. Nonetheless, the main advantage of our study is that we analyzed homogeneous data from a real-world clinical practice setting of patients who may have been excluded or may have been underrepresented in prospective studies. Our study exhibits important clinical insights into a subject that has not yet been adequately investigated.

Overall, our study demonstrates that decompressive surgery in patients with malignant space-occupying cerebellar infarction can lead to a good functional outcome, especially in cases of low premorbid mRS and low NIHSS score on admission, as well as the absence of a brainstem infarction. While our study's key results mainly coincide with results from previous studies, we exclusively considered all relevant clinical and radiological parameters in our study protocol, thus designing a potential blueprint for future prospective multicenter studies on this issue. Larger trials are required regarding this subject.

Author Contributions: E.B. as guarantor, takes full responsibility for the entire work. E.B., A.W.W., B.M., J.G. and S.W. developed the theory and designed the study. E.B., L.B. and A.W. collected and processed the clinical patient data. M.R.H.P. and T.B.-B. analyzed imaging data. V.K. performed the statistical analyses and created the figures. E.B. wrote the manuscript with support from V.K., F.H., B.H. and S.W. All authors discussed the results and contributed to the final manuscript. All authors have read and agreed to the published version of the manuscript.

Funding: This research received no external funding.

Institutional Review Board Statement: Ethical approval was granted by the Ethics Committee of the Technical University of Munich, ID: 2024-211-S-CB.

Conflicts of Interest: The authors declare no conflicts of interest.

Abbreviations

CT	computed tomography
EVD	external ventricular drain
EVT	endovascular treatment
MRI	magnetic resonance imaging
mRS	modified Rankin Scale
NIHSS	National Institutes of Health Stroke Scale
TOAST	Trial of Org 10172 in Acute Stroke Treatment
tPA	tissue plasminogen activator

References

1. Edlow, J.A.; Newman-Toker, D.E.; Savitz, S.I. Diagnosis and initial management of cerebellar infarction. *Lancet Neurol.* **2008**, *7*, 951–964. [CrossRef] [PubMed]
2. Dashyan, V.G.; Khodykin, E.A.; Nikitin, A.S.; Godkov, I.M.; Khovrin, D.V.; Sosnovsky, E.A.; Asratyan, S.A.; Sytnik, A.V.; Ochkin, S.S.; Akhmedzhanova, N.R. Zlokachestvennyĭ infarkt mozzhechka: Klinicheskoe techenie i khirurgicheskoe lechenie [Malignant cerebellar infarction: Clinical course and surgical treatment]. *S.S. Korsakov J. Neurol. Psychiatry* **2019**, *119*, 75–83. (In Russian) [CrossRef] [PubMed]

3. Baki, E.; Baumgart, L.; Kehl, V.; Hess, F.; Wolff, A.W.; Wagner, A.; Hernandez Petzsche, M.R.; Boeckh-Behrens, T.; Hemmer, B.; Meyer, B.; et al. Predictors of malignant swelling in space-occupying cerebellar infarction. *Stroke Vasc. Neurol.* **2024**, svn-2024-003360. [CrossRef] [PubMed]
4. Sims, J.R.; Gharai, L.R.; Schaefer, P.W.; Vangel, M.; Rosenthal, E.S.; Lev, M.H.; Schwamm, L.H. ABC/2 for rapid clinical estimate of infarct, perfusion, and mismatch volumes. *Neurology* **2009**, *72*, 2104–2110. [CrossRef] [PubMed] [PubMed Central]
5. Won, S.Y.; Hernández-Durán, S.; Behmanesh, B.; Bernstock, J.D.; Czabanka, M.; Dinc, N.; Dubinski, D.; Freiman, T.M.; Günther, A.; Hellmuth, K.; et al. Functional Outcomes in Conservatively vs Surgically Treated Cerebellar Infarcts. *JAMA Neurol.* **2024**, *81*, 384–393. [CrossRef] [PubMed] [PubMed Central]
6. Pfefferkorn, T.; Eppinger, U.; Linn, J.; Birnbaum, T.; Herzog, J.; Straube, A.; Dichgans, M.; Grau, S. Long-term outcome after suboccipital decompressive craniectomy for malignant cerebellar infarction. *Stroke* **2009**, *40*, 3045–3050. [CrossRef] [PubMed]
7. Jüttler, E.; Schweickert, S.; Ringleb, P.A.; Huttner, H.B.; Köhrmann, M.; Aschoff, A. Long-term outcome after surgical treatment for space-occupying cerebellar infarction: Experience in 56 patients. *Stroke* **2009**, *40*, 3060–3066. [CrossRef] [PubMed]
8. Tsitsopoulos, P.P.; Tobieson, L.; Enblad, P.; Marklund, N. Surgical treatment of patients with unilateral cerebellar infarcts: Clinical outcome and prognostic factors. *Acta Neurochir.* **2011**, *153*, 2075–2083. [CrossRef] [PubMed]
9. Rangaraju, S.; Haussen, D.; Nogueira, R.G.; Nahab, F.; Frankel, M. Comparison of 3-Month Stroke Disability and Quality of Life across Modified Rankin Scale Categories. *Interv. Neurol.* **2017**, *6*, 36–41. [CrossRef] [PubMed] [PubMed Central]
10. Johnston, K.C.; Li, J.Y.; Lyden, P.D.; Hanson, S.K.; Feasby, T.E.; Adams, R.J.; Faught, R.E., Jr.; Haley, E.C., Jr.; RANTTAS Investigators. Medical and neurological complications of ischemic stroke: Experience from the RANTTAS trial. *Stroke* **1998**, *29*, 447–453. [CrossRef] [PubMed]
11. Rollo, E.; Brunetti, V.; Scala, I.; Callea, A.; Marotta, J.; Vollono, C.; Frisullo, G.; Broccolini, A.; Calabresi, P.; Della Marca, G. Impact of delirium on the outcome of stroke: A prospective, observational, cohort study. *J. Neurol.* **2022**, *269*, 6467–6475. [CrossRef] [PubMed] [PubMed Central]
12. Oldenbeuving, A.W.; de Kort, P.L.; Jansen, B.P.; Algra, A.; Kappelle, L.J.; Roks, G. Delirium in the acute phase after stroke: Incidence, risk factors, and outcome. *Neurology* **2011**, *76*, 993–999. [CrossRef] [PubMed]

neurology international

MDPI

Article

Evaluating the Usability of a Remote Ischemic Conditioning Device for Pre-Hospital Stroke Management: Insights from Paramedic Simulations

Bogna Anna Drozdowska [1,*], Kaden Lam [1,2], Cody Doolan [1], Efrem Violato [3] and Aravind Ganesh [1]

1 Department of Clinical Neurosciences, Cumming School of Medicine, University of Calgary, Calgary, AB T2N 2T9, Canada; kaden.lam@ucalgary.ca (K.L.); cody.doolan@ucalgary.ca (C.D.); aganesh@ucalgary.ca (A.G.)
2 Graduate Science Education, Cumming School of Medicine, University of Calgary, Calgary, AB T2N 4N1, Canada
3 Centre for Advanced Medical Simulation, Northern Alberta Institute of Technology, Edmonton, AB T5G 2R1, Canada; efremv@nait.ca
* Correspondence: bognaanna.drozdowska@ucalgary.ca

Abstract: Background/Objectives: In acute stroke, often-prolonged hospital transport times present an opportunity for early interventions to salvage brain tissue. Remote ischemic conditioning (RIC), where brief cycles of ischemia–reperfusion in a limb are induced to protect the brain, is a promising treatment for this setting. We assessed the usability of a novel RIC system in a simulated emergency response scenario. Methods: Paramedics were asked to use the RIC device in an emergency stroke care and ambulance transport simulation, overseen by a confederate. Feedback on device use was collected through questionnaires, including the System Usability Scale (SUS) and the NASA Task Load Index (NASA-TLX), and a semi-structured interview. Questionnaire responses were summarized using descriptive statistics; interview transcripts were analyzed thematically. Results: Nine paramedics (including the confederate) participated, with a mean of 10.0 $\pm$ 10.3 years of professional experience. Questionnaire responses indicated high device usability (mean SUS score: 85.3 $\pm$ 12.9 out of 100) and low task-related demands, effort, and frustration (mean NASA-TLX domain scores: $\leq$3.9 out of 20). Seven paramedics stated they would use the device in daily practice. They expressed concerns related to display screen clarity, interference with standard procedures, cable management, device fragility, and patient discomfort. Suggested improvements included adding indicators of device performance and refining the cuff design. Conclusions: While the device was considered easy to use, paramedics also identified important areas of improvement. With a small, localized study sample, our findings are primarily applicable to the refinement of the RICovery system for use in future clinical trials in the same healthcare setting. However, feedback on the importance of mitigating potential interference of newly introduced procedures with those already established, robustness of equipment, and effective paramedic–patient communication may also help inform the design of other pre-hospital interventions.

Keywords: remote ischemic conditioning; stroke; neuroprotection; pre-hospital care; usability testing; ambulance; emergency medical services; medical device; paramedic; simulation

1. Introduction

In acute stroke, "time is brain" [1]. As stroke progresses, neural tissue is rapidly and irreversibly lost, rendering the effectiveness of acute treatments highly dependent on their timely administration. To meet the need for urgent medical care in the interest of improving patient prognosis, pre-hospital "code stroke" protocols have been widely endorsed, comprising the highest priority dispatch of emergency medical services (EMSs), pre-hospital notification, and rapid transfer to the closest stroke-ready center [2,3]. As supported by current evidence, implementation of "code stroke" systems is associated with shorter onset-to-admission times and door-to-needle times, as well as a higher rate of reperfusion treatment administration [3–6]. However, the beneficial impact of protocol improvements is contingent on the availability of resources and proximity to stroke centers.

Many patients continue to incur pre-hospital delays, particularly in countries like Australia and Canada, where due to geographic realities hospital transport may take an hour or more [7–9]. To address this issue, there is currently only one form of pre-hospital treatment available, which is providing thrombolysis—intravenous administration of a clot-dissolving drug—in mobile stroke units (MSUs) [10,11]. However, widespread implementation of MSUs is hindered by costs and local regulatory barriers [12]. Moreover, in the pre-hospital setting, there are no alternative acute treatments available for patients who are ineligible for thrombolysis.

In this context, remote ischemic conditioning (RIC) presents as a promising, inexpensive, non-invasive and viably administered treatment option [13–15]. It involves inducing brief cycles of ischemia–reperfusion through inflation and deflation of a cuff around a limb to protect a remote, vital organ, including the brain [14,16]. Although the mechanisms underlying the neuroprotective effect of RIC are not yet fully understood, animal model studies indicate the activation of humoral, neural, and genetic and inflammatory pathways, in turn promoting cell survival and repair, while inhibiting apoptosis and inflammation [16]. Randomized controlled trials (RCTs) in acute stroke have consistently demonstrated the safety of RIC [13,17,18], with some of the available evidence pertaining to hemorrhagic events. These findings support the potential for early RIC treatment initiation, prior to obtaining results of brain imaging to differentiate between stroke types.

To date, however, research has predominantly focused on RIC use in ischemic stroke, providing some encouraging findings regarding its effect on patient outcomes. Among these is the large, multi-center RICAMIS trial, where compared to usual care, RIC repeatedly administered during a hospital stay was reported to significantly increase the likelihood of excellent functional outcomes [19]. While providing a specific example of RIC use in a pre-hospital setting, a Danish proof-of-concept trial found that after adjustment for baseline perfusion and diffusion lesion severity in a post hoc analysis, RIC delivered during ambulance transport lowered the tissue risk of infarction [20]. However, no significant effect was observed for the primary endpoint of penumbral salvage. Building on this study, the RESIST trial involved initiating RIC in the ambulance and repeating treatment sessions over seven days [18,21]. Significant improvement in functional status with RIC use was revealed in a post hoc, sub-group analysis of patients with stroke due to cerebral small vessel disease and good treatment adherence, yet not in the primary analysis involving all participant data. These mixed results of past studies are met with continuing research interest in understanding the true effects of RIC in a pre-hospital setting, as exemplified by the recently conducted REMOTE-CAT trial [22].

Joining the pursuit to advance investigations into RIC application through a clinical trial program [23], our team recognized that efforts to gain high-quality evidence on this treatment may be compromised by important limitations of existing RIC devices. Identified concerns included (i) a lack of technical support; (ii) the inability to remotely track treatment quality or patient compliance during an intervention, and in turn to facilitate it, as devices needed to be shipped to the lead site for manual upload of data; and (iii) requiring separate sham units, which significantly increases the number of devices needed for a controlled trial. The latter also interferes with blinding; for instance, in trials like RESIST that have used sham comparators, the sham device has been visibly different (e.g., different color). With a view to address these limitations, our team collaborated with an industry partner to develop a novel RIC device, named "RICovery" (patent application: US 17/317771), seeking input from practicing stroke neurologists and persons with lived experience.

Looking ahead to the potential future adoption of RIC treatment, it is relevant to point out that there are no published data on formally involving patients and healthcare professionals in the development or refinement of existing devices. One of the primary purposes of medical technologies is to meet user needs, and incorporating user evaluations in the design process is recognized as critical to ensuring good usability of a new product in the intended context [24,25]. This translates to increased productivity, safety, user acceptance and satisfaction with introduced technologies, and a reduction in use-related errors and need for user training and support [25–27].

For usability assessments to help achieve such a development outcome, it is necessary to have real users of a device perform real-life tasks, and record what participants do and say, to then address any uncovered issues [28]. In view of this, the present study aimed to gain insights on the usability of our novel, miniaturized RIC system through its use by paramedics in a simulated emergency stroke care and transport scenario. Received feedback was intended to inform the refinement of the RICovery system and relevant operational procedures in preparation for planned feasibility and efficacy trials.

2. Materials and Methods

2.1. Study Design, Setting, and Participants

We performed a simulated use study at the Center for Advanced Medical Simulation (CAMS) at the Northern Alberta Institute of Technology (NAIT) in Edmonton, Canada. CAMS is an interactive and interdisciplinary clinical simulation facility that replicates real-world scenarios in clinical, community and home care environments. Testing was conducted over one day in October 2023. The study was approved by the NAIT Research Ethics Board (NAIT REB# 2023-12; approval date: 4 October 2023).

Potential participants were identified from a list of healthcare professionals, interested in participating in research, maintained by NAIT. To gather data from representative end-users, participants of varying knowledge and experience levels, including rural and urban experience, were targeted for recruitment, including practicum-level Primary Care Paramedic (PCP) students, practicing PCPs, and practicing Advanced Care Paramedics (ACPs). Within the practicing PCP and ACP groups, a range of years of experience were sought for inclusion. Participants were invited to participate via email. The sample size for this study was informed by usability consulting project insights, published by the Nielsen Norman Group (experts in usability heuristics) [29,30]. These projects have indicated that a sample of five to eight users provides the optimal return on investment in terms of actionable usability findings discovered to expense incurred. Therefore, we aimed to include eight paramedic users in this study.

All participants gave written, informed consent. Prior to the simulation, paramedics were provided with a standardized briefing on operating the RIC device, including cuff placement and activation of the unit. We intentionally provided minimal information regarding the device to allow an understanding of the ease and intuitiveness of its use. We further provided participants with an overview of what the simulation involves and their role (see Appendix A).

2.2. The RICovery Device Prototype

The RICovery device was developed by SnapDx Inc. and manufactured by FUEL Biomedical, both Calgary-based companies. The tested RICovery prototype, measuring 151 mm × 56 mm × 33 mm, is presented in Figure 1. Its key components include an activation unit with a display screen, a cuff, and a connecting hose. The device allows for delivering standard RIC, consisting of repetitive cycles of cuff inflation to 200 mmHg for 5 min, then deflation for 5 min, for at least four cycles. When in use, the current cycle is indicated on the display screen. As an additional feature, the cuff inflation pressures can be programmed to lower values, so as to suit the tolerance thresholds of individual patients. Treatment records are securely and remotely transmitted through a central hub, allowing quality control and compliance monitoring. Devices are also remotely programmable for "blind" randomization as true or sham RIC. The device is charged through a standard USB cable.

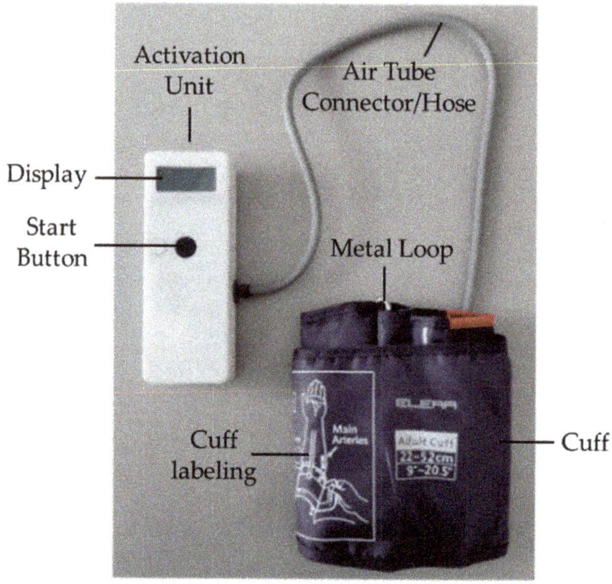

Figure 1. The tested RICovery device prototype.

2.3. Simulation Scenario

The simulated scenario, detailed in Appendix A, required the transport of a stroke patient by ambulance to a hospital. Participants were randomly assigned to one of two conditions—one with a short (22 min) transport time and one with a long (45 min) transport time. Other than duration, the two scenarios were identical. As paramedics work in teams, a confederate, independent of the research group, accompanied participants during each simulation to maintain fidelity. The confederate did not participate in the use of the RIC device nor did they provide additional instructions regarding device operation, and only acted in response to participant requests. During the simulation, the device was kept in the paramedics' primary kit. Participants were free to use it in whatever manner they considered appropriate.

A description of the situation was provided to participants at the beginning of each simulation. Briefly, they were dispatched to a stroke call for an elderly patient—an iStan manikin—who was found lying on the floor. With the assistance of the confederate, they were required to transfer the patient to an ambulance simulation box—a reproduction of an actual ambulance, built on a motion platform to replicate the sensation of driving on various road conditions. Participants were alone while providing patient care in the ambulance; however, they could communicate with the confederate over a radio transceiver. Each simulation ended at the allotted time, with a researcher informing the participant they had arrived at the hospital and requesting them to remove the RIC cuff from the patient.

2.4. Data Collection and Analysis

All simulations and interviews were audio–video recorded. We used multiple evaluation measures, including observational data, questionnaires, and semi-structured interviews. The confederate contributed to the two latter forms of data collection, separately from other participants, once all simulations had been completed. The confederate data were included as we recognized the unique value of insights gained from involvement in every simulation. For quantitative data collection, we digitally administered the following assessments using the Qualtrics survey software (Qualtrics, Provo UT):

1. The System Usability Scale (SUS)—a standardized 10-item questionnaire that provides a quantitative measure of the usability of a system or technology; each item is rated

on a scale from 1 (strongly disagree) to 5 (strongly agree), with responses producing a composite score from 0 to 100 (highest degree of usability);

2. The NASA Task Load Index (NASA-TLX)—a subjective rating scale that measures the perceived demands (mental, physical, and temporal), effort, and frustration experienced during a task, and successfulness in accomplishing it; each item is rated on a scale from 0 (very low) to 20 (very high);
3. A bespoke Device Validation Checklist—a structured instrument assessing seven specific components of the RIC device, rated on a 3-point scale—not acceptable, acceptable but improvements could be made, and acceptable; open text entry boxes were included to allow for further clarification;
4. Additional custom assessment items—five items developed based on the Post-Task Single Ease Questions [31], asking about different aspects of ease of device use on a scale from 1 (very difficult) to 5 (very easy), and two binary (yes/no) questions on liking the device and willingness to use it in daily work.

Observations of participant behavior during simulations and semi-structured interviews were conducted by one researcher, EV. The interview guide included questions regarding general impressions about the RIC device and its look and feel, the experience of using it during the scenario, any encountered challenges or difficulties, anticipated use-related risks, views on potentially using multiple devices, and thoughts about the provided training.

Quantitative data from survey responses were analyzed using descriptive statistics. The scenarios were watched in real time and multiple reviews of the simulation recordings were conducted at a later date. Interview recordings were transcribed verbatim and integrated with notes from observations. The qualitative data were analyzed thematically by BAD, KL and EV. This iterative process involved reading and re-reading transcripts, coding excerpts to capture concepts found in the data, and grouping related codes to develop themes and subthemes.

3. Results

Nine paramedics took part in the study, including four providing patient care in the long transport scenario, four in the short scenario, and one consistently serving as the confederate through all scenarios. Paramedic characteristics are summarized in Table 1, including the confederate, who was a 42-year-old male with 12 years of primarily rural professional experience.

Table 1. Participant characteristics (N = 9).

Age, mean (SD; range)	35.6 (11.3; 21–57)
Sex, N (%)	
Female	5 (55.6)
Male	4 (44.4)
Certification, N (%)	
Primary Care Paramedic	5 (55.6)
Advanced Care Paramedic	4 (44.4)
Years of professional experience, mean (SD; range)	10.0 (10.3; 0–32)
Primary Area of Work, N (%)	
Urban	5 (55.6)
Rural with or without urban	4 (44.4)
Prior usability testing experience, N(%)	
Of any medical device	2 (22.2)
Of an RIC device	2 (22.2)

3.1. Questionnaire Findings

The results of the SUS and NASA-TLX questionnaires are reported in Table 2. For items included in the Device Validation Checklist, there was no instance of a "not acceptable" response. Of the 63 total responses (7 components × 9 paramedics), "acceptable" was selected 85.7% of the time. The specific components that were indicated as "acceptable but improvements could made" included Attaching the therapy cuff to the limb (N = 4), Cuff inflation cycle (N = 1), Cuff holding at pressure (N = 1), Rest at ambient pressure (N = 1), and Removing the therapy cuff (N = 2). The two main issues mentioned by participants in the open text comments related to the use of the metal loop for placing the cuff and a lack of clarity regarding the current inflation of the cuff, including uncertainty as to whether it was in the inflation or deflation part of the cycle, and no indication of the exact pressure. The confederate, moreover, noted that accidental, premature pressing of the start button resulted in having to detach the cuff from the unit to deflate the cuff, before it could be placed on the patient's arm.

Table 2. Descriptive statistics of responses from the SUS and NASA-TLX questionnaires.

	Mean (SD) (N = 9)	Range (N = 9)
SUS	85.3 (12.0)	67.5–97.5
Would Use Frequently	4.0 (1.5)	2–5
Found Device Complex	1.2 (0.4)	1–2
Device Was Easy to Use	4.9 (0.3)	4–5
Would need Support from Technical Person	1.1 (0.3)	1–2
Functions Were Well Integrated	3.8 (1.4)	1–5
Device was Inconsistent	2.2 (0.8)	1–4
Most People would Learn to Use Device Quickly	5.0 (0.0)	5–5
Found Device Cumbersome	1.3 (0.7)	1–3
Felt Confident Using Device	3.9 (1.3)	1–5
Needed to Learn a Lot Before Using Device	1.3 (0.5)	1–2
NASA-TLX		
Mental Demand	0.9 (0.8)	0–2
Physical Demand	1.0 (0.9)	0–2
Temporal Demand	3.6 (5.3)	0–15
Effort	0.9 (1.3)	0–4
Frustration	3.9 (6.3)	0–20
Performance	13.4 (8.0)	0–20

System Usability Scale (SUS); NASA Task Load Index (NASA-TLX).

Responses from the five custom assessment items that addressed ease of understanding and using the RIC device are summarized in Figure 2. All paramedics found that understanding how to activate the device was very easy. The greatest range of responses was recorded in relation to ease of seeing if the device was working, with one participant and the confederate scoring 2/5 (where 5 is "very easy"), two participants scoring 3/5, two participants scoring 4/5, and three participants scoring 5/5. Finally, seven paramedics were "somewhat confident" or "very confident" in the functioning of the RIC device, indicated that they liked using the device, and would use it in their daily practice if it was available to them. Individual participant data for all questionnaires are presented in the Supplementary Materials (Tables S1–S4).

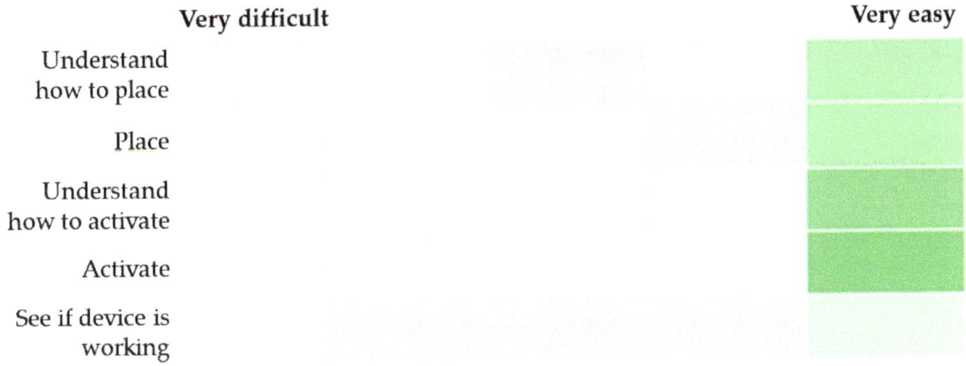

Figure 2. Heat map of responses for ease of understanding and using the RIC device. Note: Darker shade of green indicates a higher frequency of responses for a given scale point.

3.2. *Observation and Interview Findings*

Four main themes emerged from the qualitative data analysis. The first theme captured opinions and recommendations regarding the RIC device's design and way of operating. Here, our findings largely reflected the results of the questionnaires. Paramedics consistently emphasized overall ease of device use. At the same time, however, some participants indicated having difficulties placing the cuff due to the incorporated metal loop, as well as pointing to insufficient clarity of the display screen, causing uncertainty as to whether the device had been working properly. Other concerns related to the device being too fragile to withstand the 'rough' handling that is inherent to emergency response situations, and a particularly high risk of the hose becoming detached or damaged. As a solution to a number of identified issues, participants suggested enabling attachment of the activation unit to the cuff. All contributing subthemes are presented in Table 3, alongside a brief description of their key aspects, associated recommendations for improvement, and exemplar quotes.

Table 3. Detailed overview of Theme 1: RIC device design and operating.

Subtheme	Elucidation/Key Points	Related Recommendations	Exemplar Quotes
Ease of device use	Opinion that the RIC device was easy to operate and/or that providing the treatment was a low-demand task	Not applicable	*[...] in terms of actually ease of use—no issues at all. It's basically the same as a blood pressure cuff. So, we do that every day. There is literally one button to press, so it's mental load is very low. Um, I didn't feel like it interfered with my treatments.* (P1_800) *[...] it's very simple to use. It's not something I had to continually pay attention to, which I liked.* (P7_1330)
Device reliability and monitoring	Statements and observations indicating participants' uncertainty as to whether the device was functioning as intended.	Improving the clarity of the device display, including: a larger screen; use of color indicator lights (e.g., green when working correctly, red if error occurs); providing information on the cuff's inflation status.	*It was hard to see the screen. It was the only thing that I found difficult and then there wasn't anything, like, I think throughout it might have worked a couple times, but then it just kind of stopped working, so, and then I wasn't sure if it was gonna, like, supposed to work or not. I don't know.* (P2_830)
Cuff placement challenges due to metal loop	Statements and observations indicating issues with placing the cuff due to the incorporated metal loop. The loop reduced users' freedom of movement and in some situations introduced a challenge of having to unthread and rethread the cuff.	Switching to a single contiguous Velcro cuff, as is currently used by paramedics for blood pressure monitoring.	*I wish the cuff was like the ones we have on blood pressures that [...] you don't have to run the straps through a loop to go back around. Just cause then like, I had already put the IV on, I put the lead on for my ECG and then I had to, I went to just slide the cuff over, I went, 'oh, I can't do that', so I had to open the cuff, put it on, run it through that [loop].* (P8_1400)

Table 3. *Cont.*

Subtheme	Elucidation/Key Points	Related Recommendations	Exemplar Quotes
Activation unit placement	Statements and observations indicating variability and uncertainty regarding placement of the activation unit in the ambulance so as to avoid any damage to the device. With limited free surface area, some participants placed the unit on the patient, yet considered this as potentially disrespectful and disagreeable to the patient.	Enabling attachment of the activation unit to the cuff or integrating the two components. Making the device hose longer.	*I wish the cord was a little longer maybe, just cause, I had to set it on the patient or beside. Like, I feel like in the real world I wouldn't want to set it on the patient, I guess I could, but probably want to set it bedside, so the longer cord might be helpful just so it's not hanging off of them.* (P8_1400)
Cable management	Identifying the risk of the RIC device hose becoming detached from the activation unit, pinched, or severed. Monitoring for this was considered an additional attentional demand alongside watchfulness of other standard equipment cables.	Enabling attachment of the activation unit to the cuff or integrating the two components. Making the device hose shorter.	*I don't know if that would be better. It's a lot of cables running and, um, cable management is important to me, but I know, especially when I was new, My cables were all over the place and then you're getting them caught in wheels, you're getting them caught when you're lifting and lowering, um, they fall, they become, they unhook, those kind of things.* (P3_930)
Device robustness	Concerns regarding the robustness of the device. Given the circumstances in which paramedics work, accidents involving the device were viewed as inevitable.	Constructing the activation unit casing of more durable materials, making it more resistant to damage and preferably waterproof.	*We tend to drop things in puddles, in the snow, or lose them, leave them on scenes. I know, they fall out of a bag or something like that.* (P3_930)
Unintentional device activation	Identifying the risk of unintentionally pressing the activation button.	Turning the device on and off should require a longer press of the button. Adding a protective cover over the activation button.	*[. . .] that on-off button is quite sensitive, so it's easy to accidentally bump it. And then it's on, but then you turn it off by accident, as it's, you're trying to move the patient, and it's kind of just getting in the way.* (Confederate)
Device labeling	Overall opinion that the labeling on the device was simple and easy to understand, although certain issues were identified, including a nonobvious indication of the direction in which the cuff should be applied and potential to confuse the RIC device with a blood pressure monitor.	Adding a "This side to patient" label on the cuff. Adding an apparent label or distinct coloring so the device would not be confused with a blood pressure monitor. Adding step-by-step instructions on how to apply and operate the device on the back of the activation unit.	*And you see that on a blood pressure cuff as well too, it says this side to patient, and it shows like, how to put the cuff on, written on the cuff. . .* (Confederate)
Device size and weight	Overall opinion is that the device is advantageously small and light, supporting ease of transfer and storage. Some participants suggested a slightly higher weight could be beneficial, assuming an association with improved durability and being treated with more care and respect by users.	Potentially making the device slightly heavier.	*[. . .] the size is good. It would fit in lots of different kits. I was able to fit in my cargo pocket.* (Confederate) *I wasn't sure if I was going to break it if it dropped. I'm not sure if it's fragile or not.* (P5_1200)
Cuff sizes	Recognizing that the one provided cuff size may not be suitable for all patients, particularly those who are obese.	Providing multiple, different sized cuffs with the device.	*[. . .] I feel like if someone is quite overweight or edematous that's going to make it harder to put it on the lower extremity because you'll need like a bigger cuff if that's possible too to have a variety of cuff sizes.* (P1_800)

The second theme, as presented in Table 4, related to views on incorporating use of the RIC device into established workflows, including considerations on device storage, interference with standard procedures, and point of treatment initiation. Participants expressed their desire for specific guidelines on RIC use, with the absence of relevant instructions diminishing their confidence in making decisions. In determining when to begin the intervention, an assumption that RIC treatment is time-sensitive led to placing and activating the device immediately, on scene, as was the case for two paramedics. Most participants, however, initiated RIC in the ambulance—a decision reached either through applying general, established rules of conduct, where other procedures are prioritized over non-life-saving treatment, or due to practical considerations.

Table 4. Detailed overview of Theme 2: Implementation of RIC into emergency response conditions and workflow.

Subtheme	Elucidation/Key Points	Related Recommendations	Exemplar Quotes
Optimal storage	Opinions regarding where the device should be stored when not in use, which related to defining the timepoint of treatment initiation. Suggestions included storage in theprimary kit for initiating treatment on scene/as soon as possible or in the monitor case for initiating treatment in the ambulance.	Not applicable	*I'd probably keep it just down, I wouldn't put it on top of my kit, because I feel like a stroke patient, it's not something I go to every day, right, so I'd put it probably somewhere where I know where it is, but I'd have to maybe pull, maybe under the IV kit or something, just so it's, I know where it is quickly to grab it, but it's not like the first thing I'm going to grab, right.* (P8 _1400)
Decision-making around RIC initiation	Lack of specific guidelines related to uncertainty in participants' decision-making. Two users applied the device on-scene and six in the ambulance. Factors considered in making this decision included the following: Assumed time-sensitiveness of treatment (only factor favoring on-scene initiation); Prioritizing life-saving procedures, linked to avoiding delay in hospital transport; Following the Vital Signs, Oxygen, Monitor, IV, Treatment (VOMIT) protocol; Practical consideration, including risk of damaging device and/or detaching hose during stretcher transfer.	Providing evidence-based guidance as to when the RIC treatment should ideally be initiated.	*Um, I didn't, honestly, it didn't even pop into my mind, like, I know it was there to use the device, but again, my priority is getting the patient to the hospital because that's, like, the life saving treatment. […] It would have been something that I think would be nice to know, like, do we, is it an on-scene thing, or is it a, you know, in transport? Um, once we got into the back, I think it, because obviously knowing it's a study, I was like trying to prioritize getting it on sooner rather than later.* (P1_800)
Interference with other procedures	Concerns about RIC interfering with the execution or effects of other procedures, including blood pressure monitoring, establishing bilateral IV access, and delivery of IV drugs.	Providing evidence-based guidelines regarding simultaneous placement of an IV and RIC cuff, and IV drug delivery with RIC. Placing the RIC cuff on lower rather than upper limbs.	*If it needs to be on the unaffected arm, then it would be difficult to do serial blood pressures and run IVs. Because the IV also has to be on the unaffected arm.* (P6_1230)
Use of multiple RIC devices	Most participants found this feasible and were willing to use multiple devices if this translated to increased benefit to the patient. Some, however, expressed concern for excess task demands, having to activate and monitor multiple devices and manage more cables.	Enabling attachment of activations unit to the cuffs or integrating the two components. Having a single activation unit for multiple cuffs.	*I think you would definitely need more hands in the back if you're going to be applying multiple. But if this can be done en route, very easily. Um, yeah, then I don't see, I think it would just take time.* (P7_1330)

The third theme focused on the anticipated patient experience of the RIC treatment, with implications for the paramedics' conduct and the patient–paramedic relationship (Table 5). Although, overall, participants did not foresee any life- or health-threatening risks of RIC to patients, many expressed a concern that at least for some the pressure of the cuff would be difficult to tolerate. In such cases, to alleviate patient distress, avoid premature treatment termination, and maintain rapport, participants emphasized the importance of communicating with the patient about the treatment, 'coaching' them through it.

Table 5. Detailed overview of Theme 3: Anticipated patient experience of RIC and implications for the patient–paramedic relationship.

Subtheme	Elucidation/Key Points	Related Recommendations	Exemplar Quotes
Concerns over patients' tolerance of RIC	Anticipating that for some patients (particularly elderly) the cuff may cause bruising or skin tears and induce discomfort that is difficult to tolerate. This in turn could lead to increased patient fear, confusion, volatility or aggression, refusing the RIC treatment, and jeopardizing rapport between the patient and paramedic.	Placing the RIC cuff on a lower rather than upper limb (assumed to be better tolerated). Having the option to inflate the cuff to a pressure lower than 200 mgHg.	*[. . .] multiple times I've had patients that don't like the blood pressure cuff and they're pulling at it and pulling at it and pulling at it. Where they're like, ah, this hurts! And then they escalate. And then, you know, it might be something like that. That, like, the patient just getting frustrated and exploding because of the pain that it's causing. And then the medic trying to calm him down, calm him down, explain why they're doing it, and then it gets in the way of the relationship.* (Confederate)
Strategies for improving patient adherence	Emphasizing the importance of communicating with patients about the treatment and maintaining a positive relationship to help manage discomfort and improve adherence.	Displaying information on the cuff's inflation status to inform patients about the stage of treatment progress. Providing paramedics with comprehensive information on RIC to facilitate answering patient questions and providing reassurance, and conveying anticipated benefits of treatment adherence.	*So, with a, a stroke it would probably stay away with from like sedative agents and things to calm them down. So it would probably be pain management. I, having experience with, with taking blood pressures with seniors, or patients, especially with high cuff pressures, it is uncomfortable in tourniquets, so usually it's just some calming measures and pain control. [. . .] Especially when they know the risks and the benefits, so it's really an understanding, it's an education.* (P6_1230)

The fourth and final theme centered around the paramedics' need of comprehensive knowledge regarding the RIC treatment, including an understanding of the mechanisms underlying its effect and contraindications. Participants felt that being sufficiently informed was integral to confident device use and RIC treatment decision-making, including how best to embed RIC within established workflows and appropriately respond to any changes in the patient's status or external circumstances. Moreover, as indicated in Theme 3, the ability to explain the RIC process and its intended benefit was considered important in communicating with patients, and particularly key to promoting treatment adherence and maintaining rapport with those finding the cuff pressure difficult to tolerate. Participants also expressed a concern that without having evidence-based knowledge on the effects of RIC, some paramedics may view it as "just another step in their protocol" rather than an impactful treatment and may not use the device when and as intended.

> *I would have liked to know more about why we were using it. Again, that's [. . .] 50 percent of using anything is the explanation of the patient and being able to answer all those questions. So, in the training, I just would have liked more, more context into why we're using this. Are there specific cases we're using this and [. . .] what evidence is there, um, that justifies us using this in that case?* (P3_930)

> *I don't know exactly how successful these things are at reducing or helping with strokes. Um, if there, if it comes out that they are really successful, maybe that will be a higher priority on my list, but currently it's kind of an extra thing, as far as I'm concerned.* (P8_1400)

4. Discussion

We conducted a usability study in which paramedics provided feedback on a new RIC device developed for clinical trials in the pre-hospital space (among other settings). Taken together, questionnaire and interview data indicated that the RIC device was easy to use. At the same time, however, participants identified a number of issues that could be improved on to further enhance the RICovery system's usability. Concerns regarding device design and operation were primarily focused on the clarity of the display screen, risk of damage to the unit and hose, and cuff design incorporating a metal loop. Additional issues were raised around certain aspects of workflow, such as deciding on the point of treatment initiation and interference with standard procedures, as well as the patient experience of RIC, potentially entailing a need to address treatment-related distress.

Our findings provide valuable insights on the usability of the RICovery prototype, which correspond to three of four dimensions of device development distinguished by Tsai and colleagues [24]. With no feedback directly applicable to organizational characteristics, these encompass considerations regarding (i) user action, including user needs, training and personal characteristics (e.g., need to understand mechanisms of RIC to allow confident device use); (ii) technical capability, including user interface design, troubleshooting specifications, and labeling (e.g., clarity of performance indicators on the display screen); and (iii) use environment, including intended location, usage time, and necessity of specific safety and hygiene measures (e.g., stretcher transfer involving high risk of dropping device or detaching hose).

Regarding the latter, challenges inherent to emergency response situations and ambulance settings have also been recognized in previous research on system use. For example, in a study evaluating the usability of a mobile telestroke platform in the United States, it was found that the motion of the ambulance led to instability of system hardware, entailing a need to change and additionally secure its initial position [32], while in a Swedish study, the authors identified that the stress and time pressure associated with responding to a major incident posed a barrier to using a detailed triage tool [33]. The same investigation also revealed a concern that having to record information on triage cards competed with providing medical care for the patient. The issue of interference with other medical procedures was also raised in our study, particularly in relation to the cuff obstructing IV access and hindering blood pressure measurement.

Unsurprisingly, these between-study comparisons highlight that even in the similar setting of pre-hospital care, each specific system or technology entails a distinctive set of challenges. These, moreover, may need to be addressed differently depending on particular organizational characteristics [34]. As such, for the potential standard application of RIC systems in an ambulatory setting to be successful, it is necessary to expand the currently limited evidence on device use in this context.

4.1. Strengths and Limitations

To our knowledge, this is the first study to formally evaluate the usability of an RIC system in a pre-hospital setting, given that paramedic feedback had not been published from the two relevant trials conducted in Denmark [18,20]. Important strengths of our research included use of a systematic usability testing framework with structured questionnaires (including validated assessments of usability and task load), detailed feedback interviews with experts, and the fact that a party independent of the device inventors conducted the simulations in an environment that sought to closely replicate real-world transport conditions for the paramedics. Therefore, the feedback provided from participants was less likely to be biased or influenced by external stakeholders.

One limitation was that all our usability testers were from the same region and therefore our findings may not generalize to other geographical locations with different organizational characteristics, standard operating procedures, and available resources of emergency medical services. Secondly, it is important to note that we had not involved real patients at this stage of our research. While participating paramedics shared their views on the anticipated patient experience of RIC, and its potential impact on the paramedic–patient relationship and treatment adherence, these issues are yet to be formally investigated. The study limitations emphasize the need for future feasibility trials to comprehensively account for the impact of real-world, patient and environment-specific factors on the application of the RICovery system, with its usability potentially overestimated in a simulated setting.

4.2. Practical Implications of Study Findings

While some concerns raised by paramedics will require more discussion, such as addressing interference with other medical procedures, those regarding the device's design have already been incorporated into the prototype's revisions. Based on participant feedback, the blood pressure cuff has been replaced with one that does not have a metal loop and only consists of Velcro. Furthermore, the activation unit case has been upgraded to a sturdier and more robust material. Lastly, the information shown on the device screen has been updated to provide more clarity (e.g., device indicates what cycle it is currently on, the percentage of inflation achieved as inflation progresses, and the time left in the therapy cycle). A status LED (light-emitting diode) has also been added that indicates the device state (green for inflating, yellow for deflating). This revised prototype, presented in Figure 3, is now set to be tested in a pilot study with acute stroke patients in a stroke ambulance environment (clinicaltrials.gov registration: NCT05967728).

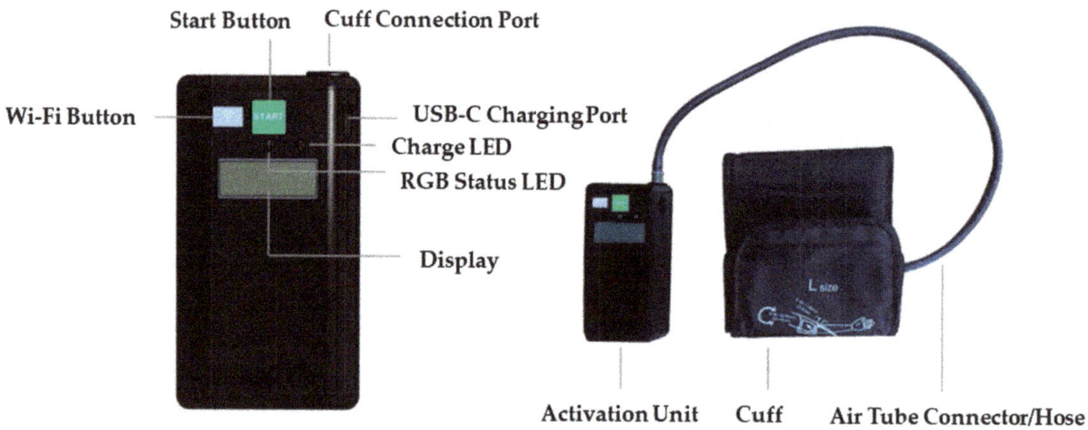

Figure 3. Revised RICovery prototype, guided by the usability testing feedback, that is set for testing in a clinical trial.

Assuming that future trials will indicate that this RIC intervention is feasible and effective in improving stroke outcomes, it appears that the use of the refined device could be implemented during patient transport without significantly increasing demands on paramedics. The ease with which the treatment can be initiated (with a single button press) and flexibility regarding where the cuff can be applied (either to the upper or lower limb) seem particularly advantageous in this context. Prior to implementing the intervention in daily practice, paramedics will require brief training on the RICovery system. Our findings suggest that successful RIC treatment delivery does not require a high level of background knowledge or experience of device use. Nonetheless, it will be important to incorporate information on how the intervention is intended to work, elaborating on the underlying mechanisms and contraindications for use, to support paramedics in confident treatment decision-making.

5. Conclusions

Through advancing the development of a novel RIC system, this study constitutes a step towards addressing an unmet need for efficacious pre-hospital stroke treatments. While the device was considered easy to use, paramedics also identified important areas for improvement. With a small, localized study sample, our findings are primarily applicable to the refinement of the RICovery system for use in future clinical trials in the same healthcare setting. However, feedback on the importance of mitigating potential interference of newly introduced procedures with those already established, robustness of equipment, and effective paramedic–patient communication may also help inform the design of other pre-hospital interventions.

Supplementary Materials: The following supporting information can be downloaded at https://www.mdpi.com/article/10.3390/neurolint16060105/s1: Table S1: Individual user responses on the System Usability Scale; Table S2: Individual user responses on the NASA Task Load Index; Table S3: Individual user responses on the Device Validation Checklist; Table S4: Individual user responses on Custom Assessment Items.

Author Contributions: Conceptualization, A.G. and C.D.; methodology, A.G., C.D. and E.V.; formal analysis, E.V., B.A.D. and K.L.; investigation, E.V. and B.A.D.; data curation, E.V., B.A.D., K.L. and C.D.; writing—original draft preparation, B.A.D. and K.L.; writing—review and editing, B.A.D., K.L., A.G., E.V. and C.D.; supervision, A.G. and C.D.; project administration, A.G., C.D., B.A.D. and E.V.; funding acquisition, A.G. and C.D. All authors have read and agreed to the published version of the manuscript.

Funding: This research was funded by a stroke catalyst grant from Brain Canada, the Canadian Stroke Consortium, and the Heart & Stroke Foundation of Canada; the University of Calgary; and the Hope for Tomorrow grant from Campus Alberta Neuroscience and the Alzheimer's Society of Alberta and Northwest Territories. B.A.D. is supported by a STROKE-COG Clinical Trials Training Platform award, co-funded by Alberta Innovates. K.L. is supported by a VAST Master's Scholarship award. A.G. received salary support through the Heart and Stroke Foundation of Canada.

Institutional Review Board Statement: The study was conducted in accordance with the Declaration of Helsinki and approved by the Institutional Research Ethics Board of the NORTHERN ALBERTA INSTITUTE OF TECHNOLOGY (NAIT REB# 2023-12; approval date: 4 October 2023).

Informed Consent Statement: Informed consent was obtained from all subjects involved in the study.

Data Availability Statement: Quantitative data collected during the study are provided in the Supplementary Materials. Interview transcripts are not publicly available due to the potential risk of identifying participants from their content but are available from the corresponding author on reasonable request.

Acknowledgments: We would like to thank FUEL Innovation Design & Manufacturing for the development and manufacturing of the tested RIC prototype and its revised version, and for ongoing technical support. We would also like to thank the CAMS Project team at the Northern Alberta Institute of Technology for organizing, hosting and conducting the usability testing of the RIC device.

Conflicts of Interest: A.G. has a patent filed (U.S. 17/317,771) for the device being studied in this project, which is being assigned to his start-up SnapDx Inc, and reports stock/stock options from SnapDx and Let's Get Proof (Collavidence Inc.). The companies had no role in the design of the study; in the collection, analyses, or interpretation of data; in the writing of the manuscript; or in the decision to publish the results. The funders had no role in the design of the study; in the collection, analyses, or interpretation of data; in the writing of the manuscript; or in the decision to publish the results. The remaining authors declare no potential conflicts of interest.

Appendix A

Pre-Simulation Briefing (or "Prebriefing") read to participants in the 'briefing' room:

1. Treat the following simulation as a real 911 ambulance call.
2. You are operating under AHS Medical Control Protocols.
3. You'll be functioning as one-half of a Basic Life Support ambulance.
4. You have been provided with a RIC device. We ask that you use the device when providing care. Please use the device in any way that you see fit.
5. What you see in the scenario is what you get. If you need more resources or information, ask for it.

The situation: You are dispatched to a "28 Charlie" (Stroke/CVA) for an elderly patient at NAIT. The caller stated that they were talking with the patient and suddenly they started to slur their speech and talk funny. The caller thinks the patient may be having a stroke. The closest hospital with a stroke center is 15 min (45 min) away. There are no stroke ambulances available. You will need to transport the patient.

Scene One: The patient is laying on the floor when EMS arrives.

The onset time was 15 min ago. The patient is a custodian at NAIT and was cleaning up a classroom and conversing with a colleague. The patient's colleague on scene witnessed the event and stated that while the patient was sweeping, they suddenly dropped their broom. The colleague helped the patient to lay down because they were worried they would fall. The patient now has left sided body weakness and slurred speech. Obvious left facial droop is noted. No trauma seen.

V/S—alert and oriented × 2 (confused to time and place), GCS 14, HR—88 bpm, BP—188/100, Sp02—95% on room air, RR—16, BGL—3.7 mmol/L, Temp—36.2 C, ECG—sinus rhythm, LAMS score—5 (facial droop—1, grip strength—2, arm strength—2).

Scene Two: The paramedics and patient are in the ambulance.

The participants should transport the patient to the ambulance. The stroke symptoms remain (LAMS 5) regardless of treatment. The patient remains confused, with slurred speech etc. The transport will last 15 or 45 min, depending on the condition. Participants will provide care as they see appropriate.

Repeat V/S—LOC—GCS 15 (a + o × 4 if D50W administered), HR—70 bpm, BP—210/100, Sp02—96% on room air, RR—16, BGL—8.2 mmol/L (if 12.5 g of D50W administered doubled if more is given), Temp—36.2 C, ECG—sinus rhythm, LAMS score—5 (facial droop—1, grip strength—2, arm strength—2).

References

1. Gomez, C.R. Editorial: Time is brain! *J. Stroke Cerebrovasc. Dis.* **1993**, *3*, 1–2. [CrossRef] [PubMed]
2. Glober, N.K.; Sporer, K.A.; Guluma, K.Z.; Serra, J.P.; Barger, J.A.; Brown, J.F.; Gilbert, G.H.; Koenig, K.L.; Rudnick, E.M.; Salvucci, A.A. Acute Stroke: Current Evidence-based Recommendations for Prehospital Care. *West J. Emerg. Med.* **2016**, *17*, 104–128. [CrossRef] [PubMed]
3. Kobayashi, A.; Czlonkowska, A.; Ford, G.A.; Fonseca, A.C.; Luijckx, G.J.; Korv, J.; Perez de la Ossa, N.; Price, C.; Russell, D.; Tsiskaridze, A.; et al. European Academy of Neurology and European Stroke Organization consensus statement and practical guidance for pre-hospital management of stroke. *Eur. J. Neurol.* **2018**, *25*, 425–433. [CrossRef] [PubMed]
4. Baldereschi, M.; Piccardi, B.; Di Carlo, A.; Lucente, G.; Guidetti, D.; Consoli, D.; Provinciali, L.; Toni, D.; Sacchetti, M.L.; Polizzi, B.M.; et al. Relevance of prehospital stroke code activation for acute treatment measures in stroke care: A review. *Cerebrovasc. Dis.* **2012**, *34*, 182–190. [CrossRef] [PubMed]
5. Sangari, A.; Akhoundzadeh, K.; Vahedian, M.; Sharifipour, E. Effect of pre-hospital notification on delays and neurological outcomes in acute ischemic stroke. *Australas. Emerg. Care* **2022**, *25*, 172–175. [CrossRef]
6. Varjoranta, T.; Raatiniemi, L.; Majamaa, K.; Martikainen, M.; Liisanantti, J.H. Prehospital and hospital delays for stroke patients treated with thrombolysis: A retrospective study from mixed rural-urban area in Northern Finland. *Australas. Emerg. Care* **2019**, *22*, 76–80. [CrossRef]
7. Holodinsky, J.K.; Williamson, T.S.; Demchuk, A.M.; Zhao, H.; Zhu, L.; Francis, M.J.; Goyal, M.; Hill, M.D.; Kamal, N. Modeling Stroke Patient Transport for All Patients with Suspected Large-Vessel Occlusion. *JAMA Neurol.* **2018**, *75*, 1477–1486. [CrossRef]
8. Fassbender, K.; Walter, S.; Grunwald, I.Q.; Merzou, F.; Mathur, S.; Lesmeister, M.; Liu, Y.; Bertsch, T.; Grotta, J.C. Prehospital stroke management in the thrombectomy era. *Lancet Neurol.* **2020**, *19*, 601–610. [CrossRef]

9. Gardiner, F.W.; Coleman, M.; Teoh, N.; Harwood, A.; Coffee, N.T.; Gale, L.; Bishop, L.; Laverty, M. Aeromedical retrievals of people for mental health care and the low level of clinical support in rural and remote Australia. *Med. J. Aust.* **2019**, *211*, 351–356. [CrossRef]
10. Fassbender, K.; Walter, S.; Liu, Y.; Muehlhauser, F.; Ragoschke, A.; Kuehl, S.; Mielke, O. "Mobile stroke unit" for hyperacute stroke treatment. *Stroke* **2003**, *34*, e44. [CrossRef]
11. Calderon, V.J.; Kasturiarachi, B.M.; Lin, E.; Bansal, V.; Zaidat, O.O. Review of the Mobile Stroke Unit Experience Worldwide. *Interv. Neurol.* **2018**, *7*, 347–358. [CrossRef] [PubMed]
12. Navi, B.B.; Audebert, H.J.; Alexandrov, A.W.; Cadilhac, D.A.; Grotta, J.C.; PRESTO (Prehospital Stroke Treatment Organization) Writing Group. Mobile Stroke Units: Evidence, Gaps, and Next Steps. *Stroke* **2022**, *53*, 2103–2113. [CrossRef] [PubMed]
13. Li, Q.; Guo, J.; Chen, H.S.; Blauenfeldt, R.A.; Hess, D.C.; Pico, F.; Khatri, P.; Campbell, B.C.V.; Feng, X.; Abdalkader, M.; et al. Remote Ischemic Conditioning with Medical Management or Reperfusion Therapy for Acute Ischemic Stroke: A Systematic Review and Meta-Analysis. *Neurology* **2024**, *102*, e207983. [CrossRef] [PubMed]
14. Hess, D.C.; Khan, M.B.; Hoda, N.; Morgan, J.C. Remote ischemic conditioning: A treatment for vascular cognitive impairment. *Brain Circ.* **2015**, *1*, 133–139. [CrossRef]
15. Heusch, G.; Bøtker, H.E.; Przyklenk, K.; Redington, A.; Yellon, D. Remote ischemic conditioning. *J. Am. Coll. Cardiol.* **2015**, *65*, 177–195. [CrossRef]
16. Stokfisz, K.; Ledakowicz-Polak, A.; Zagorski, M.; Zielinska, M. Ischaemic preconditioning—Current knowledge and potential future applications after 30 years of experience. *Adv. Med. Sci.* **2017**, *62*, 307–316. [CrossRef]
17. Zhao, W.; Jiang, F.; Li, S.; Liu, G.; Wu, C.; Wang, Y.; Ren, C.; Zhang, J.; Gu, F.; Zhang, Q.; et al. Safety and efficacy of remote ischemic conditioning for the treatment of intracerebral hemorrhage: A proof-of-concept randomized controlled trial. *Int. J. Stroke* **2022**, *17*, 425–433. [CrossRef]
18. Blauenfeldt, R.A.; Hjort, N.; Valentin, J.B.; Homburg, A.M.; Modrau, B.; Sandal, B.F.; Gude, M.F.; Hougaard, K.D.; Damgaard, D.; Poulsen, M.; et al. Remote Ischemic Conditioning for Acute Stroke: The RESIST Randomized Clinical Trial. *JAMA* **2023**, *330*, 1236–1246. [CrossRef]
19. Chen, H.S.; Cui, Y.; Li, X.Q.; Wang, X.H.; Ma, Y.T.; Zhao, Y.; Han, J.; Deng, C.Q.; Hong, M.; Bao, Y.; et al. Effect of Remote Ischemic Conditioning vs Usual Care on Neurologic Function in Patients with Acute Moderate Ischemic Stroke: The RICAMIS Randomized Clinical Trial. *JAMA* **2022**, *328*, 627–636. [CrossRef]
20. Hougaard, K.D.; Hjort, N.; Zeidler, D.; Sorensen, L.; Norgaard, A.; Hansen, T.M.; von Weitzel-Mudersbach, P.; Simonsen, C.Z., Damgaard, D.; Gottrup, H.; et al. Remote ischemic perconditioning as an adjunct therapy to thrombolysis in patients with acute ischemic stroke: A randomized trial. *Stroke* **2014**, *45*, 159–167. [CrossRef]
21. Blauenfeldt, R.A.; Mortensen, J.K.; Hjort, N.; Valentin, J.B.; Homburg, A.M.; Modrau, B.; Sandal, B.F.; Gude, M.F.; Berhndtz, A.B.; Johnsen, S.P.; et al. Effect of Remote Ischemic Conditioning in Ischemic Stroke Subtypes: A Post Hoc Subgroup Analysis From the RESIST Trial. *Stroke* **2024**, *55*, 874–879. [CrossRef]
22. Purroy, F.; Arque, G.; Mauri, G.; García-Vázquez, C.; Vicente-Pascual, M.; Pereira, C.; Vazquez-Justes, D.; Torres-Querol, C.; Vena, A.; Abilleira, S.; et al. REMOTE Ischemic Perconditioning Among Acute Ischemic Stroke Patients in Catalonia: REMOTE-CAT PROJECT. *Front. Neurol.* **2020**, *11*, 569696. [CrossRef]
23. Ganesh, A.; Barber, P.; Black, S.E.; Corbett, D.; Field, T.S.; Frayne, R.; Hachinski, V.; Ismail, Z.; Mai, L.M.; McCreary, C.R.; et al. Trial of remote ischaemic preconditioning in vascular cognitive impairment (TRIC-VCI): Protocol. *BMJ Open* **2020**, *10*, e040466. [CrossRef]
24. Tsai, I.C.; Wang, C.D.; Chen, P.T. Strategies for Medical Device Development: User and Stakeholder Perceptions. *J. Healthc. Eng.* **2023**, *2023*, 6724656. [CrossRef]
25. Jaspers, M.W. A comparison of usability methods for testing interactive health technologies: Methodological aspects and empirical evidence. *Int. J. Med. Inform.* **2009**, *78*, 340–353. [CrossRef]
26. Gosbee, J.; Klancher, J.; Arnecke, B.; Wurster, H.; Scanlon, M. The Role of Usability Testing in Healthcare Organizations. *Proc. Hum. Factors Ergon. Soc. Annu. Meet.* **2001**, *45*, 1308–1311. [CrossRef]
27. Maramba, I.; Chatterjee, A.; Newman, C. Methods of usability testing in the development of eHealth applications: A scoping review. *Int. J. Med. Inform.* **2019**, *126*, 95–104. [CrossRef]
28. Corry, M.D.; Frick, T.W.; Hansen, L. User-centered design and usability testing of a web site: An illustrative case study. *Educ. Technol. Res. Dev.* **1997**, *45*, 65–76. [CrossRef]
29. Nielsen, J. *How Many Test Users in a Usability Study?* Nielsen Norman Group: Fremont, CA, USA, 2012. Available online: https://www.nngroup.com/articles/how-many-test-users/ (accessed on 28 June 2024).
30. Nielsen, J. Heuristic evaluation. In *Usability Inspection Methods*; Nielsen, J., Mack, R.L., Eds.; John Wiley & Sons: New York, NY, USA, 1994; pp. 25–62.
31. Sauro, J.; Dumas, J.S. Comparison of three one-question, post-task usability questionnaires. In Proceedings of the SIGCHI Conference on Human Factors in Computing Systems, Boston, MA, USA, 4–9 April 2009.
32. Chapman Smith, S.N.; Brown, P.C.; Waits, K.H.; Wong, J.S.; Bhatti, M.S.; Toqeer, Q.; Ricks, J.V.; Stockner, M.L.; Habtamu, T.; Seelam, J.; et al. Development and Evaluation of a User-Centered Mobile Telestroke Platform. *Telemed. e-Health* **2019**, *25*, 638–648. [CrossRef]

33. Rådestad, M.; Lennquist Montán, K.; Rüter, A.; Castrén, M.; Svensson, L.; Gryth, D.; Fossum, B. Attitudes Towards and Experience of the Use of Triage Tags in Major Incidents: A Mixed Method Study. *Prehosp. Disaster. Med.* **2016**, *31*, 376–385. [CrossRef]
34. Sujan, M.A.; Embrey, D.; Huang, H. On the application of Human Reliability Analysis in healthcare: Opportunities and challenges. *Reliab. Eng. Syst. Saf.* **2020**, *194*, 106189. [CrossRef]

neurology international

MDPI

Review

Application of Muscle Synergies for Gait Rehabilitation After Stroke: Implications for Future Research

Jaehyuk Lee [1,†], Kimyung Kim [2,†], Youngchae Cho [2] and Hyeongdong Kim [2,*]

1 Smart Technology Laboratory, Kongju National University, Cheonan-si 31080, Republic of Korea; jhl7589@kongju.ac.kr

2 Department of Physical Therapy, School of Health and Environmental Science, College of Health Science, Korea University, Seoul 02841, Republic of Korea; bus238@naver.com (K.K.); yccho21c@hotmail.com (Y.C.)

* Correspondence: hdkimx0286@korea.ac.kr

† These authors contributed equally to this work.

Abstract: Background/Objective: Muscle synergy analysis based on machine learning has significantly advanced our understanding of the mechanisms underlying the central nervous system motor control of gait and has identified abnormal gait synergies in stroke patients through various analytical approaches. However, discrepancies in experimental conditions and computational methods have limited the clinical application of these findings. This review seeks to integrate the results of existing studies on the features of muscle synergies in stroke-related gait abnormalities and provide clinical and research insights into gait rehabilitation. Methods: A systematic search of Web of Science, PubMed, and Scopus was conducted, yielding 10 full-text articles for inclusion. Results: By comprehensively reviewing the consistencies and differences in the study outcomes, we emphasize the need to segment the gait cycle into specific phases (e.g., weight acceptance, push-off, foot clearance, and leg deceleration) during the treatment process of gait rehabilitation and to develop rehabilitation protocols aimed at restoring normal synergy patterns in each gait phase and fractionating reduced synergies. Conclusions: Future research should focus on validating these protocols to improve clinical outcomes and introducing indicators to assess abnormalities in the temporal features of muscle synergies.

Keywords: gait; muscle synergy; motor module; rehabilitation; stroke

1. Introduction

According to the 2021 mortality statistics in South Korea, stroke is classified as a cerebrovascular disease and ranks as the fourth leading cause of death, with 44.0 deaths per 100,000 in the population in 2021 [1]. As a single disease, it is the second leading cause of death domestically. Despite the completion of standard rehabilitation processes, 50–60% of stroke survivors still experience motor impairments, and at least 50% of patients report partial limitations in their activities of daily living [2–5]. Therefore, the recovery of gait function during the rehabilitation phase after a stroke is one of the most crucial goals of treatment, significantly influencing prognosis, quality of life, and the patient's return to society [6–9]. Over the past few decades, extensive research has been conducted on stroke gait rehabilitation, leading to remarkable advancements in gait rehabilitation methods based on quantitative and objective biomechanical analyses [10–14].

Gait is a daily movement that inevitably requires the central nervous system (CNS) to efficiently manage the myriad possibilities arising from redundant degrees of freedom in the musculoskeletal system [15–18]. According to numerous studies, patients with stroke exhibit pathological muscle activation patterns that deviate from normal gait patterns, such as hemiparesis, changes in muscle tone, and foot drop [19–23]. These deviations affect muscle synergies (MSs), defined as the coordinated activations of muscle groups to perform movement, which are increasingly being addressed in clinical studies [24–28].

MS analysis is used to investigate how the CNS regulates coordinated activation and has been consistently applied in gait analysis [29–36]. In patients with abnormal gait patterns, such as those with stroke, MSs are expressed differently compared to healthy individuals, ultimately leading to inefficient movement [24,37–46]. Many previous studies have reported abnormal MSs manifesting during gait in patients with stroke [46–48]. Recently, these findings have been applied to the development of electrical stimulation therapies, robotic prosthetics, and rehabilitation devices. Although a complete understanding of the neural mechanisms underlying gait remains unclear, MS analysis contributes to the fundamental understanding of abnormal gait through electromyography (EMG), which is widely used across medical centers because of its measurement practicality and time efficiency.

MS analysis addresses the degrees of freedom for problems occurring in neural and multi-joint musculoskeletal systems by distinguishing between spatial and temporal features through matrix decomposition and unsupervised machine learning algorithms [49–52]. The spatial features refer to the weight coefficients of each muscle in the synergies. These coefficients represent how strongly each muscle contributes to a particular MS. The spatial features of MS capture the distribution and contribution of different muscles to the overall movement pattern. Temporal features pertain to the activation profile of MSs, which describes how MSs experience changes over time during movement.

Various techniques, such as principal component analysis (PCA), independent component analysis (ICA), factor analysis (FA), mixed ICA/PCA, non-negative matrix factorization (NNMF), and autoencoders, have been used to analyze MS [53–59]. Owing to the use of various analytical techniques, understanding each method is essential for understanding consistency across studies, and this can simultaneously act as a barrier to effectively communicating research findings to clinicians. In addition, the findings vary across studies owing to the experimental conditions, synergy extraction criteria, and the number of muscles measured. Overall, most studies have consistently reported that stroke patients exhibit fewer MSs than healthy controls (e.g., healthy individuals typically exhibit three–five synergies, whereas stroke patients show three or fewer synergies) [24,40], although few studies have found no differences [41,44]. The reduction in MSs observed in stroke patients is often interpreted as a pathological phenomenon due to synergy merging, which leads to simplified motor control [60,61]. While many studies have primarily focused on MS during straight walking in patients with stroke, a recent finding reported that both the temporal and spatial features of MS during curved walking differ significantly from those in healthy controls, depending on the walking direction [42].

The coexistence of consistency and diversity in previous findings, along with variations in terminology among researchers (e.g., some studies refer to MSs as motor modules), highlight the need for a comprehensive view that could assist clinicians in designing rehabilitation strategies while also providing researchers with insights into the direction of future studies. Therefore, this study aimed to present state-of-the-art rehabilitation by providing a comprehensive review of MS using a systematic methodology to address motor control issues across each phase of gait.

2. Materials and Methods

Although this review adopted a comprehensive approach to explore the current literature on gait MS analysis and rehabilitation strategies in stroke patients, we followed a structured methodology inspired by the PRISMA guidelines to enhance the transparency and reproducibility of the review process [62,63]. By employing this hybrid methodology, we adhered to a structured framework that enhanced clarity and objectivity in the reporting of findings. It is important to note that while this study focused on the clinical interpretation of research outcomes rather than a qualitative assessment of each study's quality, the methodology employed aimed to ensure a thorough and systematic review. This approach maximizes the benefits of the systematic review format and provides clear and consistent information to readers.

2.1. Search Strategy

A literature search was conducted using electronic databases such as PubMed, Scopus, and Web of Science, with publication dates ranging from January 2010 to May 2024. The search was performed using a query combining the following keywords, as agreed upon by the authors of this study: the disease of interest ("Stroke"); the area of rehabilitation interest ("Gait" OR "Walk"); and the research methods of interest ("muscle synerg*" OR "motor module*"). The search query was constructed as ("Stroke") AND ("Gait" OR "Walk") AND ("muscle synerg*" OR "motor module*").

2.2. Study Eligibility Criteria

According to the search strategy, full-text journal articles related to MSs in ischemic stroke patients, including the keywords in the title and abstract, were reviewed. To comprehensively interpret and discuss the results based on the credibility and objectivity of experimental studies, the following types of papers were excluded from the review: (1) duplicate papers; (2) conference proceedings, book chapters, reports, letters, and review papers; (3) papers without Journal Citation Reports (Impact Factor, IF); (4) papers not written in English; (5) papers without gait task; and (6) papers without muscle synergy extraction. Finally, full-text papers that made it difficult to directly compare the pathological synergy between stroke patients and healthy controls were excluded.

2.3. Extraction of Study Characteristics

In the interpretation of MS research, it is essential to consider demographic information such as the characteristics of the participants (e.g., age), the measured muscles, the synergy extraction algorithms (e.g., NNMF), criteria (e.g., the variance accounted for to determine the minimum number of synergies), experimental gait tasks (e.g., overground walking), and the methods used for EMG signal processing (i.e., types of filters).

2.4. Integration of MS Features in Stroke Patients Compared to Controls

Despite the differences in experimental conditions and analytical techniques for synergy extraction across studies and the resulting variability in gait synergy patterns among stroke patients, this study aimed to integrate and compare synergy patterns in stroke patients. For temporal features, there was no visual methodology to integrate the results across studies. However, for spatial features, we visually compared the dominance of muscle weights within the synergies activated during the four gait phases (weight acceptance, push-off, foot clearance, and leg deceleration) and summarized these comparisons in a table for a clear presentation.

2.5. Extraction of Implications for Future Research

The four authors of this study conducted a cross-review to ensure that the results were not overinterpreted by deriving implications that reflected the interpretations of the researchers of each paper as accurately as possible. If a study was limited to reporting phenomena without providing detailed interpretations, the results were discussed until a consensus was reached among the authors. For the intervention studies, only the results of MS in stroke patients before the intervention were included in the review.

3. Results

3.1. Selected Studies

The final search strategy retrieved 243 studies from the three databases. A total of 113 duplicates were identified, leaving 130 studies for screening. No additional studies were found through manual reference list screening. The studies were first screened based on the title and abstract, excluding different parameters such as population, intervention, and outcomes. A total of 38 full texts were included for eligibility screening. Then, after reading the remaining 38 full texts, studies that met the exclusion criteria were removed. Eventually,

10 studies met all the inclusion and exclusion criteria and were included. Figure 1 shows a flowchart with a more detailed overview.

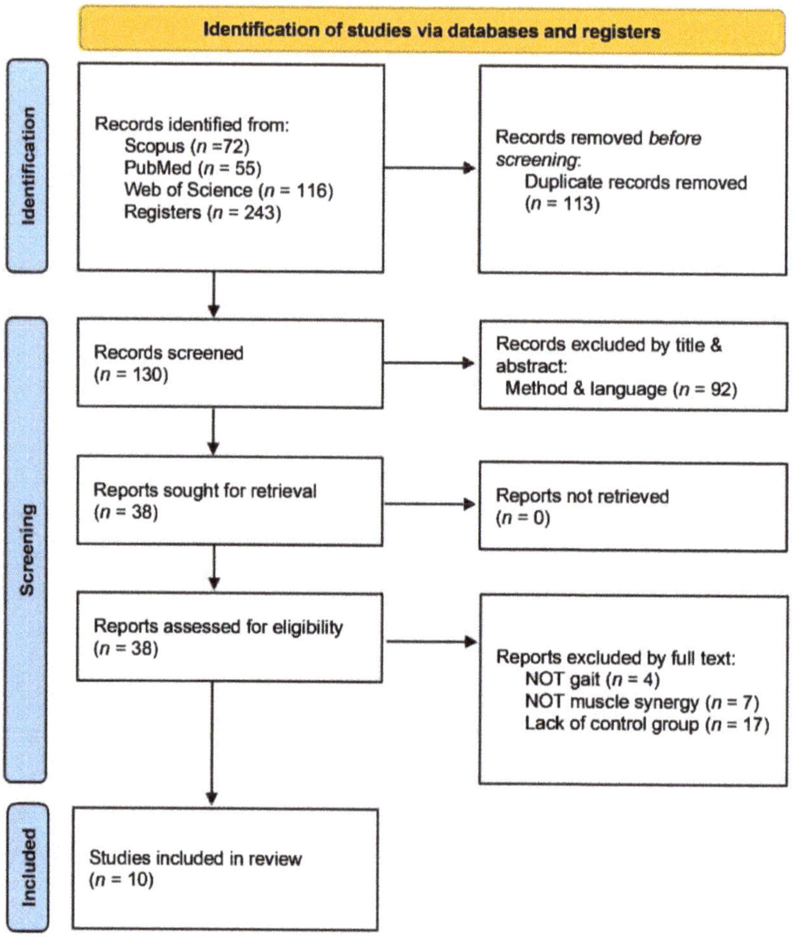

Figure 1. PRISMA flowchart for study inclusion/exclusion.

3.2. Study Characteristics

Most studies on MS in stroke patients' gait have focused on chronic patients more than 6 months post-stroke in Table 1 (8 of 10 studies) [24,38,40–42,44–46]. The gait task was predominantly conducted using overground walking (7 of 10 studies) [39–45], and the most commonly employed MS extraction technique was NNMF (8 of 10 studies) [24,39–41,43–46]. In terms of extraction criteria, the total variance accounted for (VAF) was 90% or higher in seven out of ten studies [24,39–41,44–46]. Most signal filtering methods initially removed artifacts using a band-pass filter (20–750 Hz), followed by rectification and then a low-pass filter (4–40 Hz) to create the final linear envelope before extraction [24,39–46].

Table 1. Characteristics of studies selected for the review.

Reference	Participants	Time Since Stroke	Gait Task	Extraction Algorithm	Extraction Criteria	Signal Processing
Clark, D.J. et al. (2010) [24]	55 patients, 59.5 ± 11.7 y M 35, F 20 Rt. 21, Lt. 34	57.8 ± 64.8 months	Treadmill	NNMF	Total VAF: >90%	High-pass-filtered (40 Hz), demeaned, rectified, and low-pass (4 Hz) Butterworth filter
Coscia, M. et al. (2015) [38]	12 patients, 58.5 ± 16.4 years M 9, F 3 Rt. 5, Lt. 7	54.6 ± 56.2 months	Treadmill, Overground Walking	FA	The number of retained synergies was identified using the criterion of the eigenvalue > 1; 3 synergies were extracted	Rectified and low-pass-filtered (10 Hz) Butterworth filter
Ebihara, A. et al., (2024) [39]	1 patient 49 years M 1 Lt. 1	33 days	Overground Walking	NNMF	Total VAF: >90%	Rectified and low-pass (40 Hz) Butterworth filter
Ferrante, S. et al., (2016) [40]	2 patients 67, 64 years M 2 Rt. 1, Lt. 1	C1: 11 years C2: 9 months	Overground Walking	NNMF	Total VAF: >90%	Band-pass-filtered (40–400 Hz), rectified, and low-pass (5 Hz) Butterworth filter
Gizzi, L. et al. (2011) [41]	10 patients 45.9 ± 16.5 years M 7, F3 Rt. 8, Lt. 2	12.0 ± 4.73 months	Overground Walking	NNMF	Total VAF: >80%	Band-pass filtered (20–400 Hz), rectified and low-pass-filtered (10 Hz) Butterworth filter
Lee, J. (2024) [42]	13 patients 63.2 ± 8.3 years M 7, F 6 Rt. 9, Lt. 4	5.0 ± 0.8 months	Overground Walking	Autoencoder	Fixed synergy extraction at 4	Band pass filter (20–750 Hz), a high-pass filter (35 Hz), rectified, and low-pass filter (5 Hz) Butterworth filter
Lim, J. et al. (2021) [43]	2 patients 62, 60 years M 2 Rt. 1, Lt. 1	C1: 10 months C2: 2 months	Overground Walking	NNMF	Fixed synergy extraction at 4	Band-pass filter (40–400 Hz), rectified, and low-pass filter (5 Hz) Butterworth filter
Routson, R.L. et al. (2013) [44]	22 patients 57.3 ± 13.2 M 15, F 7 Rt. 8, Lt. 14	19.0 ± 13.0 months	Treadmill	NNMF	Total VAF: higher than 90%	High-pass filter (40 Hz), demeaned, and low-pass filter (10 Hz) Butterworth filter

Table 1. *Cont.*

Reference	Participants	Time Since Stroke	Gait Task	Extraction Algorithm	Extraction Criteria	Signal Processing
Young, D.R. et al., (2022) [45]	2 groups, 8 patients each 58.1 ± 7.95 years M 14, F 2 Rt. 8, Lt. 8	HFG: 64.9 ± 38.5 months LFG: 58.1 ± 55.3 months	Overground walking	NNMF	Total VAF: >90%	Rectified and demeaned band-pass filter (10–450 Hz) and low-pass filter (7 Hz) Butterworth
Zhu, F. et al. (2021) [46]	10 patients 59.3 ± 6.8 years M 8, F 2 Rt. 5, Lt. 5	44.7 ± 35.2 months	Treadmill	NNMF	Total VAF: >90%	Band-pass filter (20–250 Hz), rectified, and low-pass filter (4 Hz) Butterworth

Abbreviations: C, case; F, female; HFG, height-functioning group; LFG, low-functioning group; Lt., left; M, male; NNMF, non-negative matrix factorization; Rt., right; VAF, variance accounted for.

3.3. Number of MS in the Paretic Side of Limb by Stroke

For this analysis, two studies were excluded because the authors predetermined the number of MSs before extraction, which made it difficult to examine pathological changes [42,43]. The findings of the eight remaining studies are summarized in Table 2. Five identified a pathological reduction in MSs, which is referred to as the merging phenomenon [24,39,40,45,46].

Table 2. The number of synergies cited from the findings reported in eight studies.

Reference	Numbers of Synergy in Stroke	Numbers of Synergy in Controls	Merging
Clark, D.J. et al. (2010) [24]	2.7 synergies for the paretic leg, and 3.5 synergies for the non-paretic leg	3.6 synergies for the healthy right leg, and 3.7 synergies for the healthy left leg	Identified
Coscia, M. et al. (2015) [38]	3 synergies for the paretic leg	3 synergies for the healthy leg	None
Ebihara, A. et al. (2024) [39]	2-3 synergies for the paretic leg	3 synergies for the non-paretic leg	Identified
Ferrante, S. et al. (2016) [40]	3 synergies for the paretic leg	4 synergies for the healthy dominant leg	Identified
Gizzi, L. et al. (2011) [41]	4 synergies for the paretic leg, and 4 for the non-paretic leg	4 synergies for the healthy left leg	None
Routson, R.L. et al. (2013) [44]	4 synergies for the bilateral legs	4 synergies for healthy bilateral legs	None
Young, D.R. et al. (2022) [45]	1.38 synergies for the bilateral legs	2 synergies for healthy bilateral legs	Identified
Zhu, F. et al. (2021) [46]	3.1 synergies for the paretic leg, and 3.8 synergies for the non-paretic leg	4.45 synergies for bilateral legs	Identified

3.4. MS Features in the Paretic Side of Limb by Stroke

Table 3 presents a comparison of the spatial features extracted from the MS analysis. A visual comparison of the spatial features of MS was possible in 10 studies; however, differences in temporal features were difficult to compare clearly. In terms of spatial features, the altered synergies in stroke patients showed significant changes in the rectus femoris (RF) during the weight acceptance phase [24,38–40,43,44,46], the tibialis anterior (TA) during the foot clearance phase [40–43,46], and hamstrings during the leg deceleration phase [38,40–42,45,46]. The calf muscles, which play key roles in the push-off phase, showed little or no change.

Table 3. The comparison of spatial features of synergies in stroke compared to the controls.

Reference	Study Design	Classification	Involved Primary Muscles												
			GM	Gm	TFL	RF	ADD	VM	ST	BF	LG	MG	SO	PEL	TA
Clark, D.J. et al. (2010) [24]	Cross-sectional (vs healthy)	None	S^{WA}_{WA}			$S^{WA,FC}_{WA}$		S^{WA}_{WA}	S^{LD}_{LD}	S^{LD}_{LD}		S^{PO}_{PO}	S^{PO}_{PO}		S^{FC}_{FC}
Coscia, M. et al. (2015) [38]	Cross-sectional (vs healthy)	None	S^{WA}_{WA}	S^{WA}_{WA}	S^{WA}_{WA}	$S^{WA, FC,LD}_{WA}$	S^{FC}_{WA}	S^{WA}_{WA}	$S^{WA,FC,LD}_{WA}$	$S^{WA, FC}_{WA}$		S^{PO}_{PO}	S^{PO}_{PO}	S^{PO}_{PO}	S^{FC}_{FC}
Ebihara, A. et al., (2024) [39]	Case–control (vs non-paretic)	C1	S^{WA}_{WA}			S^{FC}_{WA}	S^{WA}_{WA}		S^{LD}_{LD}		S^{PO}_{PO}	S^{PO}_{PO}	S^{PO}_{PO}		S^{FC}_{FC}
Ferrante, S. et al., (2016) [40]	Case–control (vs healthy)	C1	$S^{WA}_{WA,FC}$			$S^{FC}_{WA,LD}$	-	S^{WA}_{LD}	S^{LD}_{LD}	S^{LD}_{LD}	-	S^{PO}_{PO}	S^{PO}_{PO}		$S^{FC}_{WA,FC}$
		C2	S^{WA}_{WA}			S^{FC}_{WA}		S^{WA}_{WA}	$S^{LD}_{FC,LD}$	$S^{LD}_{FC,LD}$		S^{PO}_{PO}	S^{PO}_{PO}		$S^{FC}_{FC,LD}$
Gizzi, L. et al., (2011) [41]	Cross-sectional (vs healthy)	None	S^{WA}_{WA}			S^{FC}_{FC}		S^{FC}_{FC}		S^{LD}_{FC}		S^{PO}_{PO}	S^{PO}_{PO}		S^{FC}_{LD}
Lee, J. (2024) [42]	Cross-sectional (vs healthy)	ICG	S^{WA}_{WA}			$S^{WA,FC}_{WA,FC}$	S^{FC}_{FC}	S^{WA}_{WA}	$S^{LD}_{PO, FC, LD}$	$S^{WA,LD}_{WA}$		S^{PO}_{PO}			S^{WA}_{FC}
		OCG	S^{WA}_{WA}			$S^{WA,FC}_{WA,FC}$	S^{FC}_{FC}	S^{WA}_{WA}	S^{LD}_{LD}	S^{LD}_{WA}		S^{PO}_{PO}			S^{WA}_{FC}
Lim, J. et al., (2021) [43]	Case–control (vs healthy)	C1	S^{WA}_{WA}			$S^{WA}_{WA, FC}$	S^{FC}_{FC}	S^{WA}_{WA}	S^{LD}_{LD}	S^{LD}_{LD}		S^{PO}_{PO}			$S^{WA, FC}_{FC, LD}$
		C2	S^{WA}_{WA}			S^{WA}_{PO}	S^{FC}_{FC}	S^{WA}_{PO}	S^{LD}_{LD}	S^{LD}_{LD}		S^{PO}_{PO}			$S^{WA, FC}_{PO, FC}$
Routson, R.L. et al. (2013) [44]	Cross-sectional (vs healthy)	None	S^{WA}_{WA}			$S^{WA, FC}_{WA}$		S^{WA}_{WA}	S^{LD}_{LD}	S^{LD}_{LD}		S^{PO}_{PO}	S^{PO}_{PO}		S^{FC}_{FC}
Young, D.R. et al., (2022) [45]	Cross-sectional (vs healthy)	HFG	$S^{WA,LD}_{WA,LD}$	$S^{WA,LD}_{WA,LD}$		$S^{WA,LD}_{WA,LD}$		$S^{WA,LD}_{WA,LD}$		$S^{WA,FC, LD}_{WA,LD}$		S^{PO}_{PO}	S^{PO}_{PO}		$S^{WA, FC,LD}_{WA, FC,LD}$
		LFG													
Zhu, F. et al., (2021) [46]	Cross-sectional (vs. non-paretic)	None	$S^{WA,PO,LD}_{PO, FC, LD}$			$S^{WA}_{WA,PO}$		$S^{WA}_{WA,P}$	$S^{FC, LD}_{WA, LD}$	$S^{FC, LD}_{WA, LD}$		$S^{WA}_{WA, PC}$	$S^{WA}_{WA, PO}$		$S^{WA,FC, LD}_{FC}$

The areas shaded in light gray within the table represent the spatial features of the stroke patients in the study, which differed from those of the controls. S refers to the synergy, with the superscript indicating the gait phase in which the muscle prominently contributed to the synergy in controls and the gait phase in which the muscle contributed to the synergy in stroke patients. Abbreviations: ADD, adductor; BF, biceps fermoris; C, case; FC, foot clearance; GM, gluteus maximus; Gm, gluteus medius; HFG, height functioning group; ICG, inside curved walking; LD, leg deceleration; LFG, low-functioning group; LG, lateral gastrocnemius; MG, medial gastrocnemius; OCG, outside curved walking; PEL, peroneus longus; PO, push-off; RF, rectus femoris; SO, soleus; ST, semitendinosus; TA, tibialis anterior; TFL, tensor fasciae latae; VM, vastus medialis; WA, weight acceptance.

4. Discussion

Stroke recovery is predominantly achieved within the first three months after stroke onset, and the effectiveness of stroke rehabilitation gradually diminishes over time [64–66]. Therefore, specialized rehabilitation plays a crucial role during the acute phase of stroke [67]. Gait operates via a complex mechanism involving the integration of various sensory stimuli and motor signals from the CNS. However, considering that stroke patients often face difficulties in generating motor signals due to brain cell damage, the key is to maximize the sensory stimulation necessary for gait to promote the spontaneous recovery of brain cells and induce reversibility.

From this perspective, although MS analysis based on computational algorithms can provide valuable clinical insights into the mechanisms of motor control, there are limitations in applying these findings clinically because of the complexity of extraction principles and variability in results based on research methodologies. Therefore, this study aimed to enhance the clinical practicality and accessibility of MS in stroke gait rehabilitation and to suggest future research directions through a focused analysis and discussion of previous findings. Prior to conducting this review, we categorized the primary MSs based on a synthesis of studies as follows: synergy 1 (S1) for weight acceptance, synergy 2 (S2) for push-off, synergy 3 (S3) for foot clearance, and synergy 4 (S4) for leg deceleration.

4.1. Number of MSs in Stroke Gait

Most MS studies compared the number of synergies because they provide a straightforward quantitative measure of motor control complexity. Although there is variability in the reported number of synergies owing to the study's methodology, such as the measured muscles, gait task, extraction algorithms, and criteria, the reduction in the number of synergies observed in stroke patients is interpreted as a synergy-merging phenomenon [24,39,40,45,46,68]. A decrease in the number of MSs is a prominent indicator of impaired motor control in stroke patients and is observed to be restored when these synergies are fractionated [69]. It has been reported that this reduction is significantly correlated with spatiotemporal gait parameters, such as walking speed, step length, and step width. [25,70]. In addition to the number of synergies, another critical metric is the VAF by a single synergy (VAF1), which represents the proportion of data variance explained by individual synergy. In patients with stroke, a higher VAF1 value is typically associated with fewer synergies, indicating the reduced ability of an individual synergy to explain the total data variance in muscle activities compared with healthy individuals [24,39,40,45,46]. Furthermore, some studies have reported differences in the number of synergies between paretic and non-paretic limbs [24,39,46]. Therefore, these results suggest that clinicians should provide task-oriented training for gait rehabilitation by dividing the gait cycle into stages and encouraging the fractionated use of key muscles at each stage. Furthermore, such training needs to be applied independently not only to the paretic limb but also to the non-paretic limb because some studies have reported that the number of synergies in the non-paretic limb is adjusted over time to match the decreased motor control abilities of the paretic limb through an adaptive mechanism. Thus, it would be appropriate to prevent the motor control simplification of the non-paretic limb during early rehabilitation while simultaneously setting a goal to restore the complexity of the simplified motor control in the paretic limb using task-oriented approaches [71,72]. Rehabilitation training is currently feasible because of the development of specialized gait rehabilitation robots and modality equipment that were not available in the past [73,74]. For example, one study reported that when the MS analysis results were applied to electrical stimulation therapy for gait rehabilitation, gait synergies were successfully restored [43].

4.2. Spatial Features of MS in Stroke Gait

This study is the first attempt to synthesize previous findings related to the spatial features of MS in patients with stroke. In doing so, we intended to facilitate a deeper understanding of the pathological patterns of muscle weakness in patients with stroke. As a result of comparing reports on spatial features, the key muscles that functioned differently from the controls were the RF, hamstring (semitendinosus [ST], biceps fermoris [BF]), and TA. For instance, the RF normally functions in S1 and S3; however, in stroke patients, it was observed to function in other synergies, such as S2 and S4, or to be excessively activated in a particular synergy [24,38–40,43,44,46]. In recent years, task-oriented rehabilitation has been widely adopted in stroke rehabilitation [75–77]. It has been shown to aid in the recovery of gait parameters as well as muscle synergy restoration [78,79]. However, task-oriented rehabilitation typically focuses on recovering overall movement patterns rather than on the recovery of individual synergies. Most stroke patients, however, experience pronounced impairments in specific muscle synergies rather than across all synergies. Therefore, it is necessary to target the recovery of individual synergies by specifically rehabilitating the muscles primarily involved in each synergy and restoring the relationships between muscles within a single synergy. This suggests that clinicians should provide rehabilitation protocols that focus not only on the individual strengthening of the RF, ST, BF, and TA muscles but also on activating the muscle groups that work within the same synergy during the corresponding gait phase in which issues arise. For instance, rehabilitation therapists should consider methods such as inducing simultaneous muscle contractions through palpation or using the patient's posture and positioning to facilitate co-contraction.

4.3. Temporal Features of MS in Stroke Gait

In synergy research, temporal features have been relatively less emphasized than spatial features, which present challenges in comparing stroke patients and controls. However, temporal features should also be considered as major factors influencing gait performance [80]. In contrast to spatial features, consistently reported changes in temporal features across studies primarily occurred in S2, which plays a crucial role in propulsion during walking and is closely related to walking speed [41–44]. It is important to note that the temporal features of synergies, particularly when activation begins and reaches its peak, significantly affect the patient's gait ability. An excessively early onset of specific synergies has been reported in patients with stroke, which causes the merging of spatial features and prevents synergies from fully performing their required functions during the gait cycle [24,40,42,45,46]. Specifically, our experimental results showed that when stroke patients performed curved walking, the activation peak timing of the temporal features varied with direction [42]. For curved walking with the paretic leg on the inside (ICW), there was a delay in the activation timing of S1 and S4, whereas for curved walking with the paretic leg on the outside (OCW), there was a significant delay in the activation timing of S1 and S2. Additionally, other studies have indicated that temporal features that should be clearly expressed sometimes manifest ambiguously, showing lower similarity to healthy individuals [41,43,45]. For example, a study that divided stroke patients into a high-functioning group (HFG) and a low-functioning group (LFG) and compared their MSs with those of healthy controls found that, based on spatial features alone, the LFG did not exhibit abnormal patterns compared with the HFG or healthy controls. However, temporal features revealed that the activation phase was not clearly expressed in the LFG. To emphasize, it is crucial that future research provides various indicators to quantify the regularity and clarity of the phase in temporal features.

4.4. Future Challenges for Gait Rehabilitation for Stroke Patients Based on MS

Several studies have found that the recovery of MSs is closely related to improvements in gait performance [40,44,81]. However, few studies have reported significant changes in the number of gait synergies after rehabilitation [82–84]. This suggests that while gait function indicators, such as speed and stride length, may improve through the recovery of muscle strength, muscle tone, and compensatory movements, the actual recovery of motor control may remain limited. This highlights the need for the ongoing development of rehabilitation strategies to address these limitations.

Another important consideration identified in previous studies is the substantial variability in MSs among patients. As the advancement of rehabilitation therapy must be tailored to the individual functionalities of patients, goal-oriented, and cost-effective, there is an urgent need for future research to explore how to rapidly enhance MS recovery in the early stages of stroke when recovery potential is at its highest, beyond the use of MS alone as a biomarker. Additionally, the development of standardized protocols for muscle synergy analysis seems necessary to improve comparability among studies in future research. This research contributes to the development and validation of diverse rehabilitation strategies aimed at recovering specific synergies.

Our study has several limitations. First, we did not review papers written in languages other than English. This limits our ability to analyze valuable information from a diverse range of researchers. Future research that includes papers in all languages, including English, would likely result in a more comprehensive review. Second, this study focused exclusively on ischemic stroke patients. We made this choice to eliminate inconsistencies in results that may arise from the differing gait patterns of ischemic and hemorrhagic stroke patients. Future research should consider including studies that focus on hemorrhagic stroke patients to provide a more comprehensive understanding.

In summary, based on the findings of the MS studies discussed above, future researchers and clinicians in the field should address the following key questions across three major aspects:

1. Features of MS: for the key muscles (i.e., RF, hamstring, and TA), which are changed primarily in spatial features, how can we restore the weighted activation within specific MSs, and how can we improve the recovery of the activation timing of S2?
2. Synergy merging and fractionation: How can we rapidly achieve the unmerging of MSs and enhance motor control complexity? Could single synergy-focused rehabilitation bring about significant changes not only in the recovery of muscle synergies but also in the unmerging of MSs?
3. Gait difficulty considerations: what should be addressed in rehabilitation when considering gait conditions: straight versus curved walking or slow versus fast walking?

5. Conclusions

This study aimed to derive the clinical and research implications from 10 MS studies through a comprehensive review using a systematic methodology. In conclusion, we emphasize that effective gait rehabilitation requires addressing both the paretic and non-paretic limbs independently and utilizing task-oriented training to maximize motor control complexity. This involves breaking down the gait cycle into distinct stages and focusing on the specific needs of each stage to enhance the rehabilitation outcomes. Future studies should focus on quantifying the temporal features of MSs and on developing tailored rehabilitation strategies to enhance recovery, including individual strategies for strengthening and restoring each synergy.

Author Contributions: Conceptualization, J.L. and H.K.; methodology, K.K.; software, J.L. and Y.C.; validation, J.L. and K.K.; formal analysis, K.K.; investigation, J.L. and K.K.; resources, J.L.; data curation, Y.C.; writing—original draft preparation, J.L. and K.K.; writing—review and editing, J.L. and K.K.; visualization, K.K. and Y.C.; supervision, H.K.; project administration, H.K.; funding acquisition, J.L. All authors have read and agreed to the published version of the manuscript.

Funding: This research was supported by the Basic Science Research Program through the National Research Foundation of Korea (NRF) funded by the Ministry of Education (RS-2023-00275579).

Data Availability Statement: Data can be made available on request by the authors.

Conflicts of Interest: The authors declare no conflicts of interest.

References

1. Kim, C. Overview of Cardiac Rehabilitation and Current Situations in Korea. *Ann. Cardiopulm. Rehabil.* **2021**, *1*, 6–16. [CrossRef]
2. Jørgensen, H.S.; Nakayama, H.; Raaschou, H.O.; Olsen, T.S. Recovery of Walking Function in Stroke Patients: The Copenhagen Stroke Study. *Arch. Phys. Med. Rehabil.* **1995**, *76*, 27–32. [CrossRef] [PubMed]
3. Veerbeek, J.M.; Van Wegen, E.E.H.; Harmeling–Van Der Wel, B.C.; Kwakkel, G.; EPOS Investigators. Is Accurate Prediction of Gait in Nonambulatory Stroke Patients Possible Within 72 Hours Poststroke? The EPOS Study. *Neurorehabil. Neural Repair* **2011**, *25*, 268–274. [CrossRef] [PubMed]
4. Hendricks, H.T.; Van Limbeek, J.; Geurts, A.C.; Zwarts, M.J. Motor Recovery after Stroke: A Systematic Review of the Literature. *Arch. Phys. Med. Rehabil.* **2002**, *83*, 1629–1637. [CrossRef]
5. Schaechter, J.D. Motor Rehabilitation and Brain Plasticity after Hemiparetic Stroke. *Prog. Neurobiol.* **2004**, *73*, 61–72. [CrossRef]
6. Vitiello, N.; Oddo, C.M.; Lenzi, T.; Roccella, S.; Beccai, L.; Vecchi, F.; Carrozza, M.C.; Dario, P. Neuro-Robotics Paradigm for Intelligent Assistive Technologies. In *Intelligent Assistive Robots*; Mohammed, S., Moreno, J.C., Kong, K., Amirat, Y., Eds.; Springer Tracts in Advanced Robotics; Springer International Publishing: Cham, Switzerland, 2015; Volume 106, pp. 1–40. ISBN 978-3-319-12921-1.
7. Chaparro-Cárdenas, S.L.; Lozano-Guzmán, A.A.; Ramirez-Bautista, J.A.; Hernández-Zavala, A. A Review in Gait Rehabilitation Devices and Applied Control Techniques. *Disabil. Rehabil. Assist. Technol.* **2018**, *13*, 819–834. [CrossRef]
8. Selves, C.; Stoquart, G.; Lejeune, T. Gait Rehabilitation after Stroke: Review of the Evidence of Predictors, Clinical Outcomes and Timing for Interventions. *Acta Neurol. Belg.* **2020**, *120*, 783–790. [CrossRef]
9. Park, J.; Kim, T.-H. The Effects of Balance and Gait Function on Quality of Life of Stroke Patients. *NeuroRehabilitation* **2019**, *44*, 37–41. [CrossRef] [PubMed]
10. Baker, R. Gait Analysis Methods in Rehabilitation. *J. NeuroEng. Rehabil.* **2006**, *3*, 4. [CrossRef]
11. Mohan, D.M.; Khandoker, A.H.; Wasti, S.A.; Ismail Ibrahim Ismail Alali, S.; Jelinek, H.F.; Khalaf, K. Assessment Methods of Post-Stroke Gait: A Scoping Review of Technology-Driven Approaches to Gait Characterization and Analysis. *Front. Neurol.* **2021**, *12*, 650024. [CrossRef]

12. Bruni, M.F.; Melegari, C.; De Cola, M.C.; Bramanti, A.; Bramanti, P.; Calabrò, R.S. What Does Best Evidence Tell Us about Robotic Gait Rehabilitation in Stroke Patients: A Systematic Review and Meta-Analysis. *J. Clin. Neurosci.* **2018**, *48*, 11–17. [CrossRef] [PubMed]
13. Ferrarello, F.; Bianchi, V.A.M.; Baccini, M.; Rubbieri, G.; Mossello, E.; Cavallini, M.C.; Marchionni, N.; Di Bari, M. Tools for Observational Gait Analysis in Patients with Stroke: A Systematic Review. *Phys. Ther.* **2013**, *93*, 1673–1685. [CrossRef]
14. Beyaert, C.; Vasa, R.; Frykberg, G.E. Gait Post-Stroke: Pathophysiology and Rehabilitation Strategies. *Neurophysiol. Clin. Neurophysiol.* **2015**, *45*, 335–355. [CrossRef] [PubMed]
15. Knutsson, E.; Richards, C. Different Types of Disturbed Motor Control in Gait of Hemiparetic Patients. *Brain J. Neurol.* **1979**, *102*, 405–430. [CrossRef] [PubMed]
16. Petersen, T.H.; Willerslev-Olsen, M.; Conway, B.A.; Nielsen, J.B. The Motor Cortex Drives the Muscles during Walking in Human Subjects. *J. Physiol.* **2012**, *590*, 2443–2452. [CrossRef]
17. Schubert, M.; Curt, A.; Jensen, L.; Dietz, V. Corticospinal Input in Human Gait: Modulation of Magnetically Evoked Motor Responses. *Exp. Brain Res.* **1997**, *115*, 234–246. [CrossRef]
18. Winter, D.A. Biomechanics of Normal and Pathological Gait: Implications for Understanding Human Locomotor Control. *J. Mot. Behav.* **1989**, *21*, 337–355. [CrossRef]
19. Burridge, J.H.; Wood, D.E.; Taylor, P.N.; McLellan, D.L. Indices to Describe Different Muscle Activation Patterns, Identified during Treadmill Walking, in People with Spastic Drop-Foot. *Med. Eng. Phys.* **2001**, *23*, 427–434. [CrossRef]
20. Intiso, D.; Santilli, V.; Grasso, M.G.; Rossi, R.; Caruso, I. Rehabilitation of Walking with Electromyographic Biofeedback in Foot-Drop after Stroke. *Stroke* **1994**, *25*, 1189–1192. [CrossRef]
21. Den Otter, A.R.; Geurts, A.C.H.; Mulder, T.H.; Duysens, J. Abnormalities in the Temporal Patterning of Lower Extremity Muscle Activity in Hemiparetic Gait. *Gait Posture* **2007**, *25*, 342–352. [CrossRef]
22. Garland, S.J.; Gray, V.L.; Knorr, S. Muscle Activation Patterns and Postural Control Following Stroke. *Motor Control* **2009**, *13*, 387–411. [CrossRef] [PubMed]
23. De Quervain, I.A.K.; Simon, S.R.; Leurgans, S.U.E.; Pease, W.S.; Mcallister, D. Gait Pattern in the Early Recovery Period after Stroke. *JBJS* **1996**, *78*, 1506–1514. [CrossRef]
24. Clark, D.J.; Ting, L.H.; Zajac, F.E.; Neptune, R.R.; Kautz, S.A. Merging of Healthy Motor Modules Predicts Reduced Locomotor Performance and Muscle Coordination Complexity Post-Stroke. *J. Neurophysiol.* **2010**, *103*, 844–857. [CrossRef] [PubMed]
25. Barroso, F.O.; Torricelli, D.; Molina-Rueda, F.; Alguacil-Diego, I.M.; Cano-de-la-Cuerda, R.; Santos, C.; Moreno, J.C.; Miangolarra-Page, J.C.; Pons, J.L. Combining Muscle Synergies and Biomechanical Analysis to Assess Gait in Stroke Patients. *J. Biomech.* **2017**, *63*, 98–103. [CrossRef]
26. Van Criekinge, T.; Vermeulen, J.; Wagemans, K.; Schröder, J.; Embrechts, E.; Truijen, S.; Hallemans, A.; Saeys, W. Lower Limb Muscle Synergies during Walking after Stroke: A Systematic Review. *Disabil. Rehabil.* **2020**, *42*, 2836–2845. [CrossRef]
27. Safavynia, S.; Torres-Oviedo, G.; Ting, L. Muscle Synergies: Implications for Clinical Evaluation and Rehabilitation of Movement. *Top. Spinal Cord Inj. Rehabil.* **2011**, *17*, 16–24. [CrossRef] [PubMed]
28. Allen, J.L.; Kesar, T.M.; Ting, L.H. Motor Module Generalization across Balance and Walking Is Impaired after Stroke. *J. Neurophysiol.* **2019**, *122*, 277–289. [CrossRef] [PubMed]
29. Berger, D.J.; Gentner, R.; Edmunds, T.; Pai, D.K.; d'Avella, A. Differences in Adaptation Rates after Virtual Surgeries Provide Direct Evidence for Modularity. *J. Neurosci.* **2013**, *33*, 12384–12394. [CrossRef]
30. Cheung, V.C.K.; Piron, L.; Agostini, M.; Silvoni, S.; Turolla, A.; Bizzi, E. Stability of Muscle Synergies for Voluntary Actions after Cortical Stroke in Humans. *Proc. Natl. Acad. Sci. USA* **2009**, *106*, 19563–19568. [CrossRef]
31. d'Avella, A.; Portone, A.; Fernandez, L.; Lacquaniti, F. Control of Fast-Reaching Movements by Muscle Synergy Combinations. *J. Neurosci.* **2006**, *26*, 7791–7810. [CrossRef]
32. De Marchis, C.; Schmid, M.; Bibbo, D.; Bernabucci, I.; Conforto, S. Inter-Individual Variability of Forces and Modular Muscle Coordination in Cycling: A Study on Untrained Subjects. *Hum. Mov. Sci.* **2013**, *32*, 1480–1494. [CrossRef] [PubMed]
33. Gentner, R.; Edmunds, T.; Pai, D.K.; d'Avella, A. Robustness of Muscle Synergies during Visuomotor Adaptation. *Front. Comput. Neurosci.* **2013**, *7*, 120. [CrossRef] [PubMed]
34. Ivanenko, Y.P.; Poppele, R.E.; Lacquaniti, F. Five Basic Muscle Activation Patterns Account for Muscle Activity during Human Locomotion. *J. Physiol.* **2004**, *556*, 267–282. [CrossRef] [PubMed]
35. Roh, J.; Rymer, W.Z.; Perreault, E.J.; Yoo, S.B.; Beer, R.F. Alterations in Upper Limb Muscle Synergy Structure in Chronic Stroke Survivors. *J. Neurophysiol.* **2013**, *109*, 768–781. [CrossRef]
36. Torres-Oviedo, G.; Ting, L.H. Subject-Specific Muscle Synergies in Human Balance Control Are Consistent Across Different Biomechanical Contexts. *J. Neurophysiol.* **2010**, *103*, 3084–3098. [CrossRef]
37. Dipietro, L.; Krebs, H.I.; Fasoli, S.E.; Volpe, B.T.; Stein, J.; Bever, C.; Hogan, N. Changing Motor Synergies in Chronic Stroke. *J. Neurophysiol.* **2007**, *98*, 757–768. [CrossRef]
38. Coscia, M.; Monaco, V.; Martelloni, C.; Rossi, B.; Chisari, C.; Micera, S. Muscle Synergies and Spinal Maps Are Sensitive to the Asymmetry Induced by a Unilateral Stroke. *J. NeuroEng. Rehabil.* **2015**, *12*, 39. [CrossRef]
39. Ebihara, A.; Hirota, M.; Kumakura, Y.; Nagaoka, M. Analysis of Muscle Synergy and Gait Kinematics during Regain of Gait Function through Rehabilitation in a Monoplegic Patient. *Front. Hum. Neurosci.* **2024**, *17*, 1287675. [CrossRef]

40. Ferrante, S.; Chia Bejarano, N.; Ambrosini, E.; Nardone, A.; Turcato, A.M.; Monticone, M.; Ferrigno, G.; Pedrocchi, A. A Personalized Multi-Channel FES Controller Based on Muscle Synergies to Support Gait Rehabilitation after Stroke. *Front. Neurosci.* **2016**, *10*, 425. [CrossRef]
41. Gizzi, L.; Nielsen, J.F.; Felici, F.; Ivanenko, Y.P.; Farina, D. Impulses of Activation but Not Motor Modules Are Preserved in the Locomotion of Subacute Stroke Patients. *J. Neurophysiol.* **2011**, *106*, 202–210. [CrossRef]
42. Lee, J. Characteristics of Muscle Synergy Extracted Using an Autoencoder in Patients with Stroke during the Curved Walking in Comparison with Healthy Controls. *Gait Posture* **2024**, *107*, 225–232. [CrossRef] [PubMed]
43. Lim, J.; Lim, T.; Lee, J.; Sim, J.; Chang, H.; Yoon, B.; Jung, H. Patient-Specific Functional Electrical Stimulation Strategy Based on Muscle Synergy and Walking Posture Analysis for Gait Rehabilitation of Stroke Patients. *J. Int. Med. Res.* **2021**, *49*, 030006052110167. [CrossRef] [PubMed]
44. Routson, R.L.; Clark, D.J.; Bowden, M.G.; Kautz, S.A.; Neptune, R.R. The Influence of Locomotor Rehabilitation on Module Quality and Post-Stroke Hemiparetic Walking Performance. *Gait Posture* **2013**, *38*, 511–517. [CrossRef] [PubMed]
45. Young, D.R.; Banks, C.L.; McGuirk, T.E.; Patten, C. Evidence for Shared Neural Information between Muscle Synergies and Corticospinal Efficacy. *Sci. Rep.* **2022**, *12*, 8953. [CrossRef]
46. Zhu, F.; Kern, M.; Fowkes, E.; Afzal, T.; Contreras-Vidal, J.-L.; Francisco, G.E.; Chang, S.-H. Effects of an Exoskeleton-Assisted Gait Training on Post-Stroke Lower-Limb Muscle Coordination. *J. Neural Eng.* **2021**, *18*, 046039. [CrossRef]
47. Taborri, J.; Agostini, V.; Artemiadis, P.K.; Ghislieri, M.; Jacobs, D.A.; Roh, J.; Rossi, S. Feasibility of Muscle Synergy Outcomes in Clinics, Robotics, and Sports: A Systematic Review. *Appl. Bionics Biomech.* **2018**, *2018*, 3934698. [CrossRef]
48. Scano, A.; Mira, R.M.; Gabbrielli, G.; Molteni, F.; Terekhov, V. Whole-Body Adaptive Functional Electrical Stimulation Kinesitherapy Can Promote the Restoring of Physiological Muscle Synergies for Neurological Patients. *Sensors* **2022**, *22*, 1443. [CrossRef]
49. Hirashima, M.; Oya, T. How Does the Brain Solve Muscle Redundancy? Filling the Gap between Optimization and Muscle Synergy Hypotheses. *Neurosci. Res.* **2016**, *104*, 80–87. [CrossRef] [PubMed]
50. Federolf, P.A. A Novel Approach to Study Human Posture Control: "Principal Movements" Obtained from a Principal Component Analysis of Kinematic Marker Data. *J. Biomech.* **2016**, *49*, 364–370. [CrossRef]
51. Merkle, L.A.; Layne, C.S.; Bloomberg, J.J.; Zhang, J.J. Using Factor Analysis to Identify Neuromuscular Synergies during Treadmill Walking. *J. Neurosci. Methods* **1998**, *82*, 207–214. [CrossRef]
52. Trumbower, R.D.; Ravichandran, V.J.; Krutky, M.A.; Perreault, E.J. Altered Multijoint Reflex Coordination Is Indicative of Motor Impairment Level Following Stroke. In Proceedings of the 2008 30th Annual International Conference of the IEEE Engineering in Medicine and Biology Society, Vancouver, Canada, 20–25 August 2008; IEEE: Piscataway, NJ, USA, 2008; pp. 3558–3561.
53. Lee, D.; Seung, H.S. Algorithms for Non-Negative Matrix Factorization. *Adv. Neural Inf. Process. Syst.* **2000**, *13*, 535–541.
54. Ranganathan, R.; Krishnan, C. Extracting Synergies in Gait: Using EMG Variability to Evaluate Control Strategies. *J. Neurophysiol.* **2012**, *108*, 1537–1544. [CrossRef]
55. Tresch, M.C.; Cheung, V.C.K.; d'Avella, A. Matrix Factorization Algorithms for the Identification of Muscle Synergies: Evaluation on Simulated and Experimental Data Sets. *J. Neurophysiol.* **2006**, *95*, 2199–2212. [CrossRef]
56. Rasool, G.; Iqbal, K.; Bouaynaya, N.; White, G. Real-Time Task Discrimination for Myoelectric Control Employing Task-Specific Muscle Synergies. *IEEE Trans. Neural Syst. Rehabil. Eng.* **2015**, *24*, 98–108. [CrossRef] [PubMed]
57. Kieliba, P.; Tropea, P.; Pirondini, E.; Coscia, M.; Micera, S.; Artoni, F. How Are Muscle Synergies Affected by Electromyography Pre-Processing? *IEEE Trans. Neural Syst. Rehabil. Eng.* **2018**, *26*, 882–893. [CrossRef]
58. Steele, K.M.; Tresch, M.C.; Perreault, E.J. Consequences of Biomechanically Constrained Tasks in the Design and Interpretation of Synergy Analyses. *J. Neurophysiol.* **2015**, *113*, 2102–2113. [CrossRef]
59. Ebied, A.; Kinney-Lang, E.; Spyrou, L.; Escudero, J. Evaluation of Matrix Factorisation Approaches for Muscle Synergy Extraction. *Med. Eng. Phys.* **2018**, *57*, 51–60. [CrossRef] [PubMed]
60. d'Avella, A.; Saltiel, P.; Bizzi, E. Combinations of Muscle Synergies in the Construction of a Natural Motor Behavior. *Nat. Neurosci.* **2003**, *6*, 300–308. [CrossRef]
61. d'Avella, A.; Bizzi, E. Shared and Specific Muscle Synergies in Natural Motor Behaviors. *Proc. Natl. Acad. Sci. USA* **2005**, *102*, 3076–3081. [CrossRef]
62. Liberati, A. The PRISMA Statement for Reporting Systematic Reviews and Meta-Analyses of Studies That Evaluate Health Care Interventions: Explanation and Elaboration. *Ann. Intern. Med.* **2009**, *151*, W-65. [CrossRef]
63. Page, M.J.; McKenzie, J.E.; Bossuyt, P.M.; Boutron, I.; Hoffmann, T.C.; Mulrow, C.D.; Shamseer, L.; Tetzlaff, J.M.; Akl, E.A.; Brennan, S.E. The PRISMA 2020 Statement: An Updated Guideline for Reporting Systematic Reviews. *BMJ* **2021**, *372*, n71. [CrossRef] [PubMed]
64. Wade, D.T.; Wood, V.A.; Hewer, R.L. Recovery after Stroke–the First 3 Months. *J. Neurol. Neurosurg. Psychiatry* **1985**, *48*, 7–13. [CrossRef] [PubMed]
65. Horgan, N.F.; O'regan, M.; Cunningham, C.J.; Finn, A.M. Recovery after Stroke: A 1-Year Profile. *Disabil. Rehabil.* **2009**, *31*, 831–839. [CrossRef]
66. Bonita, R.; Beaglehole, R. Recovery of Motor Function after Stroke. *Stroke* **1988**, *19*, 1497–1500. [CrossRef] [PubMed]
67. Scafetta, N.; Marchi, D.; West, B.J. Understanding the Complexity of Human Gait Dynamics. *Chaos Interdiscip. J. Nonlinear Sci.* **2009**, *19*, 026108. [CrossRef]

68. Oliveira, A.S.; Gizzi, L.; Ketabi, S.; Farina, D.; Kersting, U.G. Modular Control of Treadmill vs Overground Running. *PLoS ONE* **2016**, *11*, e0153307. [CrossRef]
69. Roh, J.; Rymer, W.Z.; Beer, R.F. Evidence for Altered Upper Extremity Muscle Synergies in Chronic Stroke Survivors with Mild and Moderate Impairment. *Front. Hum. Neurosci.* **2015**, *9*, 124339. [CrossRef]
70. Mizuta, N.; Hasui, N.; Nishi, Y.; Higa, Y.; Matsunaga, A.; Deguchi, J.; Yamamoto, Y.; Nakatani, T.; Taguchi, J.; Morioka, S. Merged Swing-Muscle Synergies and Their Relation to Walking Characteristics in Subacute Post-Stroke Patients: An Observational Study. *PLoS ONE* **2022**, *17*, e0263613. [CrossRef]
71. Nadeau, S.; Duclos, C.; Bouyer, L.; Richards, C.L. Guiding Task-Oriented Gait Training after Stroke or Spinal Cord Injury by Means of a Biomechanical Gait Analysis. *Prog. Brain Res.* **2011**, *192*, 161–180.
72. Verma, R.; Narayan Arya, K.; Garg, R.K.; Singh, T. Task-Oriented Circuit Class Training Program with Motor Imagery for Gait Rehabilitation in Poststroke Patients: A Randomized Controlled Trial. *Top. Stroke Rehabil.* **2011**, *18*, 620–632. [CrossRef]
73. Srivastava, S.; Kao, P.-C.; Kim, S.H.; Stegall, P.; Zanotto, D.; Higginson, J.S.; Agrawal, S.K.; Scholz, J.P. Assist-as-Needed Robot-Aided Gait Training Improves Walking Function in Individuals Following Stroke. *IEEE Trans. Neural Syst. Rehabil. Eng.* **2014**, *23*, 956–963. [CrossRef] [PubMed]
74. Sale, P.; Franceschini, M.; Waldner, A.; Hesse, S. Use of the Robot Assisted Gait Therapy in Rehabilitation of Patients with Stroke and Spinal Cord Injury. *Eur. J. Phys. Rehabil. Med.* **2012**, *48*, 111–121. [PubMed]
75. Alsubiheen, A.M.; Choi, W.; Yu, W.; Lee, H. The Effect of Task-Oriented Activities Training on Upper-Limb Function, Daily Activities, and Quality of Life in Chronic Stroke Patients: A Randomized Controlled Trial. *Int. J. Environ. Res. Public Health* **2022**, *19*, 14125. [CrossRef]
76. Lee, H.-S.; Lim, J.-H.; Jeon, B.-H.; Song, C.-S. Non-Immersive Virtual Reality Rehabilitation Applied to a Task-Oriented Approach for Stroke Patients: A Randomized Controlled Trial. *Restor. Neurol. Neurosci.* **2020**, *38*, 165–172. [CrossRef]
77. Winstein, C.J.; Wolf, S.L.; Dromerick, A.W.; Lane, C.J.; Nelsen, M.A.; Lewthwaite, R.; Cen, S.Y.; Azen, S.P. Effect of a Task-Oriented Rehabilitation Program on Upper Extremity Recovery Following Motor Stroke: The ICARE Randomized Clinical Trial. *JAMA* **2016**, *315*, 571–581. [CrossRef]
78. Berger, D.J.; d'Avella, A. Towards a Myoelectrically Controlled Virtual Reality Interface for Synergy-Based Stroke Rehabilitation. In *Converging Clinical and Engineering Research on Neurorehabilitation II*; Ibáñez, J., González-Vargas, J., Azorín, J.M., Akay, M., Pons, J.L., Eds.; Biosystems & Biorobotics; Springer International Publishing: Cham, Switzerland, 2017; Volume 15, pp. 965–969. ISBN 978-3-319-46668-2.
79. Ghassemi, M.; Triandafilou, K.; Barry, A.; Stoykov, M.E.; Roth, E.; Mussa-Ivaldi, F.A.; Kamper, D.G.; Ranganathan, R. Development of an EMG-Controlled Serious Game for Rehabilitation. *IEEE Trans. Neural Syst. Rehabil. Eng.* **2019**, *27*, 283–292. [CrossRef]
80. Chvatal, S.A.; Ting, L.H. Voluntary and Reactive Recruitment of Locomotor Muscle Synergies during Perturbed Walking. *J. Neurosci.* **2012**, *32*, 12237–12250. [CrossRef]
81. Van Criekinge, T.; Hallemans, A.; Herssens, N.; Lafosse, C.; Claes, D.; De Hertogh, W.; Truijen, S.; Saeys, W. SWEAT2 Study: Effectiveness of Trunk Training on Gait and Trunk Kinematics after Stroke: A Randomized Controlled Trial. *Phys. Ther.* **2020**, *100*, 1568–1581. [CrossRef] [PubMed]
82. Ambrosini, E.; Parati, M.; Peri, E.; De Marchis, C.; Nava, C.; Pedrocchi, A.; Ferriero, G.; Ferrante, S. Changes in Leg Cycling Muscle Synergies after Training Augmented by Functional Electrical Stimulation in Subacute Stroke Survivors: A Pilot Study. *J. NeuroEng. Rehabil.* **2020**, *17*, 35. [CrossRef]
83. Ghislieri, M.; Lanotte, M.; Knaflitz, M.; Rizzi, L.; Agostini, V. Muscle Synergies in Parkinson's Disease before and after the Deep Brain Stimulation of the Bilateral Subthalamic Nucleus. *Sci. Rep.* **2023**, *13*, 6997. [CrossRef]
84. Jonsdottir, J.; Lencioni, T.; Gervasoni, E.; Crippa, A.; Anastasi, D.; Carpinella, I.; Rovaris, M.; Cattaneo, D.; Ferrarin, M. Improved Gait of Persons with Multiple Sclerosis after Rehabilitation: Effects on Lower Limb Muscle Synergies, Push-off, and Toe-Clearance. *Front. Neurol.* **2020**, *11*, 668. [CrossRef] [PubMed]

neurology international

MDPI

Article

Pretreatment Cranial Computed Tomography Perfusion Predicts Dynamic Cerebral Autoregulation Changes in Acute Hemispheric Stroke Patients Having Undergone Recanalizing Therapy: A Retrospective Study

Lehel-Barna Lakatos [1], Manuel Bolognese [1], Mareike Österreich [1], Martin Müller [1,*] and Grzegorz Marek Karwacki [2]

[1] Department of Neurology and Neurorehabilitation, Section Neuroradiology, Lucerne Cantonal Hospital, 6000 Lucerne, Switzerland; manuel.bolognese@luks.ch (M.B.); mareike.oesterreich@luks.ch (M.Ö.)
[2] Department of Radiology, Section Neuroradiology, Lucerne Cantonal Hospital, 6000 Lucerne, Switzerland
* Correspondence: martin.renatemueller@web.de; Tel.: +41-79-935-6220

Abstract: Objectives: Blood pressure (BP) management is challenging in patients with acute ischemic supratentorial stroke undergoing recanalization therapy due to the lack of established guidelines. Assessing dynamic cerebral autoregulation (dCA) may address this need, as it is a bedside technique that evaluates the transfer function phase in the very low-frequency (VLF) range (0.02–0.07 Hz) between BP and cerebral blood flow velocity (CBFV) in the middle cerebral artery. This phase is a prognostically relevant parameter, with lower values associated with poorer outcomes. This study aimed to evaluate whether early cranial computed tomography perfusion (CTP) can predict this parameter. Methods: In this retrospective study, 165 consecutive patients with hemispheric strokes who underwent recanalizing therapy were included (median age: 73 years; interquartile range (IQR) 60–80; women: 43 (26%)). The cohort comprised 91 patients treated with intravenous thrombolysis (IV-lysis) alone (median National Institute of Health Stroke Scale (NIHSS) score: 5; IQR 3–7) and 74 patients treated with mechanical thrombectomy (median NIHSS: 15; IQR 9–18). Regression analysis was performed to assess the relationship between pretreatment CTP-derived ischemic penumbra and core stroke volumes and the dCA VLF phase, as well as CBFV assessed within the first 72 h post-stroke event. Results: Pretreatment penumbra volume was a significant predictor of the VLF phase (adjusted $r^2 = 0.040$; $\beta = -0.001$, 95% confidence interval (CI): −0.0018 to −0.0002, $p = 0.02$). Core infarct volume was a stronger predictor of CBFV (adjusted $r^2 = 0.082$; $\beta = 0.205$, 95% CI: 0.0968–0.3198; $p = 0.0003$) compared to penumbra volume ($p = 0.01$). Additionally, in the low-frequency range (0.07–0.20 Hz), CBFV and BP were inversely related to the gain, an index of vascular tone. Conclusion: CTP metrics appear to correlate with the outcome-relevant VLF phase and reactive hyperemic CBFV, which interact with BP to influence vascular tone and gain. These aspects of dCA could potentially guide BP management in patients with acute stroke undergoing recanalization therapy. However, further validation is required.

Keywords: stroke; recanalizing therapy; cerebral autoregulation; computed tomography perfusion; transfer function

1. Introduction

Cranial computed tomography perfusion (CTP) is frequently used as a part of a multimodal imaging approach to evaluate patients with acute ischemic stroke (AIS). CTP is designed to differentiate between ischemic tissue at risk of brain cell death (penumbra) and the ischemic infarct core, where tissue damage is already irreversible. By applying specific clinical, time-dependent, and CTP criteria to identify a relevant mismatch between the infarct core or the clinical severity and the penumbral tissue, early recanalizing therapy,

such as intravenous thrombolysis (IV lysis) and/or mechanical thrombectomy (MT), within 24 h after the first AIS symptom onset can salvage the tissue at risk, leading to improved clinical outcomes [1–5].

Early CTP also provides prognostic information, such as predicting modified Rankin Scale (mRs) outcomes based on core infarct volume [6,7], and assessing the impact of recanalization therapy on reducing brain edema and improving mRs [8,9]. Additionally, CTP can estimate the likelihood of post-interventional hyperemia [10–12]. Ischemia impairs cerebral autoregulation (CA), limiting the ability of the brain to regulate the blood supply, which can exacerbate ischemic damage under conditions of low blood pressure [13–17]. Because the CA, as assessed through dynamic cerebral autoregulation (dCA), does not immediately recover to full strength after vessel reopening, a factor of prognostic relevance [18,19], it seems reasonable to estimate the post-interventional state of (d)CA as early as possible. This knowledge can guide BP management. However, it is currently unknown whether CTP parameters can predict post-interventional dCA. We hypothesized that the volume of the penumbral or core infarct predicts dCA impairment in patients undergoing recanalization therapy. If this is the case, CTP could serve as the starting point for (d) CA-driven BP management, incorporating repeated easy-to-perform bedside dCA assessments to adjust BP levels according to the severity of dCA impairment [17].

2. Methods

This retrospective cohort study of patients with AIS was approved by the Ethics Committee of Northwest and Central Switzerland (EKNZ; 2024–01039). This study was conducted in accordance with the Declaration of Helsinki, adhered to good clinical practice standards, and was part of a larger trial registered at https://clinicaltrials.gov/ NCT04611672, dated 12 October 2020. At our institution, all patients or their relatives provide general consent, allowing the use of routinely acquired data for retrospective studies. Data supporting the findings of this study are available from the corresponding author upon reasonable request.

2.1. Study Setting

The stroke care protocols at our institution have been previously reported in detail [20,21].

To summarize, all patients with stroke syndrome receive standardized care, with an initial focused clinical examination followed by multimodal cranial computed tomography (CT; Siemens Force, Edge, or XCeed CT, Siemens Healthineers, Forchheim, Germany). This includes non-enhanced cranial CT (NCCT), CTP postprocessed by Syngo.via (Siemens Healthineers) and Rapid CTP (RapidAI, San Mateo, CA, USA) to estimate the infarct core and penumbra, and CT angiography (CTA). If indicated, IV-lysis and/or MT are performed immediately. All patients diagnosed with stroke were transferred to the stroke unit for intensive clinical monitoring. This includes the National Institute of Health Stroke Scale (NIHSS) [22] and mRs [23] assessments upon hospital admission, as well as daily assessments while in the stroke unit and three months after the ischemic event (as the outcome measure). Blood pressure, heart rate, body temperature, blood glucose level, and oxygen saturation were closely monitored. Ultrasound examinations of all brain-supplying arteries, including dCA assessment, echocardiography, and brain magnetic resonance imaging (MRI) with diffusion-weighted imaging (DWI), T2, and susceptibility-weighted imaging (SWI) sequences, are performed within 72 h of hospitalization. Imaging is conducted on either a Siemens Vida fit (3 Tesla), Siemens Aera (1.5 Tesla), or Philips Achieva (3 Tesla) (Siemens Healthineers). After all relevant information is gathered, stroke etiology was classified based on the Trial of ORG 10172 in Acute Stroke Treatment (TOAST) classification [24]. Figure 1 provides an overview of the time-dependent data collection process.

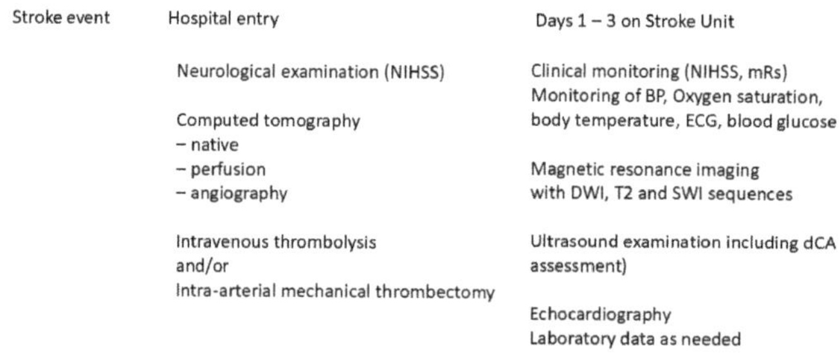

Figure 1. Flow chart of the patients' disease/hospitalization course during the first days following the stroke event and their relation to the timing of diagnostic procedures for clinical data collection.

2.2. *Cranial Computed Tomography Perfusion (CTP)*

Imaging was performed using Siemens devices (Siemens Healthineers), with detector rows ranging from 96 to 128. NCCT helical scans were acquired from the skull base to the vertex with the following scan parameters: 120 kV, 240 mA (with dose modulation when applicable), 0.6 mm collimation, 30.7 mm/s table speed, and 30.7 mm table feed per rotation. A CTA of the arterial cervical and intracranial vessels (from the aortic arch to the vertex) was performed with a prep scan at the level of the aortic arch to optimize timing of the arterial phase. A weight-adapted volume of contrast medium (1840 mg/s iodine, flow rate 6 mL/s), 100 kV, dose modulation, 0.6 mm collimation, 38.4 mm/s table speed, and 42.2. mm of table feed per rotation. CTP imaging was performed with a coverage of 10–11 cm in the Z axis, 60–65 s of total acquisition duration (one scan every 1.5 s), and 5 mm slice thickness. Parameters included 70 kV, 140 to 150 mA, 1.2 mm collimation, 38.4 mm/s table speed, 42.2 mm table feed per rotation, a weight-adapted volume of contrast medium (2240 mg/s iodine), and a recommended flow rate of 6 mL/s. After post-processing using Syngo.via and RapidAI, the tissue at risk (penumbra) was defined as having a time to maximum residue function (Tmax) prolongation of more than 6 s. Core infarction was defined as tissue with a relative cerebral blood flow <30% of the values obtained from the contralateral hemisphere.

2.3. *Dynamic Cerebral Autoregulation (dCA)*

At our institution, dCA is routinely included in the transcranial Doppler sonographic evaluation of patients with suspected ischemic cerebrovascular impairment to evaluate CA function. All investigations were performed with the patient in the supine position, with the head elevated by approximately 30°. Middle cerebral artery (MCA) cerebral blood flow velocity (CBFV) was recorded using a 2 MHz probe (MultidopX, DWL; Compumedics, Sipplingen, Germany), and blood pressure (BP) was measured simultaneously using a Finometer Pro (Finapres Medical Systems, Amsterdam, The Netherlands) for a minimum of 6 min. End-tidal pCO_2 ($ETCO_2$) concentration was measured using nostril tubes and a capnograph embedded in a TCD device. BP, CBFV, and $ETCO_2$ data were collected at a frequency of 100 Hz. Data were analyzed using MATLAB (version 2023a; MathWorks Inc., Natick, MA, USA). The data were visually inspected for artifacts, and only artifact-free periods of 5 min were used for the analysis. Each raw data time series was averaged over 1 s intervals to generate new time series representing the mean CBFV and BP. Coherence, phase, and gain between the BP and CBFV time series were extracted using their respective power auto-spectra or cross-spectra with Welch's averaged period gram method. A Hanning window with a length of 100 s, a window with a 50% overlap, and a total fast Fourier transformation data length of 300 s were used. For each patient, coherence, phase (in radians), and gain (in cm/s/mmHg) were calculated over a frequency range of 0.02–0.50 Hz. These values were then averaged within three frequency bands, namely, very low frequency range (VLF, 0.02–0.07 Hz), low frequency range (LF, 0.07–0.20 Hz), and high frequency range (HF, 0.2–0.5 Hz) [25,26]. The gain and phase usually follow opposite directions, that is, low gain is associated with high phase, and vice versa. Impaired dCA is indicated by a lower phase or higher gain in either the VLF or LF range. An illustration of this in investigation is presented in Figure 2.

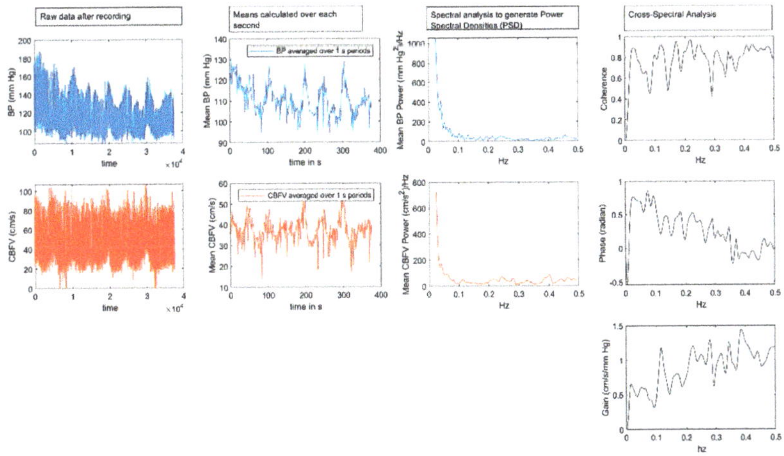

Figure 2. Illustration of performing dynamic cerebral autoregulation assessment. The first two images show the envelope curves of the blood pressure (BP) and cerebral blood flow velocity (CBFV) recordings. The data points of these time series are averaged over 1 s intervals to create new time series. From this, power spectra are generated. Cross-spectral analysis then extracts coherence, phase, and gain across the frequency range of 0–0.5 Hz. Despite some undulations, coherence is high, the phase decreases, and the gain increases. For reporting, the measured values are averaged over the following three frequency ranges: 0.02–0.07 Hz, 0.07–0.2 Hz, and 0.2–0.5 Hz. The transfer function model of dynamic cerebral autoregulation (dCA) reflects a high-pass filter behavior, that is, BP changes with a frequency of >0.2 Hz pass are transmitted through to CBFV; the exact physiological correlates in the lower frequency ranges (<0.20 Hz) are only partially understood. For example, the CO_2 regulation is primarily observed in the 0.07–0.20 Hz range, while CBFV changes in the 0.02–0.07 Hz range (corresponding to blood flow changes every 20–50 s) may reflect blood volume changes in the microcirculation [26].

2.4. Patients

This study included consecutive patients treated in our stroke unit between 1 January 2020 and 31 April 2022. Patients who underwent dCA assessments were retrospectively reviewed. The inclusion criteria for this study were ≥18 years of age, absence of pregnancy, diagnosis of a characteristic hemispheric syndrome, confirmed as a definitive supratentorial ischemic stroke in the MCA territory through initial multimodal imaging and later validated by DWI imaging, and high-quality bilateral dCA assessments (stroke-affected hemisphere (AH); stroke-unaffected hemisphere (UH)). Exclusion criteria included the final diagnosis of a stroke mimic, a primary intracranial hemorrhage, a transient ischemic attack, and a cerebral sinus or vein thrombosis.

Results are reported separately for the AH and UH to account for hemispheric differences in dCA function.

2.5. Statistical Analyses

All data analyses were performed using the MATLAB Statistical Toolbox (MathWorks Inc.). Data are presented as mean ± standard deviation (SD) for normally distributed variables and as median with their interquartile range (IQR) for non-normally distributed variables. Most continuous variables were not normally distributed; therefore, we used the nonparametric Kruskal–Wallis test for between-group comparisons. Fisher's exact test or chi-square statistics were applied for comparing categorical variables. Univariate and multiple linear regression analyses were performed to determine whether CTP-derived penumbra or infarct core volume could predict dCA parameters. Baseline variables that are reported to influence dCA, including age, history of arterial hypertension, cardiac left ventricular ejection fraction (LVEF%), CBFV, BP, and $ETCO_2$ (averaged over the dCA recording period), were included together with penumbra and infarct core volume. If either penumbra or infarct core volume, along with one or more baseline variables, were significantly related to dCA parameters, multiple linear regression models were constructed to further evaluate the role of these parameters. To address collinearity between penumbra and infarct core volume—representing different stages of cerebral ischemia—two separate multiple regression models were created. The first model included penumbra volume only, and the second included infarct core volume only [27]. Statistical significance was set at $p < 0.05$.

3. Results

A total of 336 patients (median age, 71 (59–78) years; 102 women (30.3%)) meeting the inclusion criteria were analyzed. Of them, 165 patients (age 73 (60–80) years; 43 women (26%)) had undergone recanalizing therapy, with 91 patients receiving IV-lysis alone and 74 undergoing MT. Stroke etiology was classified according to TOAST into cardio-embolism ($n = 64$), atherosclerotic large vessel disease ($n = 35$), lacunar stroke ($n = 17$), other defined sources (such as dissections; $n = 10$), and embolic strokes of unknown sources ($n = 39$). Other baseline characteristics are presented in Table 1. We focused our analysis on IV lysis and MT as recanalizing therapies without further subclassifying the patients with MT into those who received bridging IV lysis therapy and those who did not.

Table 1. Clinical baseline characteristics, risk factors, and computed tomography perfusion results in the acute ischemic stroke patients having undergone recanalizing therapy dichotomized into those who had received intravenous thrombolysis (IV-lysis) only, versus those who had undergone mechanical thrombectomy (MT) with or without prior bridging IV-lysis.

	IV-Lysis Only ($n = 91$)	MT Procedure ($n = 74$)	p-Value
Age (years)	71.5 (58–78)	75 (63–82)	0.16
Women/men	20/71	24/50	0.19
NIHSS at entry	5 (3–7)	15 (9–18)	0.0000
mRs at 3 months	1 (0–2)	3 (1–3)	0.0000
mRs at 3 months dichotomized ≤2 >2	 75 16	 36 38	0.0000
Arterial hypertension (n)	50	43	0.59
Diabetes mellitus (n)	25	20	1
Dyslipidemia (n)	64	60	0.72
Left ventricular ejection fraction (%)	59 (55–63)	58 (50–65)	0.57
Ischemic heart disease (n)	19	29	0.30
Estimated glomerular filtration rate (mL/min/10.72 m^2)	820.5 (67–91)	79 (59–89)	0.20
Blood pressure (mm Hg) at admission Systolic Mean Diastolic	 153 (137–172) 106 (93–117) 85 (74–96)	 147 (134–164) 103 (91–113) 81 (71–91)	 0.10 0.22 0.12

Table 1. *Cont.*

	IV-Lysis Only (n = 91)	MT Procedure (n = 74)	p-Value
Neuroimaging			
CTP—Penumbra (mL)	5 (0–28)	125 (55–162)	0.0000
CTP—Core (mL)	0 (0–0)	9 (0–34)	0.0000
MRI—infarct volume (mL) on DWI	20.8 (00.54–110.5)	150.3 (60.4–500.9)	0.0000

CTP, computed tomography; DWI, diffusion-weighted imaging; MRI, magnetic resonance imaging; mRs, modified Rankin scale; NIHSS, National Institutes of Health Stroke Scale. Values are presented as either numbers or medians and the interquartile range (IQR).

3.1. MT Procedures

Of the 74 patients, 47 received bridging therapy with prior IV lysis. Intracranial thrombectomy was exclusively performed using stent retriever systems. Vasospasm, a complication of MT, occurred in six instances. After MT, 31 patients achieved complete recanalization (thrombolysis in cerebral infarction (TICI) 3), while 43 had varying degrees of incomplete recanalization (TICI 2a, n = 7; 2b, n = 18; 2c, n = 18) [28]. If an extracranial carotid artery stenosis was ≥50% according to NASCET [29] or a carotid occlusion was the cause of the stroke and this artery obstruction was traversable for thrombectomy maneuver, these obstructions were permanently recanalized by stent placement.

Patients undergoing MT had a larger median penumbra and infarct core (on both CTP and MRI) and exhibited more severe neurological impairment at hospital admission and at 3 months post-stroke.

3.2. Dynamic Cerebral Autoregulation (dCA) Assessment

The earliest dCA assessment post-stroke onset was performed at varying times in 114 patients within the first 24 h after, 37 patients between 24 and 48 h, and 14 patients between 48 and 72 h. The minimum and maximum mean BP over the recording periods were 57 mmHg and 128 mmHg, respectively. In the affected hemisphere (Table 2), CBFV, VLF, and LF gains were significantly higher in patients who underwent MT than in those who underwent IV lysis. Conversely, the VLF and LF phases were significantly lower in the MT group. In the unaffected hemisphere, only the gain and phase in the LF range differed between the two groups.

Table 2. Comparison of transfer function estimates after recanalization in the middle cerebral artery of the stroke-affected hemisphere between the patients who had received intravenous thrombolysis (IV-lysis) alone versus patients who had undergone the mechanical thrombectomy (MT) procedure.

	IV-Lysis Only (n = 91)	MT Procedures (n = 74)	p-Value
Mean arterial blood pressure (mm Hg) over recording period	98 (86–104)	89 (79–102)	0.06
Heart rate	68 (60–80)	70 (62–79)	0.41
$ETCO_2$ (mm Hg) over recording period	39.5 (38.1–40.5)	39.2 (37.6–40.8)	0.40
	Affected Hemisphere		
CBFV (cm/s)	46 (41–55)	57 (48–75)	0.0000
Pulsatility index	0.95 (0.85–1.12)	0.98 (0.86–1.08)	0.45

Table 2. *Cont.*

	IV-Lysis Only (*n* = 91)	MT Procedures (*n* = 74)	*p*-Value
Coherence			
VLF	0.57 (0.48–0.64)	0.62 (0.53–0.70)	0.04
LF	0.63 (0.53–0.77)	0.66 (0.53–0.78)	0.92
HF	0.68 (0.58–0.76)	0.64 (0.55–0.70)	0.15
Gain (mmHg/cm/s)			
VLF	0.25 (0.15–0.38)	0.36 (0.23–0.55	0.001
LF	0.39 (0.28–0.52)	0.49 (0.32–0.66)	0.01
HF	0.51 (0.38–0.64)	0.56 (0.40–0.86)	0.06
Phase (radian)			
VLF	0.83 (0.59–1.06)	0.67 (0.46–0.85)	0.002
LF	0.68 (0.50–0.91)	0.53 (0.36–0.76)	0.001
HF	0.25 (0.06–0.49)	0.19 (0.03–0.39)	0.16
	Unaffected Hemisphere		
CBFV (cm/s)	47 (41–56)	49 (42–50)	0.32
Pulsatility Index	0.93 (0.84–1.11)	0.97 (0.85–1.10)	0.46
Coherence			
VLF	0.57 (0.50–0.65)	0.59 (0.48–0.67)	0.95
LF	0.62 (0.52–0.75)	0.61 (0.52–0.76)	0.71
HF	0.62 (0.52–0.77)	0.64 (0.49–0.72)	0.77
Gain (mm Hg/cm/s)			
VLF	0.26 (0.15–0.41)	0.33 (0.17–0.48)	0.17
LF	0.44 (0.26–0.56)	0.55 (0.37–0.77)	0.01
HF	0.48 (0.38–0.68)	0.62 (0.47–0.80)	0.01
Phase (radian)			
VLF	0.80 (0.61–1.04)	0.82 (0.52–1.08)	0.86
LF	0.64 (0.49–0.81)	0.56 (0.39–0.76)	0.045
HF	0.27 (0.04–0.48)	0.19 (0–0.32)	0.13

CBFV, cerebral blood flow velocity; $ETCO_2$, end-tidal carbon dioxide; HF, high frequency; LF, low frequency; VLF, very low frequency.

3.3. Linear Regression Analysis

3.3.1. CBFV in the Affected Hemisphere

In patients treated with MT, the CBVF in the AH was significantly higher compared to both patients receiving IV lysis alone and UH. Univariate linear regression analysis (Table 3) revealed that CBFV in the AH was significantly associated with penumbra size ($p = 0.01$) and core infarct volume ($p = 0.003$; Figure 3). These findings suggest that increased CBFV in the AH corresponds to reactive hyperemic blood flow.

Table 3. Univariate linear regression to predict cerebral blood flow velocity in the middle cerebral artery of the affected hemisphere.

Variable	
Left ventricular ejection fraction (LVEF, %)	$\beta = 0.019$ (95%CI −0.2202 to 0.2583), $r^2 = 0.0002$, $F(2,163) = 0.02$, $p = 0.80$
Mean BP (mm Hg) over the recording period	$\beta = -0.116$ (95%CI −0.3001 to 0.0664), $r^2 = 0.009$, $F(2,163) = 1.58$, $p = 0.20$
Mean $ETCO_2$ (mm Hg) over the recording period	$\beta = -0.044$ (95%CI −1.0060 to 0.9163), $r^2 = 0.001$, $F(2,163) = 0.008$, $p = 0.92$
Penumbra (mL^3)	$\beta = 0.047$ (95%CI 0.0108 to 0.0840), $r^2 = 0.043$, ($F(2,163) = 6.54$, $p = 0.01$
Core (mL^3)	$\beta = 0.208$ (95%CI 0.0968 to 0.3198), $r^2 = 0.085$, $F(2,163) = 13.64$, $p = 0.0003$
Age (years)	$\beta = -0.168$ (95%CI −0.3447 to 0.00859), $r^2 = 0.021$, $F(2,163) = 3.53$, $p = 0.06$
Arterial hypertension	$\beta = -3.063$ (95%CI −8.4308 to 2.3030), $r^2 = 0.007$, $F(2,163) = 1.27$, $p = 0.26$

β, beta coefficient; BP, blood pressure; $ETCO_2$, end-tidal carbon dioxide.

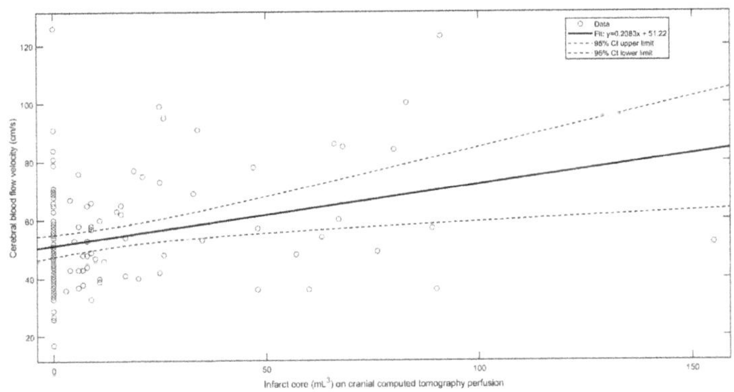

Figure 3. Linear regression model of cerebral blood flow velocity and ischemic infarct core volume on cranial computed tomography perfusion.

3.3.2. VLF Phase in the Affected Hemisphere

In the univariate linear regression analysis (Table 4), it was observed that the CBFV, penumbra, and infarct core volume were significantly related to the VLF phase.

Table 4. Univariate linear regression analysis to predict very low-frequency and low-frequency phases in the middle cerebral artery of the affected hemisphere.

Variable	Very Low Frequency Range
CBFV (cm/s)	$\beta = -0.004$ (95%CI −0.0074 to −0.0008), $r^2 = 0.03$, $F(2,163) = 5.98$, $p = 0.01$
LVEF (%)	$\beta = 0.0032$ (95%CI −0.0021 to 0.0084), $r^2 = 0.009$, $F(2,163) = 1.43$, $p = 0.23$
Mean BP (mm Hg) over recording period	$\beta = 0.002$ (95%CI −0.0021 to 0.0060), $r^2 = 0.005$, $F(2,163) = 0.88$, $p = 0.34$

Table 4. *Cont.*

Variable	Very Low Frequency Range
Mean $ETCO_2$ (mm Hg) over recording period	$\beta = 0.003$ (95%CI −0.0176 to 0.0241), $r^2 = 0.0007$, $F(2,163) = 0.09$, $p = 0.75$
Penumbra (mL^3)	$\beta = -0.001$ (95%CI −0.0018 to −0.0002), $r^2 = 0.042$, $F(2,163) = 6.76$, $p = 0.01$
Core (mL^3)	$\beta = -0.003$ (95%CI −0.0054 to −0.0005), $r^2 = 0.038$, $F(2,163) = 5.62$, $p = 0.01$
Age (years)	$\beta = 0.0002$ (95%CI −0.0037 to 0.0041), $r^2 = 0.0001$, $F(2,163) = 0.009$, $p = 0.92$
Arterial hypertension	$\beta = 0.05$ (95%CI −0.0653 to 0.1713), $r^2 = 0.005$, $F(2,163) = 0.78$, $p = 0.37$
Variable	Low Frequency Range
CBFV (cm/s)	$\beta = -0.005$ (95%CI −0.0058 to 0.0002), $r^2 = 0.02$, $F(2,163) = 3.30$, $p = 0.07$
LVEF (%)	$\beta = 0.004$ (95%CI −0.0001 to 0.0089), $r^2 = 0.023$, $F(2,163) = 3.67$, $p = 0.06$
Mean BP (mm Hg) over recording period	$\beta = 0.004$ (95%CI 0.0004 to 0.0075), $r^2 = 0.029$, $F(2,163) = 4.80$, $p = 0.02$
Mean $ETCO_2$ (mm Hg) over recording period	$\beta = 0.015$ (95%CI −0.0011 to 0.0329), $r^2 = 0.022$, $F(2,163) = 3.39$, $p = 0.06$
Penumbra (mL^3)	$\beta = -0.0009$ (95%CI −0.0016 to −0.0002), $r^2 = 0.040$, $F(2,163) = 6.01$, $p = 0.01$
Core (mL^3)	$\beta = -0.0008$ (95%CI −0.0030 to 0.0014), $r^2 = 0.003$, $F(2,163) = 0.55$, $p = 0.45$
Age (years)	$\beta = 0.001$ (95%CI −0.0018 to 0.0051), $r^2 = 0.005$, $F(2,163) = 0.86$, $p = 0.35$
Arterial hypertension	$\beta = 0.051$ (95%CI −0.0503 to 0.1591), $r^2 = 0.006$, $F(2,163) = 1.05$, $p = 0.30$

β, beta coefficient; BP, blood pressure; CBFV, cerebral blood flow velocity; $ETCO_2$, end-tidal carbon dioxide; LVEF, left ventricular ejection fraction.

In the multiple regression model with CBFV and penumbra or infarct core volume, the infarct core volume ($p = 0.01$) and penumbra volume ($p = 0.004$; Figure 4) remained significantly related to the VLF phase in the AH.

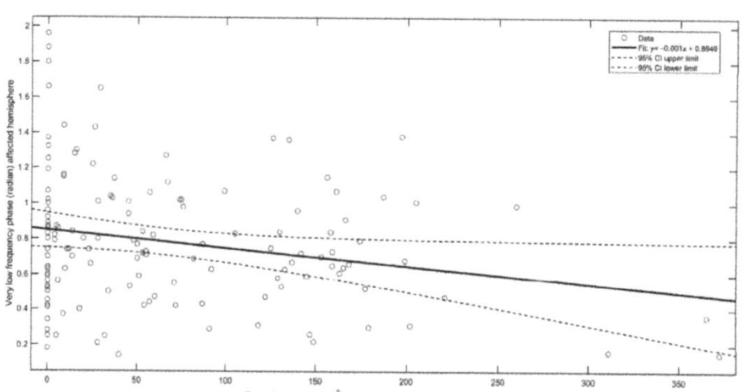

Figure 4. Linear regression model of very low frequency phase and ischemic penumbra on cranial computed tomography perfusion.

3.3.3. LF Phase in the Affected Hemisphere

Univariate linear regression analysis (Table 4) identified significant associations between LF phase in the AH and several independent variables, including CBFV, LVEF, mean BP, and mean $ETCO_2$ over the recording period, age, arterial hypertension, penumbra size, and core infarct volume. In the multiple linear regression models with mean BP over the recording period and penumbra as the independent variables, the LF phase in the AH was significantly associated with penumbra ($p = 0.04$) and showed a trend-level association with mean BP ($p = 0.06$).

3.3.4. VLF Gain in the Affected Hemisphere

Univariate linear regression analysis (Table 5) revealed a significant association between VLF gain in the AH and several independent variables, including CBFV, LVEF, mean BP, and mean $ETCO_2$ over the recording period, age, arterial hypertension, and penumbra or core infarct volume. In the various multiple linear regression analysis models with CBFV, mean BP over the recording period, and either penumbra or infarct core volume as independent variables, only CBFV remained significantly related to VLF gain in the AH ($p = 0.0000$).

Table 5. Univariate linear regression analysis to predict very low-frequency and low-frequency gain in the middle cerebral artery of the affected hemisphere.

Variable	Very Low Frequency Range
CBFV (cm/s)	$\beta = 0.006$ (95%CI 0.0050 to 0.0087), $r^2 = 0.24$, F(2,163) = 51.64, $p = 0.0000$
LVEF (%)	$\beta = -0.001$ (95%CI −0.0050 to 0.0018), $r^2 = 0.006$, F(2,163) = 0.91, $p = 0.34$
Mean BP (mm Hg) over recoding period	$\beta = -0.002$ (95%CI −0.0052 to −0.0000), $r^2 = 0.02$, F(2,163) = 3.97, $p = 0.04$
Mean $ETCO_2$ (mm Hg) over recording period	$\beta = 0.002$ (95%CI −0.0112 to 0.0156), $r^2 = 0.007$, F(2,163) = 0.10, $p = 0.74$
Penumbra (mL^3)	$\beta = 0.0005$ (95%CI 0.0000 to 0.0010), $r^2 = 0.02$, F(2,163) = 3.94, $p = 0.04$
Core (mL^3)	$\beta = 0.002$ (95%CI 0.0011 to 0.0041), $r^2 = 0.07$, F(2,163) = 11.73, $p = 0.0008$
Age (years)	$\beta = -0.001$ (95%CI −0.0040 to 0.0010), $r^2 = 0.008$, F(2,163) = 1.38, $p = 0.24$
Arterial hypertension	$\beta = -0.02$ (95%CI −0.1006 to 0.0508), $r^2 = 0.002$, F(2,163) = 0.42, $p = 0.51$
Variable	Low frequency range
CBFV (cm/s)	$\beta = 0.003$ (95%CI 0.0034 to 0.0074), $r^2 = 0.15$, (F(2,163) = 29.05, $p = 0.0000$
LVEF (%)	$\beta = -0.003$ (95%CI −0.0065 to 0.0000), $r^2 = 0.02$, F(2,163) = 3.84, $p = 0.05$
Mean BP (mm Hg) over recording period	$\beta = -0.004$ (95%CI −0.0073 to −0.0024), $r^2 = 0.08$, F(2,163) = 15.42, $p = 0.0001$
Mean $ETCO_2$ (mm Hg) over recording period	$\beta = -0.01$ (95%CI −0.0234 to 0.0026), $r^2 = 0.01$, F(2,163) = 2.51, $p = 0.11$
Penumbra (mL^3)	$\beta = 0.0004$ (95%CI −0.0001 to 0.0008), $r^2 = 0.01$, F(2,163) = 2.16, $p = 0.14$
Core (mL^3)	$\beta = 0.0001$ (95%CI −0.0014 to 0.0016), $r^2 = 0.0001$, F(2,163) = 0.01, $p = 0.90$

Table 5. *Cont.*

Variable	Very Low Frequency Range
Age (years)	$\beta = 0.001$ (95%CI −0.0014 to 0.0036), $r^2 = 0.005$, $F(2,163) = 0.79$, $p = 0.37$
Hypertension	$\beta = 0.006$ (95%CI −0.0678 to 0.0807), $r^2 = 0.002$, $F(2,163) = 0.02$, $p = 0.86$

β: beta coefficient; BP: blood pressure; CBFV: cerebral blood flow velocity; $ETCO_2$: end-tidal carbon dioxide; LVEF: left ventricular ejection fraction.

3.3.5. LF Gain in the Affected Hemisphere

Univariate linear regression analysis (Table 5) revealed significant associations between LF gain in the AH and several independent variables, including CBFV, LVEF, and mean BP, whereas penumbra and infarct core volume did not show any association. In the multiple linear regression model with CBFV, LVEF, and mean BP over the recording period as the independent variables and LF gain in the AH as the dependent variable, CBFV ($p = 0.0000$) and mean BP over the recording period ($p = 0.0000$) remained significantly related to LF gain.

3.3.6. Unaffected Hemisphere

In the unaffected hemisphere, the VLF phase was not associated with any of the predefined independent variables (Table 6).

Table 6. Univariate linear regression analysis to predict very low-frequency phase in the middle cerebral artery of the unaffected hemisphere.

	Variable
CBFV (cm/s)	$\beta = -0.002$ (95%CI −0.0061 to 0.0018), $r^2 = 0.009$, $F(2,163) = 1.13$, $p = 0.24$
LVEF (%)	$\beta = 0.002$ (95%CI −0.0036 to 0.0088), $r^2 = 0.005$, $F(2,163) = 0.66$, $p = 0.41$
Mean BP (mm Hg) over recording period	$\beta = 0.003$ (95%CI −0.0010 to 0.0085), $r^2 = 0.01$, $F(2,163) = 2.49$, $p = 0.11$
Mean $ETCO_2$ (mm Hg) over recording period	$\beta = -0.009$ (95%CI −0.0362 to 0.0164), $r^2 = 0.004$, $F(2,163) = 0.55$, $p = 0.45$
Penumbra (mL^3)	$\beta = 0.000$ (95%CI −0.001 to 0.001), $r^2 = 0.000$, $F(2,163) = 0.000$, $p = 0.98$
Core (mL^3)	$\beta = -0.001$ (95%CI −0.0044 to 0.0016), $r^2 = 0.007$, $F(2,72) = 0.87$, $p = 0.35$
Age (years)	$\beta = 0.0004$ (95%CI −0.0041 to 0.0048), $r^2 = 0.0002$, $F(2,72) = 0.02$, $p = 0.86$
Arterial hypertension	$\beta = -0.026$ (95%CI −0.1646 to 0.1111), $r^2 = 0.001$, $F(2,72) = 0.14$, $p = 0.70$

β, beta coefficient; BP, blood pressure; CBFV, cerebral blood flow velocity; $ETCO_2$, end-tidal carbon dioxide; LVEF, left ventricular ejection fraction.

The LF phase was related to mean BP over the recording period only ($p = 0.0002$, Table 7).

Table 7. Univariate linear regression analysis to predict low-frequency phase in the middle cerebral artery of the unaffected hemisphere.

	Variable
CBFV (cm/s)	$\beta = -0.006$ (95%CI -0.0060 to 0.0010), $r^2 = 0.01$, F(2,163) = 1.97, $p = 0.16$
LVEF (%)	$\beta = 0.004$ (95%CI -0.0009 to 0.0102), $r^2 = 0.02$, F(2,163) = 2.78, $p = 0.09$
Mean BP (mm Hg) over recording period	$\beta = 0.007$ (95%CI 0.0038 to 0.0118), $r^2 = 0.10$, F(2,163) = 14.8, $p = 0.0002$
Mean $ETCO_2$ (mm Hg) over recording period	$\beta = -0.001$ (95%CI -0.0241 to 0.0221), $r^2 = 0.000$, F(2,163) = 0.007, $p = 0.93$
Penumbra (mL^3)	$\beta = 0.000$ (95%CI -0.0008 to 0.0009), $r^2 = 0.0001$, F(2,163) = 0.006, $p = 0.93$
Core (mL^3)	$\beta = 0.001$ (95%CI -0.0013 to 0.0039), $r^2 = 0.009$, F(2,163) = 1.03, $p = 0.31$
Age (years)	$\beta = 0.0004$ (95%CI -0.0036 to 0.0043), $r^2 = 0.0003$, F(2,163) = 0.03, $p = 0.85$
Arterial hypertension	$\beta = 0.096$ (95%CI -0.0244 to 0.2181), $r^2 = 0.01$, F(2,163) = 2.49, $p = 0.11$

β, beta coefficient; BP, blood pressure; CBFV, cerebral blood flow velocity; $ETCO_2$, end-tidal carbon dioxide; LVEF, left ventricular ejection fraction.

In univariate linear regression analysis, VLF gain was related to mean BP over the recording period, infarct core volume, and age (Table 8).

Table 8. Univariate linear regression analysis to predict very low-frequency gain in the middle cerebral artery of the unaffected hemisphere.

	Variable
CBFV (cm/s)	$\beta = 0.002$ (95%CI -0.0005 to 0.0049), $r^2 = 0.02$, F(2,163) = 2.65, $p = 0.10$
LVEF (%)	$\beta = -0.007$ (95%CI -0.0051 to 0.0036), $r^2 = 0.000$, F(2,163) = 0.11, $p = 0.73$
Mean BP (mm Hg) over recording period	$\beta = -0.003$ (95%CI -0.0064 to 0.0001), $r^2 = 0.02$, F(2,163) = 3.71, $p = 0.05$
Mean $ETCO_2$ (mm Hg) over recording period	$\beta = -0.0003$ (95%CI -0.0188 to 0.0182), $r^2 = 0.000$, F(2,163) = 0.001, $p = 0.97$
Penumbra (mL^3)	$\beta = 0.0005$ (95%CI -0.0002 to 0.0011), $r^2 = 0.01$, F (2,72) = 2.00, $p = 0.15$
Core (mL^3)	$\beta = 0.002$ (95%CI 0.0005 to 0.0044), $r^2 = 0.05$, F(2,163) = 5.91, $p = 0.01$
Age (year)	$\beta = -0.004$ (95%CI -0.0072 to -0.0013), $r^2 = 0.06$, F(2,163) = 8.07, $p = 0.005$
Arterial hypertension	$\beta = -0.04$ (95%CI -0.1424 to 0.0463), $r^2 = 0.008$, F(2,163) = 1.01, $p = 0.31$

β, beta coefficient; BP, blood pressure; CBFV, cerebral blood flow velocity; $ETCO_2$, end-tidal carbon dioxide; LVEF, left ventricular ejection fraction.

In the multiple linear regression model, mean BP over the recording period, age, and infarct core volume were the independent variables, and VLF gain in the UH was the dependent variable. Age ($p = 0.03$) and mean BP over the recording period ($p = 0.03$) remained significantly related to VLF gain in the UH.

Univariate linear regression analysis revealed that LF gain in the UH was significantly related to CBFV, mean BP over the recording period, infarct core volume, and penumbra, with LVEF showing a trend-level relationship (Table 9). In the multiple linear regression models with CBFV, mean BP over the recording period, and either penumbra or infarct core volume as the independent variables and LF gain in the UH as the dependent variable, mean BP ($p = 0.003$), CBFV ($p = 0.008$), core ($p = 0.01$), and penumbra ($p = 0.01$) remained significantly related to LF gain in the UH.

Table 9. Univariate linear regression analysis to predict low-frequency gain in the middle cerebral artery of the unaffected hemisphere.

	Variable
CBFV (cm/s)	$\beta = 0.004$ (95%CI 0.0012 to 0.0070), $r^2 = 0.06$, $F(2,163) = 7.92$, $p = 0.005$
LVEF (%)	$\beta = -0.004$ (95%CI −0.0087 to 0.0004), $r^2 = 0.02$, $F(2,163) = 3.23$, $p = 0.07$
Mean BP (mm Hg) over recording period	$\beta = -0.005$ (95%CI −0.0090 to −0.0022), $r^2 = 0.07$, $F(2,163) = 10.6$, $p = 0.001$
Mean $ETCO_2$ (mm Hg) over recording period	$\beta = 0.0005$ (95%CI −0.0188 to 0.0199), $r^2 = 0.000$, $F(2,163) = 0.003$, $p = 0.95$
Penumbra (mL^3)	$\beta = 0.001$ (95%CI 0.0004 to 0.0018), $r^2 = 0.07$, $F(2,163) = 9.37$, $p = 0.002$
Core (mL^3)	$\beta = 0.003$ (95%CI 0.0013 to 0.0057), $r^2 = 0.08$, $F(2,163) = 10.07$, $p = 0.001$
Age (years)	$\beta = -0.001$ (95%CI −0.0044 to 0.0022), $r^2 = 0.003$, $F(2,163) = 0.44$, $p = 0.50$
Arterial hypertension	$\beta = 0.01$ (95%CI −0.0863 to 0.1192), $r^2 = 0.008$, $F(2,163) = 0.10$, $p = 0.75$

β, beta coefficient; BP, blood pressure; CBFV, cerebral blood flow velocity; $ETCO_2$, end-tidal carbon dioxide; LVEF, left ventricular ejection fraction.

4. Discussion

The two patient groups that received recanalization therapy (IV lysis vs. MT) exhibited notable differences. These clinical and CTP examination differences are most likely due to the grouping of our patients into those without a large vessel occlusion who received IV lysis only and those with large vessel occlusion who underwent MT, regardless of whether IV lysis was performed prior to MT. Regarding the cerebral hemodynamic responses, the IV lysis group exhibited, on average, more favorable outcomes, with higher VLF and LF phases and lower VLF and LF gain values in the AH. These findings are likely attributable to the absence of a large vessel occlusion in this group.

Overall, our main findings indicate that early CTP-detected ischemic parameters are associated with the future cerebral hemodynamic parameter, including phase, blood flow velocity, and gain.

When the VLF phase was assessed within the first 72 h after a stroke, a higher phase was associated with better outcomes, as defined by an mRs score < 2, and vice versa [19]. Our results indicated that the predicted VLF phase values increased as the volume of ischemic tissue (both penumbra and infarct core) decreased. This inverse relationship aligns with previous reports indicating that the dCA is more disrupted with larger brain infarct volumes [18,30]. Although the relationship observed in our patient cohort was not very strong, this study provides the first evidence that early CTP-derived ischemic parameters are associated with the dCA state in the affected hemisphere within the 48–72 h.

In the affected AH, VLF gain was related to CBFV, age, and mean BP over the recording period in the UH. In AH, it is plausible to assume that the elevated CBFV values in thrombectomy patients corresponded to a reactive hyperemic CBF that increased vascular gain (positively directed β-coefficient). In the UH group, VLF gain decreased with age, which is a noteworthy finding, as it contradicts trends observed in healthy, normal individuals [25,31]. Whether this observed age-related decrease in VLF gain represents a physiological response to a stroke event requires further research.

The LF phase in the AH was strongly associated with the penumbra and showed a trend towards significance with mean BP. The mean BP recorded for our patients was within the autoregulatory range usually considered normal of approximately 50–150 mm Hg [32]. Within this range, it is plausible that the LF phase varies, being lower at the lower end of BP and higher at the upper end. Recently, Wang and Payne [32] suggested that the CA may function adequately only within a narrow range of 80–100 mmHg of mean BP. Hence, a substantial number of our patients could have been at risk of secondary ischemia at the time of the dCA assessment. According to our results, the penumbra could be the first indicator of such a risk. In the UH, the LF phase showed a strong relationship with mean BP over the recording period but no association with CTP parameters (not even in the univariate analysis). Considering that LF phase is lower than our normative data (0.74 ± 0.20 radians) [25], further research is needed to clarify the clinical relevance of such lower phase values in the light of Wang and Payne's findings.

In the AH, LF gain was positively correlated with CBFV, suggesting that hyperemic CBF contributes to increased LF gain. Mean BP was negatively related to LF gain, indicating that an increase in BP leads to a decrease in gain. This could be due to the greater vasodilation of ischemia-damaged vessels, resulting in decreased gain, counterbalanced by the gain-enhancing effects of hyperemic CBFV. In the UH, LF gain is positively correlated with ischemic challenge in the brain. The gain in the UH was also negatively correlated with BP, as in the AH. This is also noteworthy, as the similarity in the BP–gain relationship between the two hemispheres, combined with the dependency of LF gain on ischemic factors and reactive hyperemic blood flow, suggests that ischemia triggers analogous autoregulatory processes in both hemispheres at the LF range.

Study Limitations

Although significant, the relatively low R^2 values indicate that the explanatory power of early CTP on future cerebral hemodynamics is weak, suggesting that other factors must play a role. The major limitation of this retrospective study is that our analysis assumed stable conditions between the data collected at different time points or days, disregarding potential dynamic changes, such as the initiation of restorative processes, which would have caused early dCA recovery. Another limitation is the BP dependency of dCA, that is, post-interventional BP management and the natural decline in autonomic stimulation after the acute phase of a stroke or therapeutic interventions could significantly alter dCA. Additionally, cardiac output and systemic vascular resistance may change over time because they depend on the autonomic state of patients. Thus, the reasons for the low R^2 values are not clear, and additional studies are required to determine the factors that were not measured in this study. Furthermore, we discuss whether standardized time points of dCA assessments would increase the predictive power. In our analysis, the following three clinically different patient groups were combined: those with IV-lysis only, those with MT only, and those who received IV-lysis and MT. The outcomes of dCA, especially in the IV-lysis/MT group, may not be the only effect of IV-lysis plus MT. When analyzing the relationship between ischemic tissue (e.g., the infarct core and penumbra) and VLF phase separated for each group, these relationships were no longer significant, underscoring the impact of sample size and variability. Finally, dCA assessment techniques are vulnerable to intraindividual differences [33–35].

Another limiting factor in our study was the use of CTP [27,36,37]. Cardiac output and the exact timing of contrast agent application are the reasons for inter-individual differences that can be accentuated by the patient's head movement. The latter also affects the data acquisition regions for the calculation of the penumbra and infarct core volume. Different

CTP protocols may generate variations in results, hindering the rapid generalization of monocentric findings. The dCA evaluates the CBFV response to BP changes over 2–50 s, with the VLF examining time periods between 15 and 50 s. The CTP data acquisition time used to calculate cerebral blood volume spans 20–40 s after contrast agent application [27] or up to 60–90 s (our protocol and Fainardi et al. [33]). This raises the question of whether dCA in the VLF is more likely to investigate pressure-dependent CBV changes or CBF changes contaminated by volume changes.

5. Conclusions

The assessment of cerebral autoregulation using dynamic cerebral autoregulation (dCA) can predict patient outcomes within 72 h after a stroke event [19]. In this retrospective study, we identified, for the first time, a weak but notable association between the volume of the ischemic tissue, as determined by the initial CTP to predict dCA impairment, and the occurrence of reactive hyperemic CBFV. Along with BP, CBFV determines vascular tone, which might be important for secondary hemorrhage. However, before these findings can translate into clinical applications, additional research is necessary to confirm and improve the predictive ability of CTP.

Author Contributions: Conceptualization, M.Ö.; validation, G.M.K.; formal analysis, L.-B.L., M.M. and G.M.K.; investigation, L.-B.L., M.B. and M.Ö.; data curation, M.B.; writing—original draft preparation, L.-B.L.; writing—review and editing, M.M. All authors have read and agreed to the published version of the manuscript.

Funding: This research received no external funding.

Institutional Review Board Statement: This retrospective cohort study of patients with acute stroke was approved by the Ethics Committee of Northwest and Central Switzerland number EKNZ; 2024-01039.

Informed Consent Statement: Informed consent was obtained from all subjects involved in the study.

Data Availability Statement: The corresponding author can provide all data on reasonable request.

Conflicts of Interest: The authors declare no conflicts of interest.

Abbreviations

AIS	acute ischemic stroke
BP	arterial blood pressure
CBFV	cerebral blood flow velocity in the middle cerebral artery
CTP	computed tomography perfusion
NIHSS	National Institute of Health Stroke Scale
mRs	modified Rankin scale
MT	mechanical thrombectomy
MRI	magnetic resonance imaging with diffusion-weighted imaging (DWI); T2 sequences and susceptibility-weighted imaging (SWI)
IQR	interquartile range
$ETCO_2$	end-tidal carbon dioxide tension
LVEF (%)	left ventricular ejection fraction (percentage)
VLF	very low frequencies (0.02–0.07 Hz)
LF	low frequencies (0.07–0.20 Hz)
HF	high frequencies (>0.20–0.50 Hz)

References

1. Saver, J.L.; Goyal, M.; Bonafe, A.; Diener, H.-C.; Levy, E.I.; Pereira, V.M.; Albers, G.W.; Cognard, C.; Cohen, D.J.; Hacke, W.; et al. Stent-retriever thrombectomy after intravenous t-PA vs. t-PA alone in stroke. *N. Engl. J. Med.* **2015**, *372*, 2285–2295. [CrossRef] [PubMed]
2. Campbell, B.C.; Mitchell, P.J.; Kleinig, T.J.; Dewey, H.M.; Churilov, L.; Yassi, N.; Yan, B.; Dowling, R.J.; Parsons, M.W.; Oxley, T.J.; et al. Endovascular therapy for ischemic stroke with perfusion-imaging selection. *N. Engl. J. Med.* **2015**, *372*, 1009–1018. [CrossRef] [PubMed]

3. Goyal, M.; Menon, B.K.; Van Zwam, W.H.; Dippel, D.W.J.; Mitchell, P.J.; Demchuk, A.M.; Dávalos, A.; Majoie, C.B.L.M.; Van Der Lugt, A.; De Miquel, M.A.; et al. Endovascular thrombectomy after large-vessel ischaemic stroke: A meta-analysis of individual patient data from five randomised trials. *Lancet* **2016**, *387*, 1723–1731. [CrossRef] [PubMed]
4. Albers, G.W.; Marks, M.P.; Kemp, S.; Christensen, S.; Tsai, J.P.; Ortega-Gutierrez, S.; McTaggart, R.A.; Torbey, M.T.; Kim-Tenser, M.; Leslie-Mazwi, T.; et al. Thrombectomy for stroke at 6 to 16 h with selection by perfusion imaging. *N. Engl. J. Med.* **2018**, *378*, 708–718. [CrossRef] [PubMed] [PubMed Central]
5. Nogueira, R.G.; Jadhav, A.P.; Haussen, D.C.; Bonafe, A.; Budzik, R.F.; Bhuva, P.; Yavagal, D.R.; Ribo, M.; Cognard, C.; Hanel, R.A.; et al. Thrombectomy 6 to 24 h after stroke with a mismatch between deficit and infarct. *N. Engl. J. Med.* **2018**, *378*, 11–21. [CrossRef] [PubMed]
6. Zhu, G.; Michel, P.; Aghaebrahim, A.; Patrie, J.T.; Xin, W.; Eskandari, A.; Zhang, W.; Wintermark, M. Prediction of recanalization trumps prediction of tissue fate: The penumbra: A dual-edged sword. *Stroke* **2013**, *44*, 1014–1019. [CrossRef] [PubMed]
7. Seners, P.; Oppenheim, C.; Turc, G.; Albucher, J.; Guenego, A.; Raposo, N.; Christensen, S.; Calvière, L.; Viguier, A.; Darcourt, J.; et al. Perfusion imaging and clinical outcome in acute ischemic stroke with large core. *Ann. Neurol.* **2021**, *90*, 417–427. [CrossRef] [PubMed]
8. Lansberg, M.G.; Christensen, S.; Kemp, S.; Mlynash, M.; Mishra, N.; Federau, C.; Tsai, J.P.; Kim, S.; Nogueria, R.G.; Jovin, T.; et al. Computed tomographic perfusion to Predict Response to Recanalization in ischemic stroke. *Ann. Neurol.* **2017**, *81*, 849–856. [CrossRef] [PubMed] [PubMed Central]
9. Broocks, G.; Kemmling, A.; Kniep, H.; Meyer, L.; Faizy, T.D.; Hanning, U.; Rimmele, L.D.; Klapproth, S.; Schön, G.; Zeleňák, K.; et al. Edema Reduction versus Penumbra Salvage: Investigating Treatment Effects of Mechanical Thrombectomy in ischemic Stroke. *Ann. Neurol.* **2023**, *95*, 137–145. [CrossRef] [PubMed]
10. Kneihsl, M.; Niederkorn, K.; Deutschmann, H.; Enzinger, C.; Poltrum, B.; Fischer, R.; Thaler, D.; Hermetter, C.; Wünsch, G.; Fazekas, F.; et al. Increased middle cerebral artery mean blood flow velocity index after stroke thrombectomy indicates increased risk for intracranial hemorrhage. *J. Neurointerv. Surg.* **2018**, *10*, 882–887. [CrossRef] [PubMed]
11. Lu, S.; Cao, Y.; Su, C.; Xu, X.; Zhao, L.; Jia, Z.; Liu, Q.; Hsu, Y.; Liu, S.; Shi, H.; et al. Hyperperfusion on arterial spin labeling MRI predicts the 90-day functional outcome after mechanical thrombectomy in ischemic stroke. *J. Magn. Reson. Imaging* **2021**, *53*, 1815–1822. [CrossRef] [PubMed]
12. Luijten, S.P.; Bos, D.; van Doormaal, P. J.; Goyal, M.; Dijkhuizen, R.M.; Dippel, D.W.; Roozenbeek, B.; van der Lugt, A.; Warnert, E.A. Cerebral blood flow quantification with multi-delay arterial spin labeling in ischemic stroke and the association with early neurological outcome. *NeuroImage Clin.* **2023**, *37*, 103340. [CrossRef] [PubMed] [PubMed Central]
13. Anadani, M.; Orabi, M.Y.; Alawieh, A.; Goyal, N.; Alexandrov, A.V.; Petersen, N.; Kodali, S.; Maier, I.L.; Psychogios, M.-N.; Swisher, C.B.; et al. Blood pressure and outcome after mechanical thrombectomy with successful revascularization. *Stroke* **2019**, *50*, 2448–2454. [CrossRef] [PubMed]
14. Jafari, M.; Desai, A.; Damani, R. Blood pressure management after mechanical thrombectomy in stroke patients. *J. Neurol. Sci.* **2020**, *418*, 117140. [CrossRef] [PubMed]
15. Morris, N.A.; Jindal, G.; Chaturvedi, S. Intensive blood pressure control after mechanical thrombectomy for acute ischemic stroke. *Stroke* **2023**, *54*, 1457–1461. [CrossRef] [PubMed]
16. Zhou, J.; Samara, H.; Ebrahim, A.; Kinariwala, J.; Mohamed, W. Blood pressure variability and short-term outcomes after mechanical thrombectomy in patients with acute ischemic stroke. *J. Stroke Cerebrovasc. Dis.* **2023**, *32*, 107197. [CrossRef] [PubMed]
17. Brasil, S.; Caldas, J.; Beishon, L.; Minhas, J.S.; Nogueira, R.C. Cerebral autoregulation monitoring: A guide while navigating in troubled waters. *Neurocrit. Care* **2023**, *39*, 736–737. [CrossRef] [PubMed]
18. Nogueira, R.C.; Aries, M.; Minhas, J.S.; HPetersen, N.; Xiong, L.; Kainerstorfer, J.M.; Castro, P. Review of studies on dynamic cerebral autoregulation in the acute phase of stroke and the relationship with clinical outcome. *J. Cereb. Blood Flow Metab.* **2022**, *42*, 430–453. [CrossRef]
19. Beishon, L.; Vasilopoulos, T.; Salinet, A.S.; Levis, B.; Barnes, S.; Hills, E.; Ramesh, P.; Gkargkoula, P.; Minhas, J.S.; Castro, P.; et al. Individual patient data meta-analysis of dynamic cerebral autoregulation and functional outcome after ischemic stroke. *Stroke* **2024**, *55*, 1235–1244. [CrossRef] [PubMed]
20. Lakatos, L.B.; Shin, D.C.; Müller, M.; Österreich, M.; Marmarelis, V.; Bolognese, M. Impaired dynamic cerebral autoregulation measured in the middle cerebral artery in patients with vertebrobasilar ischemia is associated with autonomic failure. *J. Stroke Cerebrovasc. Dis.* **2024**, *33*, 107454. [CrossRef] [PubMed] [PubMed Central]
21. Lakatos, L.-B.; Bolognese, M.; Österreich, M.; Weichsel, L.; Müller, M.M. Pre-Stroke antihypertensive therapy affects stroke severity and 3-month outcome of ischemic MCA-territory stroke. *Diseases* **2024**, *12*, 53. [CrossRef] [PubMed] [PubMed Central]
22. Lyden, P.; Brott, T.; Tilley, B.; Welch, K.M.; Mascha, E.J.; Levine, S.; Haley, E.C.; Grotta, J.; Marler, J. Improved reliability of the NIH Stroke Scale using video training. NINDS TPA Stroke Study Group. *Stroke* **1994**, *25*, 2220–2226. [CrossRef] [PubMed]
23. van Swieten, J.C.; Koudstaal, P.J.; Visser, M.C.; Schouten, H.J.; van Gijn, J. Interobserver agreement for the assessment of handicap in stroke patients. *Stroke* **1988**, *19*, 604–607. [CrossRef] [PubMed]
24. Adams, H.P., Jr.; Bendixen, B.H.; Kappelle, L.J.; Biller, J.; Love, B.B.; Gordon, D.L.; Marsh, E.E., 3rd. Classification of subtype of acute ischemic stroke. Definitions for use in a multicenter clinical trial. TOAST. Trial of Org 10172 in acute stroke treatment. *Stroke* **1993**, *24*, 35–41. [CrossRef] [PubMed]

25. Müller, M.; Österreich, M.; von Hessling, A.; Smith, R.S. Incomplete recovery of cerebral blood flow dynamics in sufficiently treated high blood pressure. *J. Hypertens.* **2019**, *37*, 372–379. [CrossRef] [PubMed] [PubMed Central]
26. Panerai, R.B.; Brassard, P.; Burma, J.S.; Castro, P.; Claassen, J.A.; van Lieshout, J.J.; Liu, J.; Lucas, S.J.; Minhas, J.S.; Mitsis, G.D.; et al. Transfer function analysis of dynamic cerebral autoregulation: A CARNet white paper 2022 update. *J. Cereb. Blood Flow Metab.* **2023**, *43*, 3–25. [CrossRef] [PubMed] [PubMed Central]
27. Konstas, A.A.; Goldmakher, G.V.; Lee, T.Y.; Lev, M.H. Theoretic basis and technical implementations of CT perfusion in acute ischemic stroke, part 1: Theoretic basis. *AJNR Am. J. Neuroradiol.* **2009**, *30*, 662–668. [CrossRef] [PubMed] [PubMed Central]
28. Higashida, R.T.; Furlan, A.J.; Roberts, H.; Tomsick, T.; Connors, B.; Barr, J.; Dillon, W.; Warach, S.; Broderick, J.; Tilley, B.; et al. Technology Assessment Committee of the American Society of Interventional and Therapeutic Neuroradiology, Technology Assessment Committee of the Society of Interventional Radiology. Trial design and reporting standards for intra-arterial cerebral thrombolysis for acute ischemic stroke. *Stroke* **2003**, *34*, e109–e137, Erratum in *Stroke* **2003**, *34*, e109–e137. [CrossRef] [PubMed]
29. North American Symptomatic Carotid Endarterectomy Trial Collaborators. Beneficial effect of endarterectomy in symptomatic patients with high-grade stenosis. *N. Engl. J. Med.* **1991**, *325*, 445–453. [CrossRef]
30. Reinhard, M.; Rutsch, S.; Lambeck, J.; Wihler, C.; Czosnyka, M.; Weiller, C.; Hetzel, A. Dynamic cerebral autoregulation associates with infarct size and outcome after ischemic stroke. *Acta Neurol. Scand.* **2012**, *125*, 156–162. [CrossRef]
31. Xing, C.Y.; Tarumi, T.; Meijers, R.L.; Turner, M.; Repshas, J.; Xiong, L.; Ding, K.; Vongpatanasin, W.; Yuan, L.J.; Zhang, R. Arterial pressure, heart rate, and cerebral hemodynamics across the adult life span. *Hypertension* **2017**, *69*, 712–720. [CrossRef] [PubMed] [PubMed Central]
32. Wang, Y.; Payne, S.J. Static autoregulation in humans. *J. Cereb. Blood Flow Metab.* **2024**, *44*, 1191–1207. [CrossRef] [PubMed]
33. Sanders, M.L.; Claassen, J.A.H.R.; Aries, M.; Bor-Seng-Shu, E.; Caicedo, A.; Chacon, M.; Gommer, E.D.; Van Huffel, S.; Jara, J.L.; Kostoglou, K.; et al. Reproducibility of dynamic cerebral autoregulation parameters: A multi-centre, multi-method study. *Physiol. Meas.* **2018**, *39*, 125002. [CrossRef] [PubMed]
34. Sanders, M.L.; Elting, J.W.J.; Panerai, R.B.; Aries, M.; Bor-Seng-Shu, E.; Caicedo, A.; Chacon, M.; Gommer, E.D.; Van Huffel, S.; Jara, J.L.; et al. Dynamic cerebral autoregulation reproducibility is affected by physiological variability. *Front. Physiol.* **2019**, *10*, 865. [CrossRef] [PubMed] [PubMed Central]
35. Elting, J.W.; Sanders, M.L.; Panerai, R.B.; Aries, M.; Bor-Seng-Shu, E.; Caicedo, A.; Chacon, M.; Gommer, E.D.; Van Huffel, S.; Jara, J.L.; et al. Assessment of dynamic cerebral autoregulation in humans: Is reproducibility dependent on blood pressure variability? *PLoS ONE* **2020**, *15*, e0227651. [CrossRef] [PubMed] [PubMed Central]
36. Fainardi, E.; Busto, G.; Rosi, A.; Scola, E.; Casetta, I.; Bernardoni, A.; Saletti, A.; Arba, F.; Nencini, P.; Limbucci, N.; et al. Tmax volumes predict final infarct size and functional outcome in ischemic stroke patients receiving endovascular treatment. *Ann. Neurol.* **2022**, *91*, 878–888. [CrossRef] [PubMed] [PubMed Central]
37. Katyal, A.; Calic, Z.; Killingsworth, M.; Bhaskar, S.M.M. Diagnostic and prognostic utility of computed tomography perfusion imaging in posterior circulation acute ischemic stroke: A systematic review and meta-analysis. *Eur. J. Neurol.* **2021**, *28*, 2657–2668. [CrossRef] [PubMed]

neurology international

Review

The Role of Somatic Mutations in Ischemic Stroke: CHIP's Impact on Vascular Health

Aiman Kinzhebay and Amankeldi A. Salybekov *

Qazaq Institute of Innovative Medicine, Regenerative Medicine Division, Cell and Gene Therapy Department, Astana 010000, Kazakhstan; a.kinzhebay@qaziim.kz
* Correspondence: amansaab@qaziim.kz

Abstract: Clonal hematopoiesis of indeterminate potential (CHIP) is increasingly recognized as a significant contributor to ischemic stroke and other cardiovascular diseases due to its association with somatic mutations in hematopoietic cells. These mutations, notably in genes like *DNMT3A*, *TET2*, and *JAK2*, induce pro-inflammatory and pro-atherosclerotic processes, promoting vascular damage and stroke risk. With the prevalence of CHIP rising with age, its presence correlates with higher mortality and morbidity rates in ischemic stroke patients. This article explores the mechanisms through which CHIP influences vascular aging and stroke, emphasizing its potential as a biomarker for early risk stratification and a target for therapeutic intervention. The findings highlight the necessity of integrating CHIP status in clinical evaluations to better predict outcomes and personalize treatment strategies in stroke management.

Keywords: stroke; CHIP; somatic mutation; vascular dysfunction; ischemic stroke; atherosclerosis

1. Introduction

The World Health Organization defines clonal hematopoiesis (CH) as a population of cells derived from a mutated multipotent stem or progenitor cell with a growth advantage, occurring without unexplained cytopenias, hematologic cancers, or other clonal disorders [1] (Figure 1A). The incidence of CH increases with age and has been linked to higher overall mortality, cardiovascular diseases (CVDs), and myeloid malignancies [2]. Clonal hematopoiesis of indeterminate potential (CHIP) refers specifically to CH with somatic mutations in myeloid malignancy-associated genes, detectable at a variant allele fraction (VAF) of $\geq$2% in blood or bone marrow, in individuals without a diagnosed hematologic disorder or unexplained cytopenia [3].

CH was initially observed in the 1960s through studies on mosaic X chromosome activity [4], and further understanding was gained with advancements in large-scale sequencing, revealing recurrent genetic mutations associated with this phenomenon. These cytogenetic studies of chronic myeloid leukemia showed that hematologic malignancies have a clonal origin [4]. In the 1990s, studies of non-random X-chromosome inactivation in healthy women suggested that clonal hematopoiesis could be a common aspect of aging, but the specific genetic lesions were unidentified, leaving the idea controversial [4]. The term CHIP achieved prominence a decade ago after the publication of a hallmark study by Jaiswal et al. [2]. The proposal to redefine premalignant and cancerous conditions and introduce the term CHIP was based on studies showing that cancer-causing mutations in clonal hematopoiesis-associated genes (e.g., *DNMT3A*, *TET2*, *ASXL1*) can also occur in

 https://doi.org/10.3390/neurolint16010017

individuals without related disorders, such as acute myeloid leukemia or myelodysplastic syndrome [3]. These mutations are detectable in people with normal blood counts and no apparent disease, but they increase the risk of hematological malignancies and higher mortality [3].

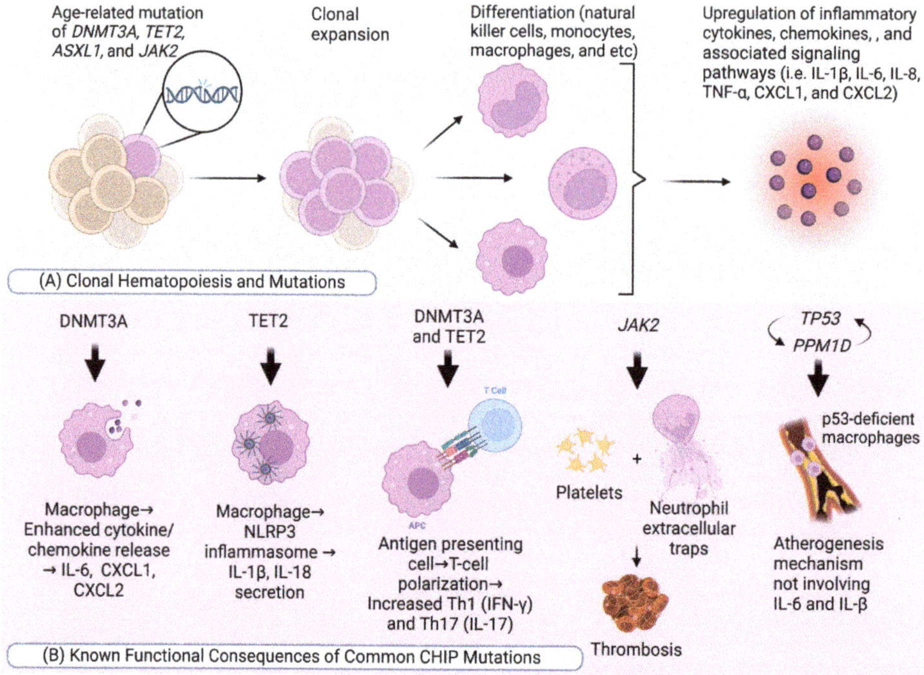

Figure 1. CHIP mutation and stroke. Created in BioRender©.

Notably, CHIP has been increasingly recognized as a major risk factor for stroke, due to the pro-inflammatory effects of mutated hematopoietic cells [5]. These mutations can accelerate vascular aging and promote atherosclerosis, two critical pathways involved in stroke pathogenesis [5]. The identification of specific CHIP mutations also holds promise for personalized medicine, providing opportunities to tailor interventions based on an individual's genetic risk profile. For instance, targeting inflammatory pathways or adjusting treatment plans according to CHIP mutation profiles is becoming a focus of precision medicine. Additionally, CHIP research has catalyzed the development of therapies targeting specific pathways, such as inflammation inhibitors or drugs designed to reduce clonal expansion. These advancements are paving the way for targeted treatments that can mitigate the risks associated with CHIP mutations, offering significant potential for improving patient outcomes. This growing evidence highlights the need for further study into how CHIP mutations contribute to stroke pathogenesis and how they can be used as predictive biomarkers.

2. Methods

To gather relevant information, guidelines, studies, and research papers associated with "somatic mutation", "clonal hematopoiesis of indeterminate potential (CHIP)", "clonal hematopoiesis", "ischemic stroke", "cardiovascular diseases", and "stroke" were identified through searches on PubMed, Google Scholar, Scopus, and Cochrane Library databases. Reference lists of selected studies were also manually reviewed. Only guidelines, studies, and research articles available in English were included in the analysis. Studies were excluded if they (i) were not in English; (ii) consisted of conference abstracts, notes, letters, case reports, or animal studies; or (iii) were duplicate entries.

3. Emerging Mechanisms of Clonal Hematopoiesis in Stroke Pathogenesis

Newer, more sensitive assays have broadened the understanding of CH. Using methods capable of detecting mutations with VAFs below 2%, CH has been identified in over half of individuals older than 70 [6]. Furthermore, whole-exome sequencing has shown that CH can affect both myeloid and lymphoid driver genes, revealing a more comprehensive spectrum of CH that includes lymphoid CH (L-CH) [6]. These findings underscore the evolving definition of CH and its dependence on the scope and sensitivity of detection methods, which have also revealed key demographic trends. For instance, detectable somatic mutations are rare in individuals younger than 40 but increase significantly with age, reflecting a strong age-related pattern in CH prevalence [6] Studies find CH mutations in 6% of individuals aged 60–69, 12% for ages 70–89, and 20% of those aged 90 and older [6]. According to another report, there is a 9.5% prevalence in individuals aged 70–79, 11.7% in those aged 80–89, and for 90–108-year-olds, it is 18.4% [2]. Despite these variations, both studies highlight the commonality of CH in aging populations and its strong association with increased risks of CHIP-associated cardiovascular and hematologic conditions, emphasizing its clinical relevance [2,6].

Initially, CH was mostly known as a risk factor for myeloid malignancies. However, the seminal 2014 article revealed the increased mortality among individuals with CHIP mutations, which could not be fully explained by the risk of hematologic cancers [2]. This increased risk was proposed to be linked to other diseases such as coronary heart disease (CHD) and ischemic stroke (IS). The exact mechanism of how CHIP mutations cause those CVDs is not fully studied, but it is an actively developing direction in research. One potential explanation is that somatic mutations in hematopoietic cells may affect immune cell function, leading to the development of conditions such as atherosclerosis and even non-cardiovascular diseases like type 2 diabetes. Both CVDs and diabetes are chronic inflammatory conditions that frequently coexist, likely due to shared underlying inflammatory processes. The presence of systemic inflammation may exacerbate both conditions, highlighting the potential role of CHIP-mediated immune dysregulation in their comorbidity [2]. The relationship between CHIP and inflammatory diseases can be likened to a "chicken-and-egg" analogy, as it remains unclear whether CHIP drives the progression of inflammatory diseases or if chronic inflammation promotes the expansion of CHIP clones [7].

In a fixed-effects meta-analysis [5], it was found that individuals with any CHIP mutation had a 14% increased risk of total stroke (95% CI, 1.03–1.27; $p = 0.01$), particularly driven by a 24% increase in the odds of hemorrhagic stroke (HS) (95% CI, 1.01–1.51]; $p = 0.04$), such as subarachnoid hemorrhage (SAH) [5]. Restricting the analysis to individuals with a variant allele fraction greater than 10% did not significantly change these associations, with a continued increased risk for total stroke (18%, [95% CI, 1.01–1.51]; $p = 0.04$). The association between CHIP and IS was weaker (11%, [95% CI, 0.98–1.25]; $p = 0.10$), with a stronger connection to small vessel stroke (SVS) [5]. The study suggests that CHIP may contribute to cerebral small vessel disease, which is linked to both IS and intracerebral hemorrhage (ICH). The discussion also addresses the potential mechanisms involving inflammatory pathways that might link CHIP to aneurysm formation and vascular fragility, particularly in older individuals. Moreover, in individuals over the age of 80, the presence of CHIP was more strongly associated with ICH (HR 1.84, $p = 0.01$). In the second cohort, there was a higher prevalence of cerebral aneurysms and nontraumatic SAH among individuals with CHIP [5].

3.1. Vascular Aging and Endothelial Dysfunction

Experimental studies have demonstrated the causal role of CHIP in cardiovascular disease, with specific mutations driving inflammation, as was mentioned above, and promoting atherogenesis (Figure 1B). For example, *TET2* deficiency in mice leads to larger atherosclerotic plaques and increased secretion of IL-1β, mediated by the NLRP3 inflammasome [8]. Similarly, $JAK2^{V617F}$ mutations enhance macrophage-driven plaque formation through the AIM2 inflammasome, leading to a more pronounced necrotic core in atherosclerotic plaques [9]. This means plaque-forming macrophages more readily engulf $JAK2^{V617F}$ erythrocytes (erythrophagocytosis), which subsequently impedes the macrophages' ability to clear apoptotic cells (efferocytosis) [10]. These studies highlight key inflammatory pathways involving IL-1β and IL-6 as central to the development of atherosclerosis in CHIP carriers (Figure 2). Additionally, CHIP mutations have been implicated in atherosclerosis across the arterial system, including PAD. Mutations in *TP53*, for example, result in the accumulation of defective macrophages, promoting atherosclerosis via mechanisms distinct from the IL-1β and IL-6 pathways. This connection between CHIP and widespread arterial disease extends beyond the coronary system to the cerebral, renal, and mesenteric vasculature [11,12]. In experiments using $Ldlr^{-/-}$ mice, the absence of *TET2* led to significantly larger atherosclerotic lesions compared to control mice, indicating that CHIP mutations enhance the severity of atherosclerosis [4,8]. Notably, these mutations did not cause leukocytosis, meaning that the effect was not due to an increased number of circulating immune cells but rather qualitative changes in immune cell function, especially myeloid cells. Further studies also demonstrated that *TET2* and *DNMT3A* loss exacerbate HF by promoting cardiac fibrosis, and hypertrophy, and reducing cardiac function. This underscores the role of altered immune cell behavior in CHIP-associated heart disease. Additionally, *JAK2* mutations, specifically $JAK2^{V617F}$, have been shown to promote thrombosis and atherosclerosis in mouse models [4,10,13]. In these cases, the formation of NETs leading to blood clot formation generally increased inflammation, and hyper-activation of platelets and coagulation pathways were contributing factors [14] (Figure 2). Even in the presence of lower cholesterol levels, mice with *JAK2* mutations developed accelerated atherosclerosis, suggesting that these mutations act as potent drivers of CVD. As was mentioned earlier in the article, direct evidence linking ASXL1 mutations to endothelial dysfunction is limited, but the pro-inflammatory environment induced by these mutations can adversely affect endothelial cells. The inflammatory milieu driven by ASXL1 mutations, marked by elevated cytokines like TNF-α, may indirectly contribute to endothelial dysfunction and increased cardiovascular risk [15].

A human study found that CHIP mutations were found in 19% of participants with coronary endothelial dysfunction (CED) and 13% in those without CED [16]. The presence of CHIP was associated with increased levels of inflammatory markers IL-6 and IL-8 in participants with CED, suggesting a link between CHIP and heightened inflammation [16]. Participants with CHIP mutations had a significantly higher risk of major adverse cardiovascular events (MACEs), particularly those with CED, even when accounting for variables like age, sex, and comorbidities [16]. The study suggests CHIP may contribute to early vascular aging and sustained cardiovascular risk, marked by inflammation in patients with endothelial dysfunction.

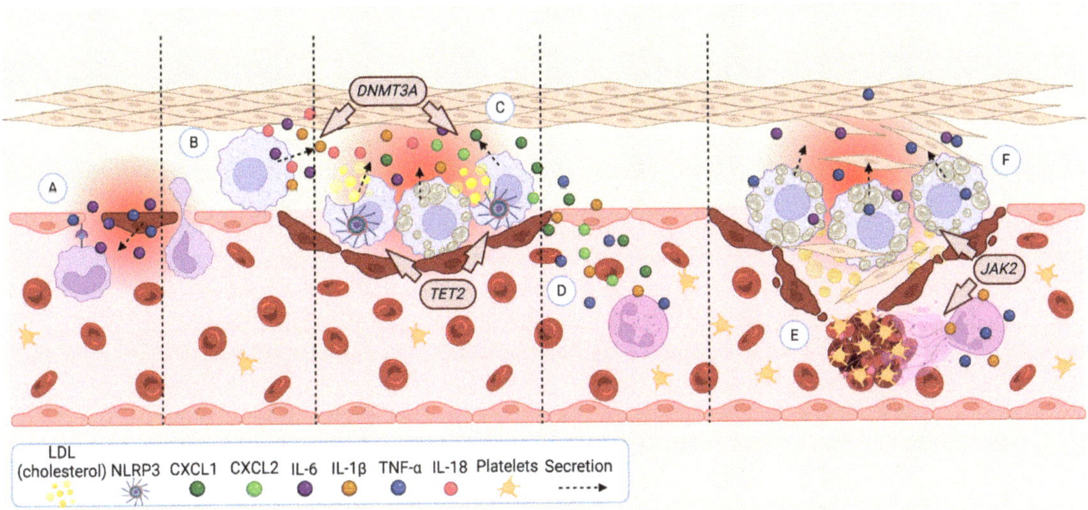

Figure 2. The general mechanism of CHIP-mediated atherogenesis and plaque rupture leading to stroke, with the inclusion of some contributing mutations. (**A**) IL-6 and TNF-α stimulate vascular permeability and expression of adhesion molecules to attract monocytes. (**B**) Monocytes exit the permeable vessel via diapedesis and turn into macrophages. (**C**) Macrophages phagocyte LDL and turn into foam cells and secrete cytokines. IL-18 activates NLRP3 inflammasome, which amplifies cytokine release. (**D**) CXCL1 and CXCL2 released by macrophages recruit additional monocytes and neutrophils to the site of inflammation. (**E**) IL-1β and TNF-α drive NET formation near the ruptured plaque leading to blood clotting. (**F**) Foam cells, smooth muscle, and other debris attracted by cytokines form a plaque that later raptures. Created in BioRender©.

3.2. Unraveling the Genetic Mutations

As humans age, a steady increase in somatic mutations occurs across various tissues, including hematopoietic stem cells (HSCs). These mutations result from processes like spontaneous deamination of 5-methylcytosine and the error-prone repair of DNA double-strand breaks. On average, each HSC acquires around 170 mutations per decade [4,17]. While most of these mutations do not affect cell fitness, occasionally, a mutation provides a selective advantage, leading to the expansion of a particular blood cell clone. There are several somatic mutations of varying frequency linked to clonal hematopoiesis [18]. Most of the instances of CHIP are caused by *DNMT3A*, *TET2*, and *ASXL1*, which are grouped as DTA genes; however, they are not the only CHIP mutations [18]. In a study of 97,691 genomes, the researchers identified 4938 CHIP mutations across 4229 individuals [18]. The median VAF of these mutations was 16%. Consistent with previous studies, over 75% of these mutations occurred in these three key DTA genes [18]. Around 15% of the mutations were found in the next five most frequently affected genes: *PPM1D*, *JAK2*, *SF3B1*, *SRSF2*, and *TP53* [18]. Significant variability in clonal fraction was observed among these eight genes [18]. For instance, the clonal fraction in peripheral blood was approximately 25% smaller for *DNMT3A* ($p = 1.3 \times 10^{-15}$) and 14% smaller for *TET2* ($p = 2.1 \times 10^{-4}$) compared to *ASXL1*, suggesting gene-specific differences in clonal selection [18]. Notably, 90% of individuals with CHIP driver mutations had only a single identified mutation [18]. CHIP prevalence was strongly correlated with age, and smoking increased the odds of developing CHIP by 18%. *JAK2* mutation carriers were the youngest, while carriers of *PPM1D*, *SF3B1*, and *SRSF2* mutations were older by 5 to 7.7 years on average [18].

DNMT3A (DNA methyltransferase 3A) encodes a methyltransferase that catalyzes DNA methylation, which is crucial for regulating gene expression [19]. Pathogenic mutations in *DNMT3A* are typically loss-of-function, including missense mutations in its catalytic domains, nonsense mutations, and indels, leading to enhanced self-renewal of HSCs [20]. These mutations promote the expression of multipotency genes and suppress differentiation genes, enabling *DNMT3A* mutations to affect various hematopoietic lineages, contributing to pro-inflammatory T-cell polarization and activation of the inflammasome complex [21]. Experimental studies using CRISPR-modified mouse models show that *DNMT3A* mutations result in aberrant inflammation [22]. *DNMT3A*-mutant HSCs outcompete wild-type cells under chronic infection conditions, leading to clonal expansion due to resistance to apoptosis and differentiation defects [22]. This inflammation-driven clonal expansion is linked to cardiovascular consequences, such as cardiac hypertrophy and fibrosis [22]. In human studies, *DNMT3A* mutations are associated with elevated expression of inflammatory cytokines and T-cell activation, contributing to cardiovascular pathologies like aortic stenosis and heart failure (HF) [23,24].

TET2 (ten-eleven translocation-2) plays a critical role in DNA demethylation by oxidizing methyl groups, functioning oppositely to *DNMT3A* [19]. Despite these biochemical differences, *TET2* mutations also enhance HSC self-renewal and lead to myeloid lineage bias [25]. *TET2*-deficient macrophages are highly pro-inflammatory, and this state accelerates atherosclerosis, as demonstrated in mouse models with increased atherosclerotic plaque burden and promotion of myocardial fibrosis, HF, and worsened cardiac outcomes in response to stress [26]. In human studies [23], *TET2* CHIP carriers have elevated circulating inflammatory markers, including non-classical monocytes that secrete TNF-α, IL-1β, and IL-8.

ASXL1 (additional sex combs-like 1) plays a key role in regulating polycomb-mediated transcriptional repression. *ASXL1* mutations drive myeloid transformation [27] but also impair HSC function in mice, leaving the precise mechanisms behind *ASXL1*-induced clonal hematopoiesis unclear. Furthermore, the relationship between *ASXL1* mutations and inflammation or atherosclerosis remains poorly understood, though it is hypothesized that the myeloid bias seen in *ASXL1* CHIP carriers may produce effects similar to those seen with *TET2* loss-of-function mutations [19].

JAK2 (Janus kinase 2) is a non-receptor tyrosine kinase that plays a crucial role in hematopoietic signaling by transmitting signals downstream of cytokine receptors [19]. It activates *TET2* via tyrosine phosphorylation in response to cytokines, thereby linking extracellular signals to epigenetic regulation of hematopoiesis [28]. $JAK2^{V617F}$ gain-of-function mutations are closely linked to myeloproliferative neoplasms like essential thrombocytopenia and myelofibrosis, which significantly elevate the risk of thrombotic events, including stroke and deep vein thrombosis ($p = 0.0003$, compared to non-CHIP) [13]. These mutations promote thrombosis through neutrophil extracellular trap formation, with studies showing increased incidence of deep vein thrombosis and pulmonary embolism in $JAK2^{V617F}$ CHIP carriers [13]. Elevated levels of pro-inflammatory cytokines, particularly IL-18 and IL-6, further exacerbate inflammation [29]. $JAK2^{V617F}$ mutations also confer the highest coronary artery disease (CAD) risk among CHIP variants, with up to a 10-fold increase [30]. In mouse models, $JAK2^{V617F}$ mutations accelerate atherosclerosis, macrophage proliferation, and oxidative damage, contributing to fibrosis and HF [31].

TP53 (tumor protein p53) and *PPM1D* (protein phosphatase 1D) are DNA repair genes. *PPM1D* plays a key role in the DNA damage response pathway and interacts in a feedback loop with the tumor suppressor *TP53* [19]. When *TP53* is activated, it induces *PPM1D* expression, which in turn downregulates apoptosis by dephosphorylating p53 [19]. *PPM1D* loss-of-function mutations are particularly associated with CH after exposure to chemotherapy agents such as cisplatin, etoposide, and doxorubicin [32]. There is an association between *TP53*-mutant CHIP and atherosclerotic diseases like CAD and peripheral artery disease (PAD) [11]. Experimental mouse models demonstrated that p53-deficient CHIP leads to larger atherosclerotic plaques with increased macrophage accumulation, suggesting that *TP53* mutations directly accelerate atherosclerosis. Unlike mutations in genes like *TET2* or *DNMT3A*, *TP53*-CHIP did not increase pro-inflammatory cytokines such as IL-1β and IL-6, indicating different mechanisms of atherogenesis.

Lastly, *SF3B1* and *SRSF2* are both crucial components of the mRNA spliceosome [19]. Defects in these genes disrupt the splicing and export of mRNAs that control translation processes [19]. Though their role in cardiovascular disease is not well-studied, elevated circulating IL-18 levels have been observed in patients with *SF3B1*-mutant CHIP [18].

When examining the association of specific CHIP driver mutations with the risk of stroke, researchers have found that mutations in *DNMT3A*, *TET2*, *ASXL1*, *JAK2*, and *TP53* were present at varying frequencies, with *TET2* being significantly associated with an increased risk of total stroke (HR, 1.93; $p = 0.006$) and IS (HR, 1.93; $p = 0.006$). For HS, both *TET2* (HR, 1.50; $p = 0.15$) and *DNMT3A* (HR, 1.44; $p = 0.03$) showed similar effect sizes, with TET2 having a slightly higher hazard ratio [5].

3.3. Epigenetics and CHIP

As mentioned above, a few of the most prevalent mutated genes of CHIP are epigenetic regulators, such as *DNMT3A*, *TET2*, *ASXL1*, and *JAK2* [19]. *DNMT3A* and *TET2* regulate opposing enzymatic functions essential for DNA methylation. *DNMT3A* promotes the production of a DNA methyltransferase enzyme, which adds methyl groups to CpG sites, altering gene expression and acting as a tumor suppressor by silencing specific genes. In contrast, *TET2* produces a DNA demethylase, which removes these methyl groups and recruits histone deacetylases to promoters, effectively reversing the epigenetic changes caused by *DNMT3A*. *ASXL1* is responsible for epigenetic modulation and regulation of chromatin-binding proteins. However, its function is relatively unknown. The NHGRI-EBI GWAS (genome-wide association study) catalog [26] and MRC-IEU EWAS (epigenome-wide association study) catalog [27] were used to search for stroke-causing genes and the corresponding epigenetic markers. The term "ischemic stroke" was employed as the primary search criterion within these databases. The identified gene sets from both catalogs were manually compared to identify common genes relevant to CHIP and IS. Between two catalogs, four overlapping genes were found in association with IS, as shown in the Table 1 below. Among these, ZFHX3 and SH2B3 have already been established as linked to stroke. Yet, the connection between the expression of the remaining genes, stroke, and other CVDs remains unclear, as discussed below. To bridge this gap, the role of each gene in the context of CVDs is reviewed and described, offering insights into their potential as future stroke biomarkers.

Table 1. Overlapping Genes Associated with Ischemic Stroke from GWAS and EWAS Catalogs.

Gene	CpG	Location	*p* Values	Study
ZFHX3	cg07786668	chr16:73092391	1.6×10^{-18}, 3.1×10^{-6}	[33–36]
	cg00614832	chr16:73092394	3.2×10^{-16}, 6.4×10^{-7}	
SH2B3	cg12885832	chr12:111843885	3.5×10^{-8}, 2.4×10^{-14}	[36–38]
DNM2	cg25487903	chr19:10828950	4.7×10^{-6}, 1.7×10^{-8}	[36,39]
SETBP1	cg20127035	chr18:42260234	9.8×10^{-7}	[36,40]
	cg00037763	chr18:42281763	3.1×10^{-6}	

ZFHX3 (zinc finger homeobox 3) has been associated with stroke, particularly atrial fibrillation (AF), which is a significant risk factor for IS [40]. Variants in the *ZFHX3* gene have been linked to increased susceptibility to AF, thus elevating the risk of stroke due to embolic events. Studies have identified single nucleotide polymorphisms (SNPs) within *ZFHX3* that are significantly associated with both AF and IS, underscoring the gene's role in the development of cerebrovascular diseases [41]. Given its involvement in cardiac conduction and AF, *ZFHX3* is considered a genetic marker for stroke risk, particularly concerning arrhythmia-induced events [40].

The *SH2B3* gene, which encodes the lymphocyte adaptor protein LNK, has been associated with an increased risk of stroke, particularly due to its role in regulating inflammation and immune responses [42]. Variants like rs3184504 in *SH2B3* are linked to pro-inflammatory pathways and the dysregulation of cytokine signaling, which can contribute to endothelial dysfunction, thrombosis, and vascular complications. LNK's negative regulation of these pathways helps maintain vascular integrity, and its deficiency or mutation may predispose individuals to ischemic stroke through enhanced inflammation and clot formation [43].

The *DNM2* gene (Dynamin 2) plays an important role in various cellular processes, including vesicle trafficking and endocytosis. While there is growing interest in *DNM2's* role in blood and hematopoietic systems, especially in clonal hematopoiesis and myelodysplastic syndromes, the direct link between *DNM2* mutations and stroke is not yet well-established in the scientific literature. There is, however, indirect evidence that *DNM2*-related mechanisms could be involved in processes relevant to stroke. *DNM2* mutations can affect blood cell formation and immune response regulation, including neutrophil migration, which could play a role in vascular inflammation—a key factor in stroke development [44]. Moreover, *DNM2* has been shown to influence endothelial cell function, which is important for maintaining vascular integrity and could, in theory, affect stroke risk.

SETBP1 has been associated with CHIP, a condition linked to an increased risk of cardiovascular diseases, including stroke. In a study investigating stroke outcomes, *SETBP1* mutations were identified in stroke patients, with CHIP carriers exhibiting distinct profiles of cardiovascular risk [45]. While the direct role of *SETBP1* in stroke pathogenesis is still being studied, its involvement in clonal hematopoiesis suggests that it may contribute to an increased risk of ischemic stroke through mechanisms involving inflammation and vascular dysfunction.

There are several studies showing this effect of CHIP on epigenetic age, such as a study by Soerensen et al. on elderly Danish twins [46]. Unlike previous research that treated CHIP as a uniform entity, this study distinguishes between mutations in specific genes like *TET2* and *DNMT3A* and their divergent effects on different types of epigenetic aging. For instance, *DNMT3A* mutations show a stronger association with intrinsic epigenetic age acceleration (IEAA), which relates to the biological aging of cells, while *TET2* mutations correlate more with extrinsic age acceleration, which reflects immune system aging. This gene-specific differentiation provides more nuanced insights into the behavior of CHIP and its influence on aging pathways. Additionally, *TET2* mutations tend to arise later in life than *DNMT3A* mutations, possibly due to the different environments needed for clonal expansion, with *TET2*-associated clones thriving in immunologically aged or senescent settings. The study [46] also highlights the potential for CHIP to be used as a tool for identifying individuals at elevated risk for adverse outcomes, such as heart disease or mortality. However, its findings diverge from those of other studies, such as Nachun et al. [47], which suggested that combining CHIP and accelerated epigenetic aging (e.g., AgeAccelHG) could help stratify high-risk individuals. This discrepancy may arise from differences in cohort sizes, age distributions, and methylation data normalization methods.

While this study did not replicate the exact results from Nachun et al., it still underscores the importance of considering age and CHIP mutations in clinical stratification models, particularly for identifying at-risk elderly individuals. As expected, the latter study also demonstrated that CHIP accelerates epigenetic aging and shows strong associations with intrinsic epigenetic aging clocks such as Horvath and IEAA, likely due to shared genetic factors, such as polymorphisms near *TERT* and *TRIM59* [47]. One of the most important findings is that combining CHIP with high scores in epigenetic aging clocks, such as Hannum and GrimAge, creates a potent risk factor for adverse health outcomes, including CHD and overall mortality. Individuals with both CHIP and high AgeAccelHG scores face a synergistic increase in risk, while CHIP without epigenetic aging shows no substantial association with poor outcomes. This context dependence highlights the influence of lifestyle factors (e.g., smoking, diet, BMI) on CHIP's impact.

3.4. Inflammatory Pathways

Most common mutated genes mentioned above are known to cause inflammation, making CHIP a pro-inflammatory condition that leads to diseases such as stroke. CHIP mutations are more commonly found in myeloid cells, natural killer cells, and monocytes, where they drive inflammatory responses [12,48]. *TET2* and *DNMT3A* loss-of-function mutations are known to cause dysregulated T-cell function, which leads to altered polarization of immune responses, promoting inflammation [49,50] (Figure 1B). *TET2* mutations are especially linked to upregulation of pro-inflammatory gene expression in macrophage cells, leading to increased production of IL-1β, IL-6, and chemokines like CXCL1 and CXCL2 [26]. This upregulation is often mediated by increased activation of the NLRP3 inflammasome, which further promotes the production of IL-1β [8]. Similarly, mutations in *DNMT3A* in macrophages exhibit comparable but distinct pro-inflammatory gene expression profiles, reinforcing the inflammatory response driven by CHIP. Shortly, CHIP mutations lead to enhanced inflammation by altering immune cell function and increasing the production of pro-inflammatory cytokines, contributing to conditions like atherosclerosis (Figure 2) and other inflammatory diseases.

Unlike the mutations described above, some CHIP mutations have no established link to stroke; however, it could be speculated based on their inflammatory actions. *ASXL1* leads to the skewed differentiation of monocytes and macrophages, resulting in increased production of pro-inflammatory cytokines such as TNF-α [15]. This upregulation is mediated through the EGR1-TNF-α axis, which enhances inflammation and contributes to disease progression [15]. $JAK2^{V617F}$ mutant cells exhibit heightened activation of the JAK-STAT1 signaling pathway. Similarly to DNMT3A and TET2 mutations, this enhances transcription of pro-inflammatory cytokines, including IL-6, IL-1β, and TNF-α. However, a thing unique to this CHIP mutation is *JAK2*-mutant neutrophils being prone to form neutrophil extracellular traps (NETs), contributing to tissue inflammation and thrombotic complications. [31]. Also, macrophages from JAK^{2V617F} mice displayed heightened inflammasome activation, marked by increased secretion of IL-1β and IL-18. [10]. The roles of *SF3B1* and *SRSF2* mutation in inflammatory pathways are now well-known; however, there were reported cases of elevated circulating IL-18 levels in patients with *SF3B1* mutation [18].

Chronic inflammatory conditions, including various pro-inflammatory and rheumatologic diseases, are associated with CHIP. For instance, systemic sclerosis (SSc) patients show a higher prevalence of CHIP (25%) compared to healthy individuals (4%), with *DNMT3A* mutations being most common, although no significant clinical differences were noted between CHIP and non-CHIP carriers [51]. Rheumatoid arthritis (RA) patients also demonstrate an elevated CHIP prevalence, especially in older patients, with *DNMT3A* and *TET2* mutations prevalent [52]. The association between CHIP and other rheumatologic conditions like systemic lupus erythematosus (SLE) is still uncertain, though these conditions carry an increased risk of myeloid neoplasms. Ulcerative colitis (UC), an inflammatory bowel disease, is also linked to *CHIP*, particularly with *DNMT3A* and *PPM1D* mutations, and is associated with an increased risk of ischemic heart disease. Elevated Th17 cells and related cytokines, important in IBD pathogenesis, are similarly observed in *DNMT3A*-CHIP carriers with severe aortic stenosis. One such mechanism is chronic, low-grade inflammation, known as "inflammaging", which is characterized by elevated levels of inflammatory markers like IL-6 and C-reactive protein (CRP) in older adults [4]. This inflammatory state is linked to an increased risk of chronic diseases of aging. Recent studies have identified clonal hematopoiesis, where blood cell clones with somatic mutations expand in older individuals, as a potential link between aging, inflammation, and cardiovascular diseases.

4. Clinical Implications of Clonal Hematopoiesis in Stroke

4.1. Risk Stratification

Stroke ranked as the second leading cause of death and long-term disability worldwide. Approximately 87% of all strokes are ischemic, while around 10% are due to intracranial hemorrhage, and 3% result from subarachnoid hemorrhage [53–55]. Incorporating CH into stroke risk stratification models as an independent predictor can enhance predictive accuracy, particularly in older adults, and facilitate more personalized preventive strategies. This approach allows for the identification of high-risk individuals earlier and promotes targeted interventions to prevent adverse outcomes.

Recent studies have incorporated CHIP in risk stratification models, particularly in predicting the progression to hematologic malignancies and understanding cardiovascular risks. For example, the Clonal Hematopoiesis Risk Score (CHRS) was developed to estimate the likelihood of CHIP/CCUS (Clonal Cytopenia of Undetermined Significance) patients progressing to myeloid malignancies [56]. This model leverages data from large cohorts, such as the UK Biobank, to identify and weight key prognostic factors like high-risk mutations (e.g., in *TP53* or *RUNX1*), clone size, and cytopenia. The CHRS then stratifies patients into low, intermediate, or high-risk groups for myeloid malignancies. A similar model could be created with a focus on stroke by adding other biomarkers and risk factors specific to the disease such as hypertension, diabetes, and cholesterol levels.

However, there are current limitations to incorporating CH into routine clinical screening. One of the main challenges is the cost and complexity of genetic testing required for CHIP detection. Technologies like NGS, while highly sensitive, are expensive and not widely accessible in all healthcare settings. Whole-genome and whole-exome sequencing offer comprehensive mutation coverage but are costly (>$1000/sample), while targeted amplicon sequencing and hybrid capture panels are more affordable ($30–$150/sample) but have technical limitations, making them more appropriate for research rather than routine clinical use [57]. This raises questions about the cost-effectiveness of routine CH screening, particularly for individuals without overt cardiovascular disease or other high-risk factors. It is also worth considering that most likely these expenses are not covered by insurance.

The clinical significance of CH, however, is not yet fully understood. While there is evidence linking CH to increased cardiovascular risk, the exact mechanisms remain unclear, and it is uncertain whether interventions targeting CH can meaningfully reduce stroke incidence or improve outcomes. In addition, not all CH mutations carry the same risk, and the presence of a CHIP mutation does not necessarily indicate an imminent cardiovascular event. This variability complicates the decision of whether to incorporate CH screening into standard practice. In light of these uncertainties, routine CHIP screening may not be necessary for the general population. Current evidence supports its potential use in high-risk individuals, such as older adults with existing cardiovascular disease or multiple risk factors, but broader screening may need further validation through large-scale studies before it can be recommended as part of standard preventive care. The decision to screen for CHIP should be weighed against the potential benefits, costs, and current gaps in our understanding of its long-term impact.

4.2. Therapeutic Targeting

Mitigating the cardiovascular and stroke risks associated with CH is an emerging area of research, focusing primarily on targeting specific mutations driving CH and anti-inflammatory treatments. FDA-approved therapies, such as IDH1 inhibitors (ivosidenib), IDH2 inhibitors (enasidenib), and *JAK2* inhibitors (ruxolitinib and fedratinib), directly target the underlying mutations [58]. Hypomethylating agents like azacitidine and decitabine have shown increased efficacy in managing *TET2*-mutant myeloid malignancies [59,60], while high-dose vitamin C can limit the expansion of hematopoietic stem and progenitor cells (HSPCs) with *TET2* deletions [61]. Mutations in the *JAK2* gene, particularly $JAK2^{V617F}$, are common in CH and are linked to increased thrombotic events. JAK-STAT pathway inhibitors, such as ruxolitinib, could potentially reduce the pro-thrombotic and pro-inflammatory effects of these mutations [14].

Since CH is strongly linked to increased systemic inflammation, which plays a key role in both stroke and CVD pathogenesis, targeting inflammatory pathways offers a promising therapeutic strategy. Studies indicate that blocking IL-6 signaling—another key inflammatory mediator—might reduce CVD risk in patients with large CHIP clones [62]. Mouse models with *TET2* or *DMMT3A* mutations exhibit elevated IL-6 levels, implicating this pro-inflammatory pathway. Canakinumab, an IL-1β inhibitor, has demonstrated efficacy in reducing cardiovascular events by lowering inflammation, as shown in the CANTOS trial, suggesting its potential benefit for CH-positive individuals [62,63]. Targeting IL-1β, upstream of IL-6, reduces CVD events, suggesting IL-6 blockade could be a promising strategy for individuals with CHIP while potentially avoiding adverse effects like elevated LDL cholesterol. Similarly, Colchicine, an anti-inflammatory agent traditionally used for gout, has been effective in reducing CVD risk in trials like COLCOT, by addressing the inflammation that may drive clonal expansion in CH [64].

5. Clonal Hematopoiesis as a Biomarker in Stroke

5.1. Prognostic Value

Current stroke risk assessment tools, like the Essen Stroke Risk Score (ESRS) and Stroke Prognostic Instrument-II (SPI-II), are only moderately effective, considering factors like age, gender, obesity, and smoking [65]. However, these tools overlook the role of CHIP, which involves mutations in genes like *TET2* and *DNMT3A* that drive chronic inflammation. This inflammation contributes to endothelial dysfunction, atherosclerosis, and increased cardiovascular disease risk, making CHIP a potential biomarker for predicting ischemic stroke risk [65]. Köhnke and Majeti suggest focusing on two high-risk groups when planning prospective trials for CHIP [58]. The first group is individuals with DNMT3A or TET2 mutations, who face elevated cardiovascular disease CVD risks, highlighting the need to determine VAF thresholds for risk assessment. Current evidence indicates that even low VAF levels may correlate with increased CVD mortality, such as VAF $\geq$ 1.15% for *DNMT3A* and $\geq$0.73% for *TET2* in IS. Secondly, in cancer patients, CH mutations like *TP53*, *PPM1D*, or *CHEK2* may increase the risk of therapy-related myeloid neoplasms (tMNs), suggesting that prospective trials should evaluate whether chemotherapy or radiotherapy should be avoided in these individuals [58]. Alternative treatment strategies, including avoiding allogeneic stem cell transplants in lymphoma patients with CH mutations, are also being considered. However, they also discuss how designing clinical trials for CH involves several challenges [58]. The low risk of CH progression in many cases means that treatment-related toxicities must be carefully balanced against the risk of disease progression. Additionally, the long latency between CH detection and events like cardiovascular disease or myeloid malignancies makes trials based on primary endpoints challenging. Surrogate markers, such as reductions in mutant VAF or decreases in mutant clones, may need to be used, though their correlation with long-term outcomes is unclear. Trials must also address difficulties like identifying asymptomatic patients and determining appropriate treatment durations. If treatments are approved, CH would become a public health issue, requiring large-scale genetic screening in primary care, demanding new infrastructure and coordination across clinical labs, primary care, and public health systems. Trials should initially focus on high-risk groups with the most clinically relevant endpoints.

Data from the previous studies [2,11] can be applied to create predictive models for stroke outcomes based on initial CHIP levels. For example, Jaiswal et al. [2] found that individuals with somatic mutations, particularly those associated with CHIP, show a higher cumulative incidence of ischemic stroke over time. This is significant, as patients with prevalent CHIP mutations are at a 2.6-fold (95% confidence interval, 1.3 to 4.8; $p = 0.003$) higher risk of IS compared to those without mutations [2]. Additional factors like hypertension, age, and diabetes further compound this risk, making CHIP a potential early marker for CVD. By incorporating variables like CHIP status (e.g., presence or absence, VAF < 10%, VAF $\geq$ 10%), age, hypertension, and diabetes, multivariable models can predict stroke incidence and progression. The studies' hazard ratios, especially for larger CHIP clones, offer insights into the vascular risks. Competing risk regression models, accounting for comorbidities, could then predict stroke outcomes, and these models would need to be validated in independent cohorts.

5.2. Monitoring Disease Progression

Monitoring CH mutations over time in stroke patients could reveal early warning signs of clonal expansion, which is linked to increased inflammation and thrombotic events. This monitoring helps predict recurrent stroke risk, as patients with CHIP often experience worse outcomes due to systemic inflammation, driven by pathways like IL-6 and IL-1β. Detecting early changes in these pathways provides insight into vascular deterioration. Additionally, CH monitoring can inform treatment strategies for related conditions such as coronary artery disease, where clonal expansion destabilizes plaques, increasing heart attack or secondary stroke risks. Tracking these changes supports precision medicine approaches: rapid clonal expansion may necessitate aggressive interventions, while stable clones might suggest conservative management. This approach parallels cancer prognostic studies, where long-term mutation tracking informs disease progression and guides personalized care.

6. Conclusions

Somatic mutations, or clonal hematopoiesis of indeterminate potential (CHIP), significantly impact ischemic stroke and cardiovascular diseases. Recent studies emphasize the need for close monitoring of patients with substantial somatic mutations due to their high risk of mortality and morbidity. Our research underscores the critical importance of risk stratification in these patients. Understanding the mechanistic role of somatic mutations in stroke and cardiovascular conditions using/integrating various multi-omics approaches is essential for improving patient outcomes.

Author Contributions: Conceptualization, A.A.S.; methodology, A.A.S. and A.K.; software, A.K. and A.A.S.; validation, A.A.S.; resources, A.A.S.; writing—original draft preparation, A.K. and A.A.S.; writing—review and editing, A.K. and A.A.S.; visualization, A.K. and A.A.S.; supervision, A.A.S.; project administration, A.A.S.; funding acquisition, A.A.S. All authors have read and agreed to the published version of the manuscript.

Funding: This study was supported by the Science Committee of the Ministry of Science and Higher Education of the Republic of Kazakhstan, grant no. AP23490482.

Data Availability Statement: Not applicable.

Acknowledgments: The authors thank BioRender for providing a platform to create compelling and insightful illustrations.

Conflicts of Interest: The authors declare that they have no known competing financial interests or personal relationships that could have appeared to influence the work reported in this paper.

References

1. Khoury, J.D.; Solary, E.; Abla, O.; Akkari, Y.; Alaggio, R.; Apperley, J.F.; Bejar, R.; Berti, E.; Busque, L.; Chan, J.K.C.; et al. The 5th edition of the World Health Organization Classification of Haematolymphoid Tumours: Myeloid and Histiocytic/Dendritic Neoplasms. *Leukemia* **2022**, *36*, 1703–1719. [CrossRef] [PubMed]
2. Jaiswal, S.; Fontanillas, P.; Flannick, J.; Manning, A.; Grauman, P.V.; Mar, B.G.; Lindsley, R.C.; Mermel, C.H.; Burtt, N.; Chavez, A.; et al. Age-Related Clonal Hematopoiesis Associated with Adverse Outcomes. *N. Engl. J. Med.* **2014**, *371*, 2488–2498. [CrossRef]
3. Steensma, D.P.; Bejar, R.; Jaiswal, S.; Lindsley, R.C.; Sekeres, M.A.; Hasserjian, R.P.; Ebert, B.L. Clonal hematopoiesis of indeterminate potential and its distinction from myelodysplastic syndromes. *Blood* **2015**, *126*, 9–16. [CrossRef] [PubMed]
4. Jaiswal, S.; Libby, P. Clonal haematopoiesis: Connecting ageing and inflammation in cardiovascular disease. *Nat. Rev. Cardiol.* **2020**, *17*, 137–144. [CrossRef]
5. Bhattacharya, R.; Zekavat, S.M.; Haessler, J.; Fornage, M.; Raffield, L.; Uddin, M.M.; Bick, A.G.; Niroula, A.; Yu, B.; Gibson, C.; et al. Clonal Hematopoiesis Is Associated With Higher Risk of Stroke. *Stroke* **2022**, *53*, 788–797. [CrossRef]
6. Winter, S.; Götze, K.S.; Hecker, J.S.; Metzeler, K.H.; Guezguez, B.; Woods, K.; Medyouf, H.; Schäffer, A.; Schmitz, M.; Wehner, R.; et al. Clonal hematopoiesis and its impact on the aging osteo-hematopoietic niche. *Leukemia* **2024**, *38*, 936–946. [CrossRef]
7. Cypris, O.; Božić, T.; Wagner, W. Chicken or Egg: Is Clonal Hematopoiesis Primarily Caused by Genetic or Epigenetic Aberrations? *Front. Genet.* **2019**, *10*, 785. [CrossRef]
8. Fuster, J.J.; MacLauchlan, S.; Zuriaga, M.A.; Polackal, M.N.; Ostriker, A.C.; Chakraborty, R.; Wu, C.-L.; Sano, S.; Muralidharan, S.; Rius, C.; et al. Clonal hematopoiesis associated with TET2 deficiency accelerates atherosclerosis development in mice. *Science* **2017**, *355*, 842–847. [CrossRef]
9. Fidler, T.P.; Xue, C.; Yalcinkaya, M.; Hardaway, B.; Abramowicz, S.; Xiao, T.; Liu, W.; Thomas, D.G.; Hajebrahimi, M.A.; Pircher, J.; et al. The AIM2 inflammasome exacerbates atherosclerosis in clonal haematopoiesis. *Nature* **2021**, *592*, 296–301. [CrossRef]
10. Wang, W.; Liu, W.; Fidler, T.; Wang, Y.; Tang, Y.; Woods, B.; Welch, C.; Cai, B.; Silvestre-Roig, C.; Ai, D.; et al. Macrophage Inflammation, Erythrophagocytosis, and Accelerated Atherosclerosis in $Jak2^{V617F}$ Mice. *Circ. Res.* **2018**, *123*, 11. [CrossRef] [PubMed]
11. Zekavat, S.M.; Viana-Huete, V.; Matesanz, N.; Jorshery, S.D.; Zuriaga, M.A.; Uddin, M.M.; Trinder, M.; Paruchuri, K.; Zorita, V.; Ferrer-Pérez, A.; et al. TP53-mediated clonal hematopoiesis confers increased risk for incident atherosclerotic disease. *Nat. Cardiovasc. Res.* **2023**, *2*, 144–158. [CrossRef] [PubMed]
12. Amancherla, K.; Wells, J.A.; Bick, A.G. Clonal hematopoiesis and vascular disease. *Semin. Immunopathol.* **2022**, *44*, 303–308. [CrossRef] [PubMed]
13. Wolach, O.; Sellar, R.S.; Martinod, K.; Cherpokova, D.; McConkey, M.; Chappell, R.J.; Silver, A.J.; Adams, D.; Castellano, C.A.; Schneider, R.K.; et al. Increased neutrophil extracellular trap formation promotes thrombosis in myeloproliferative neoplasms. *Sci. Transl. Med.* **2018**, *10*. [CrossRef] [PubMed]
14. Baldini, C.; Moriconi, F.R.; Galimberti, S.; Libby, P.; De Caterina, R. The JAK–STAT pathway: An emerging target for cardiovascular disease in rheumatoid arthritis and myeloproliferative neoplasms. *Eur. Heart J.* **2021**, *42*, 4389–4400. [CrossRef] [PubMed]
15. Shi, Z.; Liu, J.; Zhao, Y.; Yang, L.; Cai, Y.; Zhang, P.; Xu, Z.; Qin, T.; Qu, S.; Pan, L.; et al. ASXL1 mutations accelerate bone marrow fibrosis via EGR1-TNFA axis-mediated neoplastic fibrocyte generation in myeloproliferative neoplasms. *Haematologica* **2023**, *108*, 1359–1373. [CrossRef] [PubMed]

16. Akhiyat, N.; Lasho, T.; Ganji, M.; Toya, T.; Shi, C.-X.; Chen, X.; Corban, M.; Braggio, E.; Ahmad, A.; Keith, S.; et al. Clonal hematopoiesis of indeterminate potential is associated with future cardiovascular events in patients with early vascular aging. *J. Am. Coll. Cardiol.* **2022**, *79*, 1760. [CrossRef]
17. Lee-Six, H.; Øbro, N.F.; Shepherd, M.S.; Grossmann, S.; Dawson, K.; Belmonte, M.; Osborne, R.J.; Huntly, B.J.P.; Martincorena, I.; Anderson, E.; et al. Population dynamics of normal human blood inferred from somatic mutations. *Nature* **2018**, *561*, 473–478. [CrossRef] [PubMed]
18. Bick, A.G.; Weinstock, J.S.; Nandakumar, S.K.; Fulco, C.P.; Bao, E.L.; Zekavat, S.M.; Szeto, M.D.; Liao, X.; Leventhal, M.J.; Nasser, J.; et al. Inherited causes of clonal haematopoiesis in 97,691 whole genomes. *Nature* **2020**, *586*, 763–768. [CrossRef] [PubMed]
19. Marnell, C.S.; Bick, A.; Natarajan, P. Clonal hematopoiesis of indeterminate potential (CHIP): Linking somatic mutations, hematopoiesis, chronic inflammation and cardiovascular disease. *J. Mol. Cell. Cardiol.* **2021**, *161*, 98–105. [CrossRef] [PubMed]
20. Jeong, M.; Park, H.J.; Celik, H.; Ostrander, E.L.; Reyes, J.M.; Guzman, A.; Rodriguez, B.; Lei, Y.; Lee, Y.; Ding, L.; et al. Loss of Dnmt3a Immortalizes Hematopoietic Stem Cells In Vivo. *Cell Rep.* **2018**, *23*, 1–10. [CrossRef] [PubMed]
21. Challen, G.A.; Sun, D.; Jeong, M.; Luo, M.; Jelinek, J.; Berg, J.S.; Bock, C.; Vasanthakumar, A.; Gu, H.; Xi, Y.; et al. Dnmt3a is essential for hematopoietic stem cell differentiation. *Nat. Genet.* **2012**, *44*, 23–31. [CrossRef]
22. Sano, S.; Oshima, K.; Wang, Y.; Katanasaka, Y.; Sano, M.; Walsh, K. CRISPR-Mediated Gene Editing to Assess the Roles of Tet2 and Dnmt3a in Clonal Hematopoiesis and Cardiovascular Disease. *Circ. Res.* **2018**, *123*, 335–341. [CrossRef]
23. Mas-Peiro, S.; Hoffmann, J.; Fichtlscherer, S.; Dorsheimer, L.; Rieger, M.A.; Dimmeler, S.; Vasa-Nicotera, M.; Zeiher, A.M. Clonal haematopoiesis in patients with degenerative aortic valve stenosis undergoing transcatheter aortic valve implantation. *Eur. Heart J.* **2020**, *41*, 933–939. [CrossRef] [PubMed]
24. Abplanalp, W.T.; Cremer, S.; John, D.; Hoffmann, J.; Schuhmacher, B.; Merten, M.; Rieger, M.A.; Vasa-Nicotera, M.; Zeiher, A.M.; Dimmeler, S. Clonal Hematopoiesis–Driver DNMT3A Mutations Alter Immune Cells in Heart Failure. *Circ. Res.* **2021**, *128*, 216–228. [CrossRef]
25. Buscarlet, M.; Provost, S.; Zada, Y.F.; Bourgoin, V.; Mollica, L.; Dubé, M.-P.; Busque, L. Lineage restriction analyses in CHIP indicate myeloid bias for TET2 and multipotent stem cell origin for DNMT3A. *Blood* **2018**, *132*, 277–280. [CrossRef]
26. Cull, A.H.; Snetsinger, B.; Buckstein, R.; Wells, R.A.; Rauh, M.J. Tet2 restrains inflammatory gene expression in macrophages. *Exp. Hematol.* **2017**, *55*, 56–70.e13. [CrossRef] [PubMed]
27. Fujino, T.; Kitamura, T. ASXL1 mutation in clonal hematopoiesis. *Exp. Hematol.* **2020**, *83*, 74–84. [CrossRef]
28. Jeong, J.J.; Gu, X.; Nie, J.; Sundaravel, S.; Liu, H.; Kuo, W.-L.; Bhagat, T.D.; Pradhan, K.; Cao, J.; Nischal, S.; et al. Cytokine-Regulated Phosphorylation and Activation of TET2 by JAK2 in Hematopoiesis. *Cancer Discov.* **2019**, *9*, 778–795. [CrossRef] [PubMed]
29. Zhang, Q.; Zhao, K.; Shen, Q.; Han, Y.; Gu, Y.; Li, X.; Zhao, D.; Liu, Y.; Wang, C.; Zhang, X.; et al. Tet2 is required to resolve inflammation by recruiting Hdac2 to specifically repress IL-6. *Nature* **2015**, *525*, 389–393. [CrossRef] [PubMed]
30. Jaiswal, S.; Natarajan, P.; Silver, A.J.; Gibson, C.J.; Bick, A.G.; Shvartz, E.; McConkey, M.; Gupta, N.; Gabriel, S.; Ardissino, D.; et al. Clonal Hematopoiesis and Risk of Atherosclerotic Cardiovascular Disease. *N. Engl. J. Med.* **2017**, *377*, 111–121. [CrossRef] [PubMed]
31. Sano, S.; Wang, Y.; Yura, Y.; Sano, M.; Oshima, K.; Yang, Y.; Katanasaka, Y.; Min, K.-D.; Matsuura, S.; Ravid, K.; et al. JAK2-Mediated Clonal Hematopoiesis Accelerates Pathological Remodeling in Murine Heart Failure. *JACC Basic Transl. Sci.* **2019**, *4*, 684–697. [CrossRef] [PubMed]
32. Hsu, J.I.; Dayaram, T.; Tovy, A.; De Braekeleer, E.; Jeong, M.; Wang, F.; Zhang, J.; Heffernan, T.P.; Gera, S.; Kovacs, J.J.; et al. PPM1D Mutations Drive Clonal Hematopoiesis in Response to Cytotoxic Chemotherapy. *Cell Stem Cell* **2018**, *23*, 700–713.e706. [CrossRef] [PubMed]
33. Rosand, J.; Mitchell, B.D.; Ay, H.; de Bakker, P.I.W.; Gwinn, K.; Kittner, S.J.; Lindgren, A.; Meschia, J.F.; Pulit, S.L.; Sudlow, C.L.M.; et al. Loci associated with ischaemic stroke and its subtypes (SiGN): A genome-wide association study. *Lancet. Neurol.* **2016**, *15*, 174–184. [CrossRef]
34. Song, Y.; Choi, J.E.; Kwon, Y.J.; Chang, H.J.; Kim, J.O.; Park, D.H.; Park, J.M.; Kim, S.J.; Lee, J.W.; Hong, K.W. Identification of susceptibility loci for cardiovascular disease in adults with hypertension, diabetes, and dyslipidemia. *J. Transl. Med.* **2021**, *19*, 85. [CrossRef]
35. Malik, R.; Chauhan, G.; Traylor, M.; Sargurupremraj, M.; Okada, Y.; Mishra, A.; Rutten-Jacobs, L.; Giese, A.K.; Van Der Laan, S.W.; Gretarsdottir, S.; et al. Multiancestry genome-wide association study of 520,000 subjects identifies 32 loci associated with stroke and stroke subtypes. *Nat. Genet.* **2018**, *50*, 524–537. [CrossRef] [PubMed]
36. Soriano-Tárraga, C.; Lazcano, U.; Giralt-Steinhauer, E.; Avellaneda-Gómez, C.; Ois, Á.; Rodríguez-Campello, A.; Cuadrado-Godia, E.; Gómez-González, A.; Fernández-Sanlés, A.; Elosua, R.; et al. Identification of 20 novel loci associated with ischaemic stroke. Epigenome-wide association study. *Epigenetics* **2020**, *15*, 988–997. [CrossRef]

37. Temprano-Sagrera, G.; Sitlani, C.M.; Bone, W.P.; Martin-Bornez, M.; Voight, B.F.; Morrison, A.C.; Damrauer, S.M.; de Vries, P.S.; Smith, N.L.; Sabater-Lleal, M.; et al. Multi-phenotype analyses of hemostatic traits with cardiovascular events reveal novel genetic associations. *J. Thromb. Haemost.* **2022**, *20*, 1331–1349. [CrossRef]
38. Sakaue, S.; Kanai, M.; Tanigawa, Y.; Karjalainen, J.; Kurki, M.; Koshiba, S.; Narita, A.; Konuma, T.; Yamamoto, K.; Akiyama, M.; et al. A cross-population atlas of genetic associations for 220 human phenotypes. *Nat. Genet.* **2021**, *53*, 1415–1424. [CrossRef] [PubMed]
39. Jaworek, T.; Xu, H.; Gaynor, B.J.; Cole, J.W.; Rannikmae, K.; Stanne, T.M.; Tomppo, L.; Abedi, V.; Amouyel, P.; Armstrong, N.D.; et al. Contribution of Common Genetic Variants to Risk of Early-Onset Ischemic Stroke. *Neurology* **2022**, *99*, e1738–e1754. [CrossRef]
40. Jameson, H.S.; Hanley, A.; Hill, M.C.; Xiao, L.; Ye, J.; Bapat, A.; Ronzier, E.; Hall, A.W.; Hucker, W.J.; Clauss, S.; et al. Loss of the Atrial Fibrillation-Related Gene, *Zfhx3*, Results in Atrial Dilation and Arrhythmias. *Circ. Res.* **2023**, *133*, 313–329. [CrossRef] [PubMed]
41. Gudbjartsson, D.F.; Holm, H.; Gretarsdottir, S.; Thorleifsson, G.; Walters, G.B.; Thorgeirsson, G.; Gulcher, J.; Mathiesen, E.B.; Njølstad, I.; Nyrnes, A.; et al. A sequence variant in ZFHX3 on 16q22 associates with atrial fibrillation and ischemic stroke. *Nat. Genet.* **2009**, *41*, 876–878. [CrossRef]
42. Flister, M.J.; Hoffman, M.J.; Lemke, A.; Prisco, S.Z.; Rudemiller, N.; O'Meara, C.C.; Tsaih, S.-W.; Moreno, C.; Geurts, A.M.; Lazar, J.; et al. *SH2B3* Is a Genetic Determinant of Cardiac Inflammation and Fibrosis. *Circ. Cardiovasc. Genet.* **2015**, *8*, 294–304. [CrossRef] [PubMed]
43. Murphy, M.B.; Yang, Z.; Subati, T.; Farber-Eger, E.; Kim, K.; Blackwell, D.J.; Fleming, M.R.; Stark, J.M.; Van Amburg, J.C.; Woodall, K.K.; et al. LNK/SH2B3 loss of function increases susceptibility to murine and human atrial fibrillation. *Cardiovasc. Res.* **2024**, *120*, 899–913. [CrossRef]
44. Willis, A.J.; Corey, S.J.; Murga-Zamalloa, C.; Karimi, S.S.; Khaddour, K.; Quigley, J.; Eklund, E.A.; Chen, Y. Dynamin-2 deficiency causes age- and sex-dependent neutropenia and myelodysplasia in mice. *Blood Adv.* **2023**, *7*, 1418–1431. [CrossRef] [PubMed]
45. Qiu, X.; Weng, J.; Jiang, Y.; Cui, L.; Gu, H.; Jiang, Y.; Dai, Y.; Li, H.; Wang, Y.; Li, Z. Association between clonal hematopoiesis-related gene mutations and unfavorable functional outcome in patients with large-artery atherosclerotic stroke. *Eur. J. Med. Res.* **2023**, *28*, 599. [CrossRef]
46. Soerensen, M.; Tulstrup, M.; Hansen, J.W.; Weischenfeldt, J.; Grønbæk, K.; Christensen, K. Clonal Hematopoiesis and Epigenetic Age Acceleration in Elderly Danish Twins. *HemaSphere* **2022**, *6*, e768. [CrossRef] [PubMed]
47. Nachun, D.; Lu, A.T.; Bick, A.G.; Natarajan, P.; Weinstock, J.; Szeto, M.D.; Kathiresan, S.; Abecasis, G.; Taylor, K.D.; Guo, X.; et al. Clonal hematopoiesis associated with epigenetic aging and clinical outcomes. *Aging Cell* **2021**, *20*, e13366. [CrossRef]
48. Arends, C.M.; Galan-Sousa, J.; Hoyer, K.; Chan, W.; Jäger, M.; Yoshida, K.; Seemann, R.; Noerenberg, D.; Waldhueter, N.; Fleischer-Notter, H.; et al. Hematopoietic lineage distribution and evolutionary dynamics of clonal hematopoiesis. *Leukemia* **2018**, *32*, 1908–1919. [CrossRef]
49. Ladle, B.H.; Li, K.-P.; Phillips, M.J.; Pucsek, A.B.; Haile, A.; Powell, J.D.; Jaffee, E.M.; Hildeman, D.A.; Gamper, C.J. De novo DNA methylation by DNA methyltransferase 3a controls early effector $CD8^+$ T-cell fate decisions following activation. *Proc. Natl. Acad. Sci. USA* **2016**, *113*, 10631–10636. [CrossRef] [PubMed]
50. Carty, S.A.; Gohil, M.; Banks, L.B.; Cotton, R.M.; Johnson, M.E.; Stelekati, E.; Wells, A.D.; Wherry, E.J.; Koretzky, G.A.; Jordan, M.S. The Loss of TET2 Promotes CD8+ T Cell Memory Differentiation. *J. Immunol.* **2018**, *200*, 82–91. [CrossRef]
51. Ricard, L.; Hirsch, P.; Largeaud, L.; Deswarte, C.; Jachiet, V.; Mohty, M.; Rivière, S.; Malard, F.; Tenon, M.; de Vassoigne, F.; et al. Clonal haematopoiesis is increased in early onset in systemic sclerosis. *Rheumatology* **2020**, *59*, 3499–3504. [CrossRef] [PubMed]
52. Savola, P.; Lundgren, S.; Keränen, M.A.I.; Almusa, H.; Ellonen, P.; Leirisalo-Repo, M.; Kelkka, T.; Mustjoki, S. Clonal hematopoiesis in patients with rheumatoid arthritis. *Blood Cancer J.* **2018**, *8*, 69. [CrossRef] [PubMed]
53. Johnson, C.O.; Nguyen, M.; Roth, G.A.; Nichols, E.; Alam, T.; Abate, D.; Abd-Allah, F.; Abdelalim, A.; Abraha, H.N.; Abu-Rmeileh, N.M.E.; et al. Global, regional, and national burden of stroke, 1990–2016: A systematic analysis for the Global Burden of Disease Study 2016. *Lancet Neurol.* **2019**, *18*, 439–458. [CrossRef]
54. Virani, S.S.; Alonso, A.; Benjamin, E.J.; Bittencourt, M.S.; Callaway, C.W.; Carson, A.P.; Chamberlain, A.M.; Chang, A.R.; Cheng, S.; Delling, F.N.; et al. Heart Disease and Stroke Statistics—2020 Update: A Report From the American Heart Association. *Circulation* **2020**, *141*, 9. [CrossRef]
55. Saini, V.; Guada, L.; Yavagal, D.R. Global Epidemiology of Stroke and Access to Acute Ischemic Stroke Interventions. *Neurology* **2021**, *97*, S6–S16. [CrossRef]
56. Weeks, L.D.; Niroula, A.; Neuberg, D.; Wong, W.; Lindsley, R.C.; Luskin, M.R.; Berliner, N.; Stone, R.M.; DeAngelo, D.J.; Soiffer, R.J.; et al. Prediction of Risk for Myeloid Malignancy in Clonal Hematopoiesis. *NEJM Evid.* **2023**, *2*. [CrossRef]
57. Mack, T.; Vlasschaert, C.; von Beck, K.; Silver, A.J.; Heimlich, J.B.; Poisner, H.; Condon, H.R.; Ulloa, J.; Sochacki, A.L.; Spaulding, T.P.; et al. Cost-Effective and Scalable Clonal Hematopoiesis Assay Provides Insight into Clonal Dynamics. *J. Mol. Diagn.* **2024**, *26*, 563–573. [CrossRef]

58. Köhnke, T.; Majeti, R. Clonal Hematopoiesis: From Mechanisms to Clinical Intervention. *Cancer Discov.* **2021**, *11*, 2987–2997. [CrossRef]
59. Itzykson, R.; Kosmider, O.; Cluzeau, T.; Mansat-De Mas, V.; Dreyfus, F.; Beyne-Rauzy, O.; Quesnel, B.; Vey, N.; Gelsi-Boyer, V.; Raynaud, S.; et al. Impact of TET2 mutations on response rate to azacitidine in myelodysplastic syndromes and low blast count acute myeloid leukemias. *Leukemia* **2011**, *25*, 1147–1152. [CrossRef] [PubMed]
60. Bejar, R.; Lord, A.; Stevenson, K.; Bar-Natan, M.; Pérez-Ladaga, A.; Zaneveld, J.; Wang, H.; Caughey, B.; Stojanov, P.; Getz, G.; et al. TET2 mutations predict response to hypomethylating agents in myelodysplastic syndrome patients. *Blood* **2014**, *124*, 2705–2712. [CrossRef]
61. Cimmino, L.; Dolgalev, I.; Wang, Y.; Yoshimi, A.; Martin, G.H.; Wang, J.; Ng, V.; Xia, B.; Witkowski, M.T.; Mitchell-Flack, M.; et al. Restoration of TET2 Function Blocks Aberrant Self-Renewal and Leukemia Progression. *Cell* **2017**, *170*, 1079–1095.e1020. [CrossRef] [PubMed]
62. Bick, A.G.; Pirruccello, J.P.; Griffin, G.K.; Gupta, N.; Gabriel, S.; Saleheen, D.; Libby, P.; Kathiresan, S.; Natarajan, P. Genetic Interleukin 6 Signaling Deficiency Attenuates Cardiovascular Risk in Clonal Hematopoiesis. *Circulation* **2020**, *141*, 124–131. [CrossRef] [PubMed]
63. Ridker, P.M.; Everett, B.M.; Thuren, T.; MacFadyen, J.G.; Chang, W.H.; Ballantyne, C.; Fonseca, F.; Nicolau, J.; Koenig, W.; Anker, S.D.; et al. Antiinflammatory Therapy with Canakinumab for Atherosclerotic Disease. *N. Engl. J. Med.* **2017**, *377*, 1119–1131. [CrossRef] [PubMed]
64. Bouabdallaoui, N.; Tardif, J.-C.; Waters, D.D.; Pinto, F.J.; Maggioni, A.P.; Diaz, R.; Berry, C.; Koenig, W.; Lopez-Sendon, J.; Gamra, H.; et al. Time-to-treatment initiation of colchicine and cardiovascular outcomes after myocardial infarction in the Colchicine Cardiovascular Outcomes Trial (COLCOT). *Eur. Heart J.* **2020**, *41*, 4092–4099. [CrossRef]
65. Huang, Y.; Wang, Z.; Huang, Z.-X.; Liu, Z. Biomarkers and the outcomes of ischemic stroke. *Front. Mol. Neurosci.* **2023**, *16*, 1171101. [CrossRef]

MDPI AG
Grosspeteranlage 5
4052 Basel
Switzerland
Tel.: +41 61 683 77 34

Neurology International Editorial Office
E-mail: neurolint@mdpi.com
www.mdpi.com/journal/neurolint

www.ingramcontent.com/pod-product-compliance
Lightning Source LLC
Chambersburg PA
CBHW061452210926
51850CB00008B/1691